CRIMINOLOGY

This book is dedicated to my wife, Grace; my parents, Lawrence and Winifred;
my sons, Robert and Michael; my stepdaughters, Heidi and Kasey; and my grandchildren, Robbie,
Ryan, Mikey, Randy, Christopher, Stevie, Ashlyn, and Morgan; and to my great granddaughter, Kaelyn.

CRIMINOLOGY

ANTHONY WALSH

Boise State University

Los Angeles | London | New Delhi
Singapore | Washington DC

Los Angeles | London | New Delhi
Singapore | Washington DC

FOR INFORMATION:

SAGE Publications, Inc.
2455 Teller Road
Thousand Oaks, California 91320
E-mail: order@sagepub.com

SAGE Publications Ltd.
1 Oliver's Yard
55 City Road
London EC1Y 1SP
United Kingdom

SAGE Publications India Pvt. Ltd.
B 1/I 1 Mohan Cooperative Industrial Area
Mathura Road, New Delhi 110 044
India

SAGE Publications Asia-Pacific Pte. Ltd.
33 Pekin Street #02-01
Far East Square
Singapore 048763

Acquisitions Editor: Jerry Westby
Editorial Assistant: Erim Sarbuland
Production Editor: Karen Wiley
Copy Editor: Teresa Herlinger
Typesetter: C&M Digitals (P) Ltd.
Proofreader: Theresa Kay
Indexer: Jeanne Busemeyer
Cover Designer: Gail Buschman
Marketing Manager: Erica DeLuca
Permissions Editor: Karen Ehrmann

Printed in the United States of America

Library of Congress Cataloging-in-Publication Data

Walsh, Anthony, 1941-

Criminology : the essentials / Anthony Walsh.

p. cm.
Includes bibliographical references and index.

ISBN 978-1-4129-9943-4 (pbk.)

1. Criminology. 2. Criminal behavior. I. Title.

HV6025.W3654 2012 364—dc23 2011028305

This book is printed on acid-free paper.

11 12 13 14 15 10 9 8 7 6 5 4 3 2 1

Contents

Chapter 3. The Early Schools of Criminology and Modern Counterparts   41

Chapter 4. Social Structural Theories   57

Chapter 12. Property Crime   201

Chapter 13. Public Order Crime   217

Preface

There are a number of excellent criminological textbooks available to students and professors, so why this one? The reason is that the typical textbook has become inordinately expensive (often as high as $150), which is a true hardship for many students today. In addition, many of the books are filled with enormous amounts of information that cannot possibly be digested in one semester. Moreover, there is so much to try to cover that professors may be reluctant to bring in additional materials such as journal articles that they may consider very important.

By way of contrast, this book provides the essentials of criminology in a compact and affordable volume. It covers all the material that is necessary to know and omits what is merely nice to know. It does not inundate students with scores of minor facts that may turn them glassy-eyed, but it does engage them in straightforward language with the latest advances in criminology from a variety of disciplines (and it costs them one half to one third of the price charged for the more glitzy hardback texts). This book can serve as the primary text for an undergraduate course in criminology, or as the primary text for a graduate course when supplemented by additional readings available on the Sage website.

Structure of the Book

This book uses the typical outline for criminology textbook topics/sections, beginning with the definitions of crime and criminology and descriptions of how crime is measured, proceeding into theories of crime and criminality, and following with typologies. I depart from the typical textbook sequencing in one way only, that of the ordering of the theory chapters. The typical criminology textbook begins with a discussion of biological and psychological theories and proceeds to demolish concepts that others demolished decades ago such as atavism and the XYY syndrome. Having shown how wrong these concepts were, and leaving the impression that they exhaust the content of modern biological and psychological theories, they proceed to sociological theories.

Unfortunately, this is the exact opposite of the way that normal science operates. Normal science begins with observations and descriptions of phenomena on a large (macro) scale and then asks a series of "why" questions that systematically take it down to lower levels of analysis. Wholes are wonderful, meaningful things, and holistic explanations are fine as far as they go. But they only go so far before they exhaust their explanatory power and before the data require a more elementary look. Philosophers of science agree that holistic accounts describe phenomena, whereas reductionist (examining a phenomenon at a more fundamental level) accounts explain them. Scientists typically observe and describe what is on the surface of a phenomenon and then seek to dig deeper to find the fundamental mechanisms that drive the phenomenon.

In the natural sciences, useful observations go in both holistic and reductionist directions, such as from quarks to the cosmos in physics and from nucleotides to ecological systems in biology. There is no zero-sum competition between levels of analysis in these sciences, nor should there be in ours. Thus, following a discussion of the early schools, the book moves into the most holistic (social structural) theories. These theories describe elements of whole societies that are supposedly conducive to high rates of criminal behavior such as capitalism or racial heterogeneity. Because only a small proportion of people exposed to these alleged criminogenic forces commit crimes, we must move down to social process theories that talk about how individuals interpret and respond to structural forces. We then have to move to more individualistic (psychosocial) theories that focus on the traits and abilities of individuals that would lead them to arrive at different interpretations from other individuals, and finally to theories (biosocial) that try to pin down the exact mechanisms underlying these predilections.

This book is divided into 15 chapters that mirror the sections in a typical criminology textbook, each dealing with a particular type of subject matter in criminology. Each of the theory sections concludes with an evaluation of the theories based on the policy implications derivable from them. These sections are as follows:

1. Criminology, Crime, and Criminal Law: This chapter opens with a discussion of criminology, describing crime and criminality and introducing the concepts, functions, and pitfalls of criminological theory. Criminals are not defined as such until convicted in a criminal court. Thus, this chapter takes students on a brief excursion of the criminal justice system from arrest to incarceration. It also offers a brief history of the discipline, which will allow readers to understand how the science of criminology got started. The social context in which various perspectives began, at key times in certain periods of cultural and political development, is discussed, as well as the technology available for criminologists seeking to understand the quicksilver of criminal behavior. I also discuss connecting criminological theory to social policy in this chapter.

2. Measuring Crime and Criminal Behavior: This chapter describes the various ways data on the prevalence and incidence of crime are collected. It describes the strengths and weaknesses of the Uniform Crime Reports, the National Incident-Based Reporting System, and the National Crime Victimization Survey, which are measures collected by government agencies, as well as self-report studies, which are collected by criminologists. I show how comparisons of the various measures can be used to address important criminological debates, such as racial disproportionality in arrests, that cannot be addressed by a single data source. The chapter also looks at crime trends and the pitfalls of trying to interpret them. Also included is a discussion of the FBI's Financial Crimes Report, which contains data on white-collar crime not reported in the UCR.

3. The Early Schools of Criminology and Modern Counterparts: This chapter explores a basic dichotomy in criminology: classical versus positivist. The Classical School emphasizes human rationality, free will, and choice; the Positivist School emphasizes the scientific search for factors that influence how these human attributes are exercised. The chapter looks at how modern thinkers view the argument about "free will" and "determinism" in the context of criminal behavior.

4. Social Structural Theories: Social structural theories are "macro" theories that explore the behavioral effects of how society is structured on criminal behavior. They look at such things as culture, neighborhood, and social practices and how these things serve to generate crime. They do not seek to explain individuals' criminal behavior, but rather aggregate crime rates of different groups who are exposed to these factors.

5. Social Process Theories: Social process theories are "micro" theories that explore how individuals subjectively perceive the kinds of factors social structural theories identify; all people do not react similarly to similar situations. Theorists in this tradition concentrate on exploring the influence of smaller social groupings (such as peer groups and the family) on the behavior of individuals.

6. Critical Theories: Critical theories are also structural theories, but they differ on the critical stance that they have on society and on their emphasis on social conflict rather than social consensus. The capitalist mode of production is *the* cause of crime for many theorists in this tradition, although many others do not share this extreme view. Feminist theories are also discussed as part of the critical tradition. The major concerns of feminist criminology are the generalizability of traditional "male" theories to female crime, and to understand why always and everywhere females commit far less crime than males.

7. Psychosocial Theories: Although some writers have classified social process theories as "psychological" because of their emphasis on subjective interpretation, the primary difference between them and the material in this chapter is that psychological traits such as IQ, impulsiveness, and empathy are emphasized more than influences outside of the actor. This chapter also discusses the "antisocial personalities": psychopaths and sociopaths.

8. Biosocial Approaches: Biosocial perspectives are having an ever-increasing impact on criminology. This chapter examines what the disciplines of genetics, evolutionary psychology, and neuroscience have to offer our discipline. Theorists in these disciplines go to great pains to convince us that we cannot understand the role of genes, hormones, and brain structures without understanding the complementary role of the environment—there is no nature versus nurture argument here; only nature *via* nurture.

9. Developmental Theories: Developmental theories bring the disciplines of biology, psychology, and sociology together to offer a more complete understanding of antisocial behavior. This chapter looks at offending in terms of the onset, acceleration, deceleration, and desistance from it along with all the risk and protective factors for it. This is followed by discussions of major developmental theories.

10. Violent Crimes: This chapter examines the UCR Part I violent crimes (murder, rape, robbery, and aggravated assault). It also features multiple murder (mass, spree, and serial killing) and discusses how different disciplines and theories try to explain violence.

11. Terrorism: This chapter examines the very contemporary problem of terrorism, with emphasis on al-Qaeda and its influence on the recruitment of domestic terrorists. The causes and context of terrorism are addressed with reference to previously discussed theories.

12. Property Crime: This chapter examines the UCR Part I property crimes (burglary, larceny-theft, motor vehicle theft, and arson) as well as some important Part II crimes. It concentrates primarily on the subjective reasons for engaging in property crime by looking at what offenders themselves have to say about why they engage in crime. In addition, the growing area of cybercrime is discussed.

13. Public Order Crime: Public order crimes can be more harmful than many other types of crime, although they may be legal at some times and in some places. This chapter looks at the links between alcohol, drugs, and crime. It also examines prostitution and drunk driving.

14. White-Collar and Organized Crime: White-collar crime is more costly to American society than common street crime. In this chapter, I differentiate between occupational and corporate crime, and look at such issues as the similarities and differences between white-collar and street criminals. The chapter then examines organized crime and the reasons that it exists, as well as where it is most likely to occur. Theories explaining white-collar and organized crime are discussed.

15. Victimology: The final chapter discusses the neglected topic of victimology. I examine who is most likely to be victimized in terms of gender, race, age, and socioeconomic class, and find that those most likely to be perpetrators of crime are also those most likely to be victims of crime. Also explored are victimization theories and the consequences of victimization.

Acknowledgments

I would first of all like to thank the ever jovial executive editor Jerry Westby for his faith in this project from the beginning. Thanks also for the commitment of his very able colleagues and assistants Karen Wiley and Erim Sarbuland. This tireless threesome kept up a most useful dialogue between authors, publisher, and a number of excellent reviewers. The copy editor, Teresa Herlinger, spotted every errant comma, dangling participle, missing reference, and misspelled word in the manuscript, for which I am truly thankful. The production editor, Karen Wiley, made sure everything went quickly and smoothly thereafter. Thank you one and all.

I am also most grateful for the reviewers who spent considerable time providing me with the benefit of their expertise during the writing/rewriting phase of the book's production. Their input and encouragement has undoubtedly made the book better than it would otherwise have been. These expert criminologists are Gennifer Furst, William Paterson University; Scott Maggard, Old Dominion University; Heather Melton, The University of Utah; Allison Payne, Villanova University; Kelly Asmussen, Peru State College; Tracey Steele, Wright State University; Linda Tobin, Austin Community College; and Steven Egger, University of Houston.

Most of all, I would like to acknowledge the love and support of my most wonderful and drop-dead gorgeous wife, Grace Jean (aka "Grace the face"). Grace's love and support has sustained me for so long that I cannot imagine life without her; she is a real treasure and the center of my universe: Szeretlek nagyonsok, Gracie.

Criminology, Crime, and Criminal Law

In 1996, Iraqi refugees Majed Al-Timimy, 28, and Latif Al-Husani, 34, married the daughters, aged 13 and 14, of a fellow Iraqi refugee in Lincoln, Nebraska. The marriages took place according to Muslim custom and everything seemed to be going well for awhile until one of the girls ran away and the concerned father and her husband reported it to the police. It was at this point that American and Iraqi norms of legality and morality clashed head-on. Under Nebraska law, people under 17 years old cannot marry, so both grooms and the father and mother of the girls were arrested and charged with a variety of crimes from child endangerment to rape.

According to an Iraqi woman interviewed by the police (herself married at 12 in Iraq), both girls were excited and happy about the wedding. The Iraqi community was shocked that the parents of the brides faced up to 50 years in prison for their actions, as would have been earlier generations of Americans who were legally permitted to marry girls of this age. The grooms were sentenced to 4 to 6 years in prison and paroled in 2000 on the condition that they have no contact with their "wives."

Thus, something that is legally and morally permissible in one culture can be severely punished in another. Did the actions of these men constitute child sex abuse or simply unremarkable marital sex? Which culture is right? Can we really ask such a question? Is Iraqi culture "more right" than American culture given that marrying girls of that age was permissible in the United States, too, at one time? Most importantly, how can criminologists hope to study crime scientifically if what constitutes a crime is relative to time and place?

⌧ What Is Criminology?

Criminology is an interdisciplinary science that gathers and analyzes data on various aspects of criminal, delinquent, and general antisocial behavior. It is different from the discipline of criminal justice. Criminal justice is concerned with how the criminal justice system investigates, prosecutes, and controls/supervises individuals who have committed crime, while criminology wants to know *why* those individuals committed crimes. As with all scientific disciplines, the goal of criminology is to understand its subject matter and to determine how that understanding can benefit humankind. In pursuit of this understanding, criminologists ask questions such as

- Why do crime rates vary across time and from culture to culture?
- Why are some individuals more prone to committing crime than others?
- Why do crime rates vary across different ages, genders, and racial/ethnic groups?
- Why are some harmful acts criminalized and not others?
- What can we do to prevent crime?

By a *scientific* study of crime and criminal behavior, we mean that criminologists use the scientific method to try to answer the questions they ask rather than just philosophizing about them from their armchairs. The scientific method is a tool for separating truth from error by demanding evidence for any conclusions. Evidence is obtained by formulating hypotheses derived from theory that are rigorously tested with data. How this is accomplished will be addressed after we discuss the nature of crime.

⌧ What Is Crime?

The term *criminal* can and has been applied to many types of behavior, some of which nearly all of us have been guilty of at some time in our lives. We can all think of acts that we feel *ought* to be criminal but are not, or acts that should not be criminal but are. The list of things that someone or another at different times and at different places may consider to be crimes is very large, with only a few being defined as criminal by the law in the United States at this time. Despite these difficulties, we need a definition of crime in order to proceed. The most often quoted definition is that of Paul Tappan (1947), who defined **crime** as "an intentional act in violation of the criminal law committed without defense or excuse, and penalized by the state" (p. 100). A crime is thus an *act* in violation of a *criminal law* for which a *punishment* is prescribed; the person committing it must have *intended* to do so and must have done so without legally acceptable *defense* or *justification*.

Tappan's definition is strictly a legal one that reminds us that the state, and only the state, has the power to define crime. Hypothetically, a society could eradicate crime tomorrow simply by canceling all of its criminal statutes. Of course, this would not eliminate the behavior specified by the laws; in fact, the behavior would doubtless increase since the behavior could no longer be officially punished. While it is absurd to think that any society would try to solve its crime problem by eliminating its criminal statutes, legislative bodies are continually revising, adding to, and deleting from their criminal statutes.

Crime as a Moving Target

Every vice is somewhere and at some times a virtue. There are numerous examples, such as the vignette at the beginning of this chapter, of acts defined as crimes in one country being tolerated and even expected behavior in another. We might congratulate ourselves for protecting young girls from the kind of "fate" that befell the 13- and 14-year-old girls in the vignette, but in 1885, no state in the union had an age of consent above 12 (Friedman, 2005). Laws also vary within the same culture across time as well as across different cultures. Until the Harrison Narcotics Act of 1914, there were few legal restrictions in the United States on the sale, possession, or use of most drugs such as heroine and cocaine. Following the Harrison Act, many drugs became controlled substances, their possession became a crime, and a brand new class of criminals was created overnight.

Crimes pass out of existence also, even acts that had been considered crimes for centuries. Until the United States Supreme Court invalidated sodomy (anal or oral sex) statutes in *Lawrence v. Texas* (2003), it was legally punishable in many states. Likewise,

▲ **Photo 1.1** Prohibition era police officer standing by car loaded with moonshine wrecked in a chase.

burning the American flag had serious legal consequences until 1989 when the Supreme Court ruled anti-flag-burning statutes unconstitutional in *Texas v. Johnson* (1989).

What constitutes a crime, then, can be defined into or out of existence by the courts or by legislators. As long as human societies remain diverse and dynamic, there will always be a moving target of activities with the potential for nomination as crimes, as well as illegal activities nominated for decriminalization.

If what constitutes crime differs across time and place, how can criminologists hope to agree upon a scientific explanation for crime and criminal behavior? Science is about making universal statement about stable or homogeneous phenomena. Atoms, the gas laws, the laws of thermodynamics, photosynthesis, and so on, are not defined or evaluated differently by scientists around the globe according to local customs or ideological preferences. But what we call "crime" keeps moving around, and because it does, some criminologists have declared it impossible to generalize about what is and is not "real" crime.

What these criminologists are saying is that crime is a socially constructed phenomenon that lacks any "real" objective essence and is defined into existence rather than discovered. In a trivial sense, everything is socially constructed. Nature does not reveal herself to us sorted into ready-labeled packages, so humans must do it for her. *Social construction* means nothing more than humans having perceived a phenomenon, named it, and categorized it according to some classificatory rule that makes note of the similarities and differences among the things being classified. Most classification schemes are not arbitrary; if they were, we would not be able to make sense of anything. Categories have empirically meaningful referents and are used to impose order on the diversity of human experience, although arguments exist about just how coherent that order is.

Crime as a Subcategory of Social Harms

So, what *can* we say about crime? How *can* we conceive of it in ways that at least most people would agree are logical, consistent, and correspond with their view of reality? When all is said and done, crime is a subcategory of all harmful acts that range from simple things like smoking to very serious things like murder. Some harmful acts such as smoking tobacco and drinking to excess are not considered anyone's business other than the actor's if they take place in private or even in public if the person indulging in those things creates no annoyance to others.

Socially (as opposed to private) harmful acts are acts deemed to be in need of regulation (health standards, air pollution, etc.), but not by the criminal law except under exceptional circumstances. Private wrongs (such as someone reneging on a contract) are socially harmful, but not sufficiently so to require the heavy hand of the criminal law. Such wrongs are regulated by the civil law in which the wronged party (the plaintiff) rather than the state initiates legal action, and the defendant does not risk deprivation of his or her liberty if the plaintiff prevails.

Further along the continuum, we find a category of harmful acts considered so socially harmful that they come under the scope of the criminal justice system. Even here, we are still confronted with the problem of human judgment in determining what goes into this subcategory. But this is true all along the line; smoking was once actually considered rather healthy, and air pollution and unhealthy conditions were simply facts of life about which nothing could be done. Categorization always requires a series of human judgments, but that does not render the categorizations arbitrary.

The harm caused by criminal activity is financially and emotionally very costly. The emotional pain and suffering borne by crime victims is obviously impossible to quantify, but many estimates of the financial harm are available. Most estimates focus on the costs of running the criminal justice system, which includes the salaries and benefits of personnel and the maintenance costs of buildings (offices, jails, prisons, stations) and equipment (vehicles, weapons, uniforms, etc.). Added to these costs are those associated with each crime (the average cost per incident multiplied by the number of incidents as reported to the police). All these costs combined are estimates of the *direct* costs of crime.

The *indirect* costs of crime must also be considered as part of the burden. These costs include all manner of surveillance and security devices, protective devices (guns, alarms, security guards) and insurance costs, medical services, and the productivity and tax loss of incarcerated individuals. Economist David Anderson (1999) lists numerous direct and indirect costs of crime and concludes that the aggregate burden of crime in the United States (in 1997 dollars) is about $1,102 *billion,* or a per capita burden of $4,118 ($5,480 in 2009 dollars). Crime thus places a huge financial burden on everyone's shoulders, as well as a deep psychological burden on its specific victims.

Beyond Social Construction: The Stationary Core Crimes

Few people would argue that an act is not arbitrarily categorized or is not seriously harmful if it is universally condemned. That is, there is a core of offenses defined as wrong at almost all times and in almost all cultures. Some of the strongest evidence in support of the *stationary core* perspective comes from the International Criminal Police Organization (INTERPOL) (1992), headquartered in Lyon, France. INTERPOL serves as a repository for crime statistics from each of its 188 member nations. INTERPOL's data show that such acts as murder, assault, rape, and theft are considered serious crimes in every single country.

Criminologists call these universally condemned crimes **mala in se** ("inherently bad"). Crimes that are time and culture bound are described as **mala prohibita** ("bad because they are prohibited"). But how can we be sure that an act is inherently bad? The litmus test for determining a mala in se crime is that no one except under the most bizarre of circumstances would want to be victimized by one. While millions of people seek to be "victimized" by prostitutes, drug dealers, or bookies, no one wants to be murdered, raped, robbed, or have their property stolen. Being victimized by such actions evokes physiological reactions (anger, helplessness, sadness, depression, a desire for revenge) in all cultures, and would do so even if the acts were not punishable by law or custom. Mala in se crimes engage these emotions not because some legislative body has defined them as wrong, but because they hammer at our deepest instincts. Evolutionists propose that these built-in emotional mechanisms exist because mala in se crimes threatened the survival and reproductive success of our distant ancestors, and that they function to strongly motivate people to try to prevent such acts from occurring and punishing them if they do (O'Manique, 2003; Walsh, 2000).

Figure 1.1 illustrates the relationship of core crimes (mala in se) to acts that have been arbitrarily defined (mala prohibita) as crimes and all harmful acts that may potentially be criminalized. The figure is inspired by John Hagan's (1985) effort to distinguish between "real" crimes and "socially constructed" arbitrary crimes by examining the three highly interrelated concepts of *consensus* (the degree of public agreement on the seriousness of an act), the *severity* of penalties attached to an act, and the level of *harm* attached to an act.

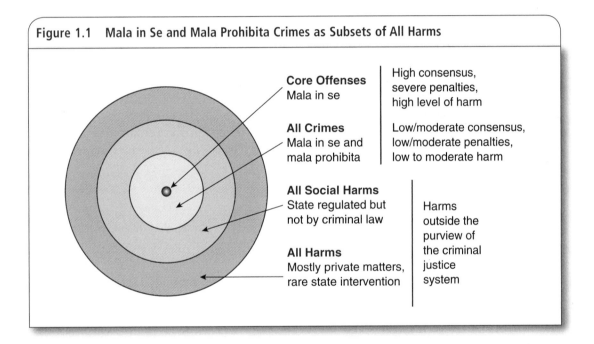

Figure 1.1   Mala in Se and Mala Prohibita Crimes as Subsets of All Harms

Core Offenses
Mala in se

High consensus,
severe penalties,
high level of harm

All Crimes
Mala in se and
mala prohibita

Low/moderate consensus,
low/moderate penalties,
low to moderate harm

All Social Harms
State regulated but
not by criminal law

Harms
outside the
purview of
the criminal
justice
system

All Harms
Mostly private matters,
rare state intervention

✄ Criminality

Perhaps we can avoid altogether the problem of defining crimes by studying individuals who commit predatory *harmful* acts, regardless of the legal status of the acts. Criminologists do this when they study criminality. **Criminality** is a clinical or scientific term rather than a legal one, and one that can be defined independently of legal definitions of crimes. Crime is an intentional act of commission or omission contrary to the law and is a property of society; *criminality* is a property of individuals that signals the willingness to commit crimes and other harmful acts. Criminality is a trait that lies on a continuum ranging from saint to sociopath, and is composed of other traits such as callousness and impulsiveness that also vary greatly among people. People can use and abuse others for personal gain regardless of whether the means used have been defined as criminal; it is the propensity to do this that defines criminality independent of the labeling of an act as a crime or of the person being legally defined as a criminal.

Defining criminality as a continuous trait acknowledges that there is no sharp line separating individuals with respect to this trait—it is not a trait that one has or does not have. Just about everyone at some point in life has committed an act or two in violation of the law. But that doesn't make us all criminals; if it did, the term would become virtually synonymous with the word *human*. The point is, we are all situated somewhere on the criminality continuum, just as our heights range from the truly short to the truly tall. Some are so extreme in height that any reasonable person would call them "tall." Likewise, a small number of individuals have violated so many criminal statutes over such a long period of time that few would question the appropriateness of calling them "criminals." Thus, both height and criminality can be thought of as existing along a continuum, even though the words we use often imply that people's heights and criminal tendencies come in more or less discrete categories (tall/short, criminal/noncriminal). In other words, just as height varies in fine gradations, so too does involvement in crime.

The Legal Making of a Criminal

No one is a criminal until he or she has been defined as such by the law, which makes it necessary to briefly discuss the process of arriving at that definition. The legal answer to the question "What is a criminal?" is that he or she is someone who has committed a crime and has been judged guilty of having done so. Before the law can properly call a person a criminal, it must go through a series of actions governed by well-defined legal rules guiding the serious business of officially labeling a person a criminal. In this section, I introduce the American criminal justice system by following the processing of felony cases from arrest to trial and beyond.

What Constitutes a Crime?

Corpus delicti is a Latin term meaning "body of the crime" and refers to the elements of an act that must be present in order to legally define it as a crime. All crimes have their own specific *elements,* which are the essential constituent parts that define the act as criminal. In addition, all crimes share a set of general elements or principles underlying and supporting the specific elements. There are five principles to be satisfied in order for a person to be "officially" labeled a criminal, but in actuality it is only necessary for the state to prove actus reus and mens rea to satisfy corpus delicti. The other principles are typically automatically proven in the course of proving actus reus and mens rea.

Actus reus means *guilty act* and refers to the principle that a person must commit some forbidden act or neglect some mandatory act before he or she can be subjected to criminal sanctions. In effect, this principle of law means that people cannot be criminally prosecuted for thinking something or being something, only for *doing* something. This prevents governments from passing laws criminalizing statuses and systems of thought they don't like. For instance, although drunken *behavior* may be a punishable crime, *being* an alcoholic cannot be punished because "being" something is a status, not an act.

Mens rea means *guilty mind* and refers to whether or not the suspect had a wrongful purpose in mind when carrying out the actus reus. For instance, although receiving stolen property is a criminal offense, if you were to buy a stolen television set from an acquaintance without knowing it had been stolen, you would have lacked mens rea, and would not be subject to prosecution. If you were to be prosecuted, the state would have to prove that you knew the television was stolen. Negligence, recklessness, and carelessness that results in some harmful consequences, even though not intended, *do not* excuse such behavior from criminal prosecution under mens rea. Conditions that may preclude prosecution under this principle are self-defense, defense of others, youthfulness (a person under 7 years of age cannot be held responsible), insanity (although being found insane does not preclude confinement), and extreme duress or coercion.

Concurrence means that the act (actus reus) and the mental state (mens rea) concur in the sense that the criminal intention actuates the criminal act. For instance, if John sets out with his tools to burglarize Mary's apartment and takes her VCR, he has fused the guilty mind with the wrongful act and has therefore committed burglary. However, assume John and Mary are friends who habitually visit each other's apartment unannounced. One day, John decides to visit Mary, finds her not at home, but walks in and suddenly decides that he could sell Mary's VCR for drug money. Although the loss to Mary is the same in both scenarios, in the latter instance John cannot be charged with burglary because he did not enter her apartment "by force or fraud," the crucial element needed to satisfy such a charge. In this case, the concurrence of guilty mind and wrongful act occurred after lawful entry, so he is only charged with theft, a less serious crime.

Causation refers to the necessity to establish a causal link between the criminal act and the harm suffered. This causal link must be proximate, not ultimate. Suppose Tony wounds Frank in a knife fight. Being macho, Frank attends to the wound himself. Three weeks later, the wound becomes severely infected and results in his death. Can Tony be charged with murder? Although the wounding led to Frank's death (the ultimate cause), Frank's disregard for the seriousness of his injury was the most proximate (or direct) cause of his death. The question the law asks in cases like this is, "What would any reasonable person do?" Most people would agree that the reasonable person would have sought medical treatment. This being the case, Tony cannot be charged with homicide; the most he could be charged with is aggravated assault.

Harm refers to the negative impact a crime has either on the victim or on the general values of the community. Although the harm caused by the criminal act is often obvious, the harm caused by many so-called victimless crimes is often less obvious. Yet some victimless crimes can cause more social harm in the long run than many crimes with obvious victims.

⊠ An Excursion Through the American Criminal Justice System

The best way to explain the process of becoming a legal criminal is to follow the processing of felony cases from arrest to trial and beyond. There are many points at which the arrested person may be shunted off the criminal justice conveyor belt via the discretionary decisions of a variety of criminal justice officials. This process will vary in some specifics from state to state, but the principles underlying the specifics are uniform. Presented here are the stages and procedures that are most common among the U.S. 50-state court systems.

Arrest. A felony suspect first enters the criminal justice system by arrest. When a person has been legally detained to answer criminal charges, he or she has been arrested. Some arrests are made on the basis of an *arrest warrant,* which is an official document signed by a judge on the basis of evidence presented by law enforcement indicating that the person named in the warrant has probably committed a crime. The warrant authorizes the police to make an arrest, although the great majority of arrests are initiated by the police without a warrant. A police officer making a warrantless arrest is held to the same legal constraints involved in making application for a warrant. To make a legal felony arrest, the officer must have probable cause. *Probable cause* means that the officer must possess a set of facts that would lead a reasonable person to

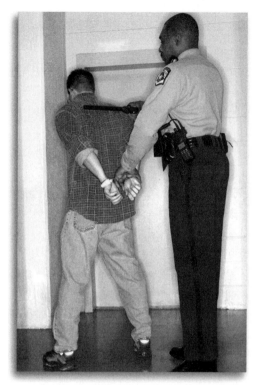

▲ **Photo 1.2**   Police officer arresting a crime suspect

conclude that the arrested person had committed a crime. Although a person can be stopped on the basis of an officer's suspicion and frisked for a weapon, he or she cannot be arrested on the basis of suspicion alone. It is only after an arrest that the Fifth Amendment right to protection against self-incrimination comes into play.

Preliminary hearing. After arrest and booking into the county jail, the suspect must be presented in court for the preliminary hearing before a magistrate or judge at the earliest opportunity. The preliminary hearing has two purposes: (1) to advise suspects of their constitutional rights and of the charges against them and (2) to set bail. The suspect may be released on monetary bail on his or her "own recognizance." If bail is denied, it is usually because of the gravity of the crime, the risk the suspect poses to the community, or the risk that the suspect might flee the court's jurisdiction. There is no constitutional right to bail. The Eighth Amendment only states that "excessive bail shall not be required." The traditional assumption has been that bail is only designed to assure the suspect's appearance at the next court hearing, and that "excessive" means that the amount set should be within the suspect's means.

Preliminary arraignment. The preliminary arraignment is a proceeding before a magistrate or judge in which three major matters must be decided: (1) whether or not a crime has actually been committed, (2) whether or not there are reasonable grounds to believe that the person before the bench committed it, and (3) whether or not the crime was committed in the jurisdiction of the court. These matters determine if the suspect's arrest and detention are legal. The onus of proving the legality of the suspect's arrest and detention is on the prosecutor, who must establish probable cause and present the court with evidence pertinent to the suspect's probable guilt. This is usually a relatively easy matter for the prosecutor since defense attorneys rarely cross-examine witnesses or introduce their own evidence at this point, their primary use of the preliminary hearing being only to discover the strength of the prosecutor's case.

The grand jury. If the prosecutor is successful, the suspect is *bound over* to a higher court for further processing. Prior to the suspect's next court appearance, prosecutors in some states must seek an indictment (a document formally charging the suspect with a specific crime or crimes) from a *grand jury*. The grand jury, so called to distinguish it from the "petit" or trial jury, is nominally an investigatory body and a buffer between the awesome power of the state and its citizens, but some see it as a historical anachronism that serves only prosecutorial purposes. The grand jury is composed of citizens chosen from voter or automobile registration lists and numbers anywhere from 7 to 23 members.

Arraignment. Armed with an indictment (or an *information* in states not requiring grand jury proceedings), the prosecutor files the case against the accused in felony court (variably called a district, superior, or common pleas court), which sets a date for arraignment. The arraignment proceeding is the first time defendants will have had the opportunity to respond to the charges against them. After the charges are read to the defendant, he or she must then enter a formal response to them, known as a *plea*. The plea alternatives are guilty, not guilty, or no contest. A guilty plea is usually the result of a plea bargain agreement concluded before the arraignment. About 90% of all felony cases in the United States are settled by plea bargains in which the state extends some benefit to defendants, such as reduced charges, in exchange for their coopera- tion. By pleading guilty, defendants give up their right to be proven guilty "beyond a reasonable doubt," their right of protection against self-incrimination, and the right to appeal.

A "not guilty" plea results in a date being set for trial; a "guilty" or "no contest" plea results in a date being set for sentencing. A "no contest" plea is one in which defendants do not admit guilt (usually to prevent that admission of guilt being used against them in a civil suit) but will not contest the matter, and the pros- ecutor "wins" by default.

The trial. A trial by a jury of one's peers is a Sixth Amendment right, and is an examination of the facts of a case by a judge (if the defendant elects to be tried by a judge alone) or by a jury for the purpose of reach- ing a judgment. The trial is an adversarial process pitting the prosecutor against the defense attorney, with each side trying to "vanquish" the other. There is no sense that each side is interested in seeking truth or justice in this totally partisan process. It is the task of the judge to ensure that both sides play by the rules. The prosecution's job is a little more difficult than the defense's since it must "prove beyond a reasonable doubt" that the accused is indeed guilty. Except in states that allow for non-unanimous jury decisions, the defense need only plant the seed of reasonable doubt in the mind of one juror to upset the prosecution's case.

Having heard the facts of the case, and having been instructed by the judge on the principles of law pertaining to it, the jury is charged with reaching a verdict. The jury's verdict may be guilty or not guilty, or if it cannot reach a verdict (a "hung" jury), the judge may declare a mistrial. A hung jury results in either dismissal of the charges by the prosecutor or a retrial. If the verdict is guilty, in most cases the judge will delay sentencing to allow time for a presentence investigation report to be prepared. It is at the point of conviction (or entering a plea of guilty) that the defendant officially becomes a criminal.

Probation. Presentence investigation reports (PSI's) are prepared by probation officers and contain a variety of information about the crime and the offender's background (criminal record, education and work history, marital status, substance abuse history, and attitude). On the basis of this information, the proba- tion officer offers a sentencing recommendation. The most important factors influencing these recommen- dations are crime seriousness and the defendant's criminal history. A judge may place the offender on probation, the most common sentence in the United States today. A probation sentence is a suspended commitment to prison, and if at any time during their probationary period offenders do not abide by the imposed probation conditions (consisting of a variety of general and offender-specific conditions), they may face revocation of probation and the imposition of the original prison sentence. Probation officers supervise and monitor offenders' behavior and assure that all conditions of probation are adhered to. Probation officers thus function as both social workers and law enforcement officers, sometimes conflicting roles that officers may find difficult to reconcile.

Incarceration. If the sentence imposed for a felony conviction is some form of incarceration, the judge has the option of sentencing the offender to a state penitentiary, a county jail, or a county work release program. The latter two options are almost invariably imposed as supplements to probation orders.

Parole. Parole is a conditional release from prison granted to inmates prior to the completion of their sentences. An inmate is granted parole by an administrative body called a parole board, which decides for or against parole based on such factors as inmate behavior while incarcerated and the urgency of the need for cell space. Once released on parole, parole officers, whose job is almost identical to that of probation officers, supervise parolees. In many states, probation and parole officers are one and the same. The primary difference between probation and parole is that probationers are under the supervision of the courts and parolees are under the supervision of the state Department of Corrections. Revocation of probation is a judicial function; revocation of parole is an executive administrative function.

A Short History of Criminology

Criminology is a young discipline, although humans have probably been theorizing about crime and its causes ever since they first made rules and observed others breaking them. What and how people thought about crime and criminals (as well as all other things) in the past was strongly influenced by the social and intellectual currents of their time. This is no less true of what and how modern criminologists think about crime and criminals. In pre-scientific days, explanations for bad behavior were often of a religious or spiritual nature such as demonic possession or the abuse of free will. From the Christian perspective, because of the legacy of Original Sin, all human beings were considered born sinners. The gift of the grace of God kept men and women on the straight and narrow, and if they deviated from this line, it was because God was no longer their guide and compass.

Others believed that the human character and personality are transparent in physical appearance. Such folk wisdom was systematized by an Italian physician named Giambattista della Porta, who developed a theory of human personality called *physiognomy* in 1558. Porta claimed that the study of physical appearance, particularly of the face, could reveal much about a person's personality and character. Thieves, for instance, were said to have large lips and sharp vision. Porta was writing during the Renaissance, a period between approximately 1450 and 1600 that saw a shift in thinking away from the pure God-centered supernaturalism of the Middle Ages and toward a more human-centered naturalism. *Renaissance* literally means "rebirth" and refers to the rediscovery of the thinking traditions of the ancient Greeks. The sciences and arts were becoming important, the printing press was invented, and Christopher Columbus "discovered" America during this period. In short, the Renaissance began to mold human thinking away from the absolute authority of received opinion and toward a way that would eventually lead to the modern scientific method.

Another major thrust toward the emergence of the modern world was the Enlightenment, a period approximately between 1650 and 1800. It might be said that the Renaissance provided a key to the human mind and the Enlightenment opened the door. Whereas the Renaissance is associated with advances in art, literature, music, and philosophy, the Enlightenment is associated with advances in mathematics, science, and the dignity and worth of the individual as exemplified by a concern for human rights. This concern led to reforms in criminal justice systems throughout Europe, a process given a major push by Cesare Beccaria's work *On Crime and Punishment,* which ushered in the Classical School. The *Classical School* emphasized human rationality and free will in its explanations for criminal behavior.

Modern criminology really began to take shape with the increasing faith that science could provide answers for everything. This period saw the harnessing of the forces of nature to build and operate the great machines that drove the Industrial Revolution, and the strides made in biology by Charles Darwin's works on the evolution of the species. Criminology saw the beginning of the *positivist school* during this period.

Theories of character, such as phrenology, abounded. The basic idea behind *phrenology* was that cognitive and personality functions are localized in the brain, and that the parts regulating the most dominant functions were bigger than parts regulating the less dominant ones. Criminals were said to have large bumps on parts of the skull thought to regulate craftiness, brutishness, moral insensibility, and so forth, and small bumps in such "localities" as intelligence, honor, and piety.

The biggest impact during this period, however, was made by Cesare Lombroso's theory of *atavism,* or the born criminal. Criminologists from this point on were obsessed with measuring, sorting, and sifting all kinds of data about criminal behavior. The main stumbling block to criminological advancement during this period was the inadequacy of its research. The intricacies of scientifically valid research design and measurement were not appreciated, and statistical techniques were truly primitive by today's standards. The early classical and positivist thinkers will be discussed at length in Chapter 3.

The so-called Progressive Era (about 1890 to 1920) ushered in new social ideologies and new ways of thinking about crime. It was an era of liberal efforts to bring about social reform as unions, women, and other disadvantaged groups struggled for recognition. Criminology largely turned away from what was disparaged as "biological determinism," which

▲ **Photo 1.3**   Charles Darwin

implied that nothing could be done to reform criminals, and toward cultural determinism. If behavior is caused by what people experience in their environments, so the optimistic argument went, then we can change their behavior by changing their environment. It was during this period that sociology became the disciplinary home of criminology. Criminology became less interested in why individuals commit crime from biological or psychological points of view and more concerned with aggregate-level data (social structures, neighborhoods, subcultures, etc.); that is, where is crime most prevalent and among what groups? It was during this period that the structural theories of crime (discussed in Chapter 4), such as the *Chicago School* of social ecology, were formulated. *Anomie strain theory* was another structural/cultural theory that emerged somewhat later (1938). This theory was doubtless influenced strongly by the American experience of the Great Depression and of the exclusion of African Americans from many areas of American society.

The period from the 1950s through the early 1970s saw considerable dissatisfaction with the strong structural approach, which many viewed as proceeding as if individuals were almost irrelevant to explaining criminal behavior. Criminological theory moved toward integrating psychology and sociology during this period and strongly emphasized the importance of socialization. Control theories were highly popular at this time, as was labeling theory; these are addressed in Chapter 5.

Because the latter part of this period was a time of great civil unrest in the United States (the anti-Vietnam War, civil rights, women's, and gay rights movements), it also saw the emergence of several theories, such as conflict theory, that were highly critical of American society. These theories extended to earlier works of Marxist criminologists, who tended to believe that the only real cause of crime was capitalism. These theories provided little that was new in terms of our understanding of "street" criminal behavior, but they did spark an interest in white-collar crime and how laws were made by the powerful and applied against the powerless. These theories are addressed in Chapter 6.

Perhaps because of a new conservative mood in the United States, theories with the classical taste for free will and rationality embedded in them reemerged in the 1980s. These were rational choice, deterrence, and routine activities theories, all of which had strong implications for criminal justice policy, and in the late 1990s/early 2000s, we witnessed a resurgence of biosocial theories. These theories view behavior as the result of biological factors interacting with the past and present environments of the actors involved. Biosocial theories have been on the periphery of criminology since its beginning but have been hampered by perceptions of being driven by an illiberal agenda and by the inability to "get inside" the mysteries of heredity and the workings of the brain. The truly spectacular advances in the observational techniques (brain scan methods, $10 cheek swabs to test DNA, etc.) in the genetic- and neurosciences over the past three decades have made these things less mysterious, and social scientists are increasingly realizing that there is nothing illiberal about recognizing the biology of human nature.

Lilly, Cullen, and Ball (2007) note that the sciences' most dramatic developments come most often from new observational techniques rather than new developments in theory. No science advances without the technology at its disposal to plumb its depths. For instance, physicists and chemists argued for centuries about the existence of atoms, and the issue was only settled when they were finally able to see them with the scanning tunneling microscope in 1981. Criminology is in a similar position today to the one chemists were in 150 years ago. The concepts, methods, and measuring devices available to us today may do for the progress of criminology what physics did for chemistry, what chemistry did for biology, and what biology is doing for psychology. Exceptionally ambitious longitudinal studies carried out over decades in concert with medical and biological scientists, such as the Dunedin Multidisciplinary Health and Development Study (Moffitt, 1993), the National Longitudinal Study of Adolescent Health (Udry, 2003), and the National Youth Survey (Menard, Mihalic, & Huizinga, 2001), are able to gather a wealth of genetic, neurological, and physiological data as well as psychological and sociological data. Integrating these hard data into criminology will no more rob it of its autonomy than physics robbed chemistry or chemistry robbed biology. For those who agree with this assessment, this is an exciting time to study criminology!

The Role of Theory in Criminology

When an FBI agent asked the Depression-era bank robber Willie Sutton why he robbed banks, Sutton replied, "Because that's where the money is." In his witty way, Sutton was offering a theory explaining bank robbery encompassing the kind of person who has learned how to take advantage of opportunities provided by convenient targets (banks) flush with a valued commodity (cash). Thus, if we put a certain kind of personality and learning together with opportunity and coveted resources, we get bank robbery. This is what theory making is all about: trying to grasp how all the known correlates of a phenomenon are linked together in non-coincidental ways to produce an effect.

Just as medical scientists want to find out what causes disease, criminologists are interested in finding out factors that cause criminal behavior. As is the case with disease, there are a variety of risk factors that may lead to criminal behavior. The first step in detecting causes is to discover **correlates,** which are factors that are related to the phenomenon of interest. To discover if two factors are correlated, we must see if they vary together; i.e., if one variable goes up or down, the other goes up or down as well, either in the same direction (a positive correlation) or in the opposite direction (a negative correlation).

Establishing causality requires more than simply establishing a correlation. Take gender, the most thoroughly documented correlate of criminal behavior ever identified. Literally thousands of studies throughout the world, some European studies going back five or six centuries, consistently reported strong gender differences in criminal behavior, and the more serious the crime the greater the difference (Ellis & Walsh, 2000). However, establishing *why* gender is such a strong correlate of crime is the real challenge, as it is with any other correlate. Trying to establish causes is the business of theory.

What Is Theory?

A **theory** is a set of logically interconnected propositions explaining how phenomena are related, and from which a number of hypotheses can be derived and tested. Theories should provide logical explanations of the phenomena they address, they should correspond with the relevant empirical facts, and they should provide practical guidance for researchers looking for further facts. This guidance takes the form of a series of statements that can be logically deduced from the assertions of the theory called **hypotheses,** which are statements about relationships between and among factors we expect to find based on the logic of our theories. Theories provide the raw material (the ideas) for generating hypotheses, and hypotheses support or fail to support theories by exposing them to empirical (based on experiment and observation) testing.

▲ **Photo 1.4**   Willie Sutton

Theories are devised to explain how a number of different correlates may actually be causally related to criminal behavior rather than simply associated with it. When we talk of causes, we do not mean that when *X* is present *Y will* occur in a completely prescribed way. We mean that when *X* is present, *Y* has a certain *probability* of occurring, and perhaps only if *X* is present along with factors *A, B,* and *C.* Criminologists have never uncovered a necessary cause (a factor that *must* be present for criminal behavior to occur, and in the absence of which criminal behavior has never occurred) or a sufficient cause (a factor that is able to produce criminal behavior without being augmented by some other factor).

Theories help us to make sense of a diversity of seemingly unrelated facts, and even tell us where to look for more facts. We all use theory every day to fit facts together. A detective confronted with a number of facts about a mysterious murder must fit them together, even though their meaning and relatedness to one another are ambiguous and perhaps even contradictory. Using years of experience, training, and good common sense, the detective constructs a theory linking those facts together so that they begin to make some sense, to begin to tell their story. An initial theory derived from the available facts then guides the detective in the search for additional facts in a series of "*if* this is true, *then* this should be true" statements. There may be many false starts as our detective misinterprets some facts, fails to uncover others, and considers some to be relevant when they are not. Good detectives, like good scientists, will adjust their theory as new facts warrant; poor detectives and poor scientists will stand by their favored theory by not looking for more facts, or by ignoring, downplaying, or hiding contrary facts that come to their attention.

How to Think About Theories

You will be a lot less concerned about the numerous theories in criminology if you realize that different theories deal with different levels of analysis. A **level of analysis** is that segment of the phenomenon of interest that is measured and analyzed. We can ask about causes of crime at the levels of whole societies, subcultures, neighborhoods, families, or individuals. If the question asks about crime rates in societies, the answers must address sociocultural differences among different societies or in the same society at different times. Conversely, if crime rates are found to be related to the degree of industrialization or racial/ethnic diversity in societies, this tells us nothing about why some people in an industrialized, heterogeneous society commit crimes and others in the same society do not. To answer questions about individuals, we need theories about individuals. Generally speaking, questions of cause and effect must be answered at the same level of analysis at which they were posed; thus, different theories are required at different levels.

Interpreting research findings is not as simple as documenting correlates of crime. There is little room for error when contrasting rates of crime between and among the various demographic variables such as age, gender, and race/ethnicity. However, theory testing looks for causal explanations rather than simple descriptions, and that's where our problems begin. For example, when we consistently find positive correlations between criminal behavior and some other factor, it is tempting to assume that something causal is going on, but as we have said previously, correlations merely *suggest* causes; they do not demonstrate them. Resisting the tendency to jump to causal conclusions from correlations is the first lesson of statistics.

▨ Ideology in Criminological Theory

In addition to criminological theorizing being linked to the social and intellectual climate of the times, it is also strongly linked to ideology. **Ideology** is a way of looking at the world, a general emotional picture of "how things should be." It is often so strongly held that it narrows the mind and inflames the passions, leading to a selective interpretation and understanding of evidence rather than an objective and rational evaluation of it. Ideology forms, shapes, and colors our concepts of crime and its causes in ways that lead to a tendency to accept or reject new evidence according to how well or poorly it fits our ideology.

According to Thomas Sowell (1987), two contrasting visions have shaped thoughts about human nature throughout history, and these visions are in constant conflict with each other. The first of these is the **constrained vision,** so called because believers in this vision view human activities as constrained by an innate human nature that is self-centered and largely unalterable. The **unconstrained vision** denies an innate human nature, viewing it as formed anew in each different culture. Believers in the unconstrained vision see human nature is perfectible, a view scoffed at by those who profess the constrained vision. A major difference between the two visions is that the constrained vision says, "This is how the world *is*," and the unconstrained vision says, "This is how the world *should be.*" For instance, unconstrained visionaries might ask what causes crime or poverty, but constrained visionaries would ask the opposite question: What causes a well-ordered society and wealth? Note that this implies that unconstrained visionaries believe that crime and poverty are deviations from the norm that must be explained, while constrained visionaries see them as historically normal and inevitable, and believe that what has to be understood are the conditions that prevent them. We will see the tension between visions constantly as we discuss the various theories in this book.

A criminological theory is at least partly shaped by ideology, and those who feel drawn to a particular theory owe a great deal of their attraction to it to the fact that they share the theory's vision (Cullen, 2005). This observation reminds us of the Indian parable of the six blind men feeling different parts of an elephant. Each man described the elephant according to the part of its anatomy he had felt, but each failed to appreciate the descriptions of the others who had felt different parts. The men fell into dispute and departed in anger, each convinced of the utter stupidity of the others. The point is that ideology often leads criminologists to "feel" only part of the criminological elephant and then to confuse the parts with the whole, and even to question the intelligence and motives (e.g., having some kind of political agenda) of others who have examined different parts of the criminological elephant.

The evidence that ideology is linked to what theories criminologists favor is strong. Cooper, Walsh, and Ellis (2010) asked a number of criminologists which theory best explained criminal behavior. As you see from Table 1.1, a total of 24 different theories are represented. Obviously, they cannot all "best explain criminal behavior," so something other than evidence led the respondents to their choices, and the best predictor was criminologists' self-reported ideology—conservative, moderate, liberal, or radical. The "$X^2 = 134.6, p < 0.001$" notation means that such a result could be found by chance less than 1 time in 1,000 similar samplings, so we can be quite confident in the finding. When reading this text, try to understand where the originators, supporters, and detractors of any particular theory being discussed are "coming from" ideologically as well as theoretically.

Table 1.1 Favored Theory Cross-Tabulated by Self-Reported Political Ideology

Theory favored*	Political ideology				
	Conservative	Moderate	Liberal	Radical	Total
Social learning (2,6)	1	22	22	5	50
Life course/developmental (n/a,11)	3	8	28	3	42
Social control (1,1)	0	14	27	1	42
Social disorganization (7,14)	0	11	26	3	40
Self-control (n/a,2)	3	6	15	0	24
Biosocial (4,12)	5	5	11	0	21
Rational choice	2	7	11	1	21
Conflict (n/a,4)	0	2	8	6	16
Critical (10,18)	0	0	8	8	16
Differential association (4,3)	1	4	10	1	16
Age-graded developmental	1	5	7	0	13
Strain (n/a,8)	0	3	9	0	12

(Continued)

Table 1.1 (Continued)	Political ideology				
Theory favored*	**Conservative**	**Moderate**	**Liberal**	**Radical**	**Total**
Dual-pathway developmental (n/a,5)	1	0	10	0	11
Routine activities (n/a,9)	1	2	8	0	11
General strain	0	2	4	1	7
Institutional anomie	0	1	5	0	6
Interactional	0	1	5	0	6
Opportunity (5,15)	1	2	2	0	5
Ecological (n/a,23)	1	1	2	0	4
Labeling (6,17)	0	1	2	1	4
Psychological	0	1	3	0	4
Classical (n/a,20)	0	3	0	0	3
Feminist (n/a,10)	0	0	2	1	3
Anomie (9,6)	0	1	1	0	2
TOTAL	20	102	226	31	379

SOURCE: Cooper, Walsh, and Ellis (2010).

$X^2 = 134.6, p < 0.001$.

*Numbers in parentheses represent ranking of theories in the Ellis and Hoffman (1990) and Walsh and Ellis (2004) surveys. Theories without ranking or designated n/a (not applicable) were not represented in those surveys.

Connecting Criminological Theory and Social Policy

Theories of crime imply that changing the conditions the theory holds responsible for causing crime can reduce it and even prevent it. I say "imply" because few theorists are explicit about the public policy implications of their work. Scientists are primarily concerned with gaining knowledge for its own sake; they are only secondarily concerned with how useful that knowledge may be to practitioners and policy makers. Conversely, policy makers are less concerned with hypothesized "causes" of a problem and more concerned with what can be done about the problem that is both politically and financially feasible.

Policy is simply a course of action designed to solve some problem that has been selected from among alternative courses of action. Solving a social problem means attempting to reduce the level of the problem currently being experienced or to enact strategies that try to prevent it from occurring in the first place. Social science findings can and have been used to help policy makers determine which course of action to follow to "do something" about the crime problem, but there are many other concerns that policy makers must consider that go beyond maintaining consistency with social science theory and data. The question of "what to do about crime" involves political and financial considerations, the urgency of other problems competing for scarce financial resources (schools, highways, environmental protection, public housing, national defense), and a host of other major and minor considerations.

Policy choices are, at bottom, value choices, and as such, only those policy recommendations that are ideologically palatable are likely to be implemented. Given all of these extra theoretical considerations, it would be unfair to base our judgment of a theory's power solely by its impact on public policy. Even if some aspects of policy are theory-based, unless all recommendations of the theory are fully implemented, the success or failure of the policy cannot be considered evidence of theoretical failure any more than a recipe can be blamed for a lousy cake if the baker neglected to include all the ingredients it calls for.

Connecting problems with solutions is a tricky business in all areas of government policy making, but nowhere is it more difficult than in the area of criminal justice. No single strategy can be expected to produce significant results, and it may sometimes make matters worse. For example, President Johnson's "War on Poverty" was supposed to have a significant impact on the crime problem by attacking what informed opinion of the time considered its "root cause." Programs and policies developed to reduce poverty did so, but reducing poverty had no effect on reducing crime; in fact, crime rose to record levels as poverty was falling (Walsh & Ellis, 2007). Another high-profile example of failed policy is the Volstead Act of 1919 that prohibited the manufacture and sale of alcohol in the United States. Although based on a true premise (that alcohol is a major factor in facilitating violent crime), it failed because it ushered in a wild period of crime as gangs fought over control of the illegal alcohol market. Policies often have effects that are unanticipated by policy makers, and these effects can be positive or negative.

Nevertheless, every theory has policy implications deducible from its primary assumptions and propositions. The deep and lasting effects of the classical theories on legal systems around the world have long been noted, but the broad generalities about human nature contained in those theories offer little specific advice on ways to change criminals or to reduce their numbers. Although I would caution against using the performance of a theory's public policy recommendations as a major criterion to evaluate its power, the fact remains that a good theory *should* offer useful practical recommendations, and we will discuss a theory's policy implications where appropriate.

SUMMARY

- Criminology is the scientific study of crime and criminals. It is an interdisciplinary/multidisciplinary study, although criminology has yet to integrate these disciplines in any comprehensive way.

- The definition of crime is problematic because acts that are defined as criminal vary across time and cultures. Many criminologists believe that because crimes are defined into existence, we cannot determine what real crimes are and what criminals are. However, there is a stationary core of crimes that are universally condemned and always have been. These are predatory crimes that cause serious harm and are defined as *mala in se,* or "inherently bad" crimes, as opposed to *mala prohibita*—"bad because they are forbidden" crimes.

- A person is not "officially" a criminal until such time as he or she has been found guilty beyond a reasonable doubt of having committed a crime. In order to prove that he or she did, the state has to prove *corpus delicti* ("the body of the crime"), which essentially means that the person committed a criminal act (*actus reus*) with full awareness that the act was wrong (*mens rea*—guilty mind). Other basic principles—concurrence, harm, and causation—are proven in the process of proving *corpus delicti.*

- The history of criminology shows that the cultural and intellectual climate of the time strongly influences how scholars think about and study crime and criminality. The Renaissance brought more secular thinking, the Enlightenment inspired more humane and rational thinking, the Industrial Revolution brought with it more scientific thinking, and the Progressive Era saw a reform-oriented criminology reminiscent of the Classical School.

- Advances in any science are also constrained by the tools available to test theories. The ever-improving concepts, methods, and techniques available from modern genetics, neuroscience, and other biological sciences should add immeasurably to criminology's knowledge base in the near future.

- Theory is the "bread and butter" of any science, including criminology. There are many contending theories seeking to explain crime and criminality. Although we do not observe such theoretical disagreement in the more established sciences, the social/behavioral sciences are young, and human behavior is extremely difficult to study.

- When judging among the various theories, we have to keep certain things in mind, including the theory's predictive accuracy, scope, simplicity, and falsifiability. We must also remember that crime and criminality can be discussed at many levels (societywide, subcultural, family, or individual), and that one theory that may do a good job of predicting crime at one level may do a poor job at another level.

- Theories can also be offered at different temporal levels. They may focus on the evolutionary history of the species (the most ultimate level), the individual's subjective appraisal of a situation (the most proximate level), or any other temporal level in between. A full account of an individual's behavior may have to take all these levels into consideration, because any behavior arises from an individual's propensities interacting with the current environmental situation as that individual perceives it. This is why we approach the study of crime and criminality from social, psychosocial, and biosocial perspectives.

- Criminologists have not traditionally done this, preferring instead to examine only aspects of criminal behavior that they find congenial to their ideology and, unfortunately, often maligning those who focus on other aspects. The main dividing line in criminology has separated conservatives (who tend to favor explanations of behavior that focus on the individual) and liberals (who tend to favor structural or cultural explanations). The theories favored by criminologists are strongly correlated with sociopolitical ideology.

- All theories have explicit or implicit recommendation for policy since they posit causes of crime or criminality. Removing those alleged causes should reduce crime if the theory is correct, but the complex nature of crime and criminality make basing policy decisions on them very risky indeed. Policy makers must consider many other issues demanding scarce resources, so the policy content of a theory should never be used to pass judgment on the usefulness of theory for criminologists.

DISCUSSION QUESTIONS

1. Which of the following 10 acts do you consider mala in se crimes, mala prohibita crimes, or no crime at all? Defend your choices.

   (a) drug possession, (b) vandalism, (c) drunk driving, (d) collaborating with the enemy, (e) sale of alcohol to minors, (f) fraud, (g) spouse abuse, (h) adult male having consensual sex with underage person, (i) prostitution, (j) homosexual behavior, (k) pornography.

2. Why is it important to consider ideology when evaluating criminologists' work? Is it possible for them to divorce their ideology from their work?

3. Policies aimed at reducing crime (think of Prohibition and the War on Poverty) rarely have the desired effect. Can you think of any good reasons why this should be so?

4. Locate the online journal *Quarterly Journal of Ideology*. Click on *archive* and find and read "Ideology: Criminology's Achilles' Heel." What does this article say about the "conflict of visions" in criminology?

USEFUL WEBSITES

Anderson, K. "Social Constructionism and Belief Causation." www.stanford.edu/group/dualist/vol8/pdfs/anderson.pdf

Classical School vs. Conflict Criminology. See the useful YouTube discussion at www.youtube.com/watch?v=FVbjzND9kDg

Conflict Criminology. http://faculty.ncwc.edu/toconnor/301/301lect13.htm

Critical Criminology. www.critcrim.org

Learning Theories of Crime. http://en.wikipedia.org/wiki/Social_learning_theory

Links to Criminological Theory. www.crimetheory.com/

CHAPTER TERMS

Actus reus	Criminology	Parole
Causation	Hypotheses	Preliminary arraignment
Concurrence	Ideology	Preliminary hearing
Constrained vision	Incarceration	Probation
Corpus delicti	Level of analysis	Theory
Correlates	Mala in se	Unconstrained vision
Crime	Mala prohibita	
Criminality	Mens rea	

CHAPTER

2

Measuring Crime
and Criminal Behavior

A weary English bobby (a popular nickname for British police officers) patrolling his foot beat on a chilly November night hears the unmistakable sounds of sexual activity from the dark entranceway of a closed greengrocer's shop. He smiles to himself and tiptoes toward the sound. When he reaches the entranceway, he switches on his flashlight and booms out the favorite line of the stereotypical bobby: "What's goin' on 'ere then?" The squeaking couple immediately come to attention and adjust their dress before the young man—obviously still in a state of arousal—stammers, "Why, nothing, constable." The officer recognizes the woman as a local "slapper" (prostitute) and he vaguely recognizes the man (more of a boy of around 17, really) as a local supermarket worker. The constable reasons that he should arrest both parties for public indecency, but that would entail about an hour of paperwork (an hour in the warm police station with a nice cup of tea sounds good, though) and lead to the profound embarrassment of the poor boy. He finally decides to give the boy some sound advice about sexually transmitted diseases and a stern warning to the woman and sends them both on their way.

This short story illustrates that official statistics are measuring police behavior as much as they are measuring crime. Sir Josiah Stamp, director of the Bank of England in the 1920s, cynically stated this criticism: "The government are very keen on amassing statistics. They collect them, raise them to the nth power, take the cube root and prepare wonderful diagrams. But you must never forget that every one of these figures comes in the first instance from the village watchmen, who just puts down what he damn pleases" (quoted in Nettler, 1984, p. 39). I don't recommend this kind of cynicism, but I do counsel that you keep a healthy skepticism about statistics as you read this chapter.

▧ Categorizing and Measuring Crime and Criminal Behavior

When attempting to understand, predict, and control any social problem, including the crime problem, the first step is to determine its extent. Gauging the extent of the problem means discovering how much of it there is, where and when it occurs most often, and among what social categories it occurs most frequently. It also helps our endeavors if we have knowledge of the patterns and trends of the problem over time. Note that we did not address "why" questions (why does crime occur, why is it increasing/decreasing, who commits it and why, and so on); such questions can only be adequately addressed after we have reliable data about the extent of the problem. However, all social statistics are suspect to some extent, and crime statistics are perhaps the most suspect of all. They have been collected from many different sources in many different ways and have passed through many sieves of judgment before being recorded.

There is a wide variety of data provided by government and private sources to help us to come to grips with America's crime problem, all with their particular strengths and weaknesses. The major data sources that we have can be grouped into three categories: official statistics, victimization survey data, and self-reported data. *Official statistics* are those derived from the routine functioning of the criminal justice system. The most basic category of official statistics comes from the calls made to police by victims or witnesses and from crimes that the police discover on patrol. Other major categories of official crime data consist of information about arrests, about convictions, and about correctional (prison, probation/parole) populations.

▧ The Uniform Crime Reports: Counting Crime Officially

▲ **Photo 2.1**   The J. Edgar Hoover building, headquarters of the FBI, in Washington, D.C. Annual Uniform Crime Reports are compiled by the FBI after local, county, and state criminal justice agencies send in their annual crime data.

The primary source of official crime statistics in the United States is the annual **Uniform Crime Reports (UCR)** compiled by the Federal Bureau of Investigation (FBI). The UCR reports crimes known to the nation's police and sheriff's departments and the number of arrests made by these agencies; federal crimes are not included. Offenses known to the police are recorded whether or not an arrest is made or if an arrested person is subsequently prosecuted and convicted. Participation in the UCR reporting program is voluntary, and thus all agencies do not participate. This is unfortunate for anyone hoping for comprehensive crime data. In 2009, law enforcement agencies active in the UCR program represented more than 295 million U. S. inhabitants—96.3% of the total population (FBI, 2010). This means that crimes committed in the jurisdictions of agencies representing about 4% of the population (about 12.5 million people) were not included in the UCR data.

The UCR reports the number of each type of crime reported to the police as well as the rate of occurrence. The rate of a given crime is the actual number of reported crimes standardized by some unit of the population. We expect the raw number of crimes to increase as the population increases, so comparing the number of crimes reported today with the number reported 30 years ago, or the number of crimes reported in New York with the number reported in Wyoming, tells us little without considering population differences. For instance, California reported 1,972 murders to the FBI in 2009, and Louisiana reported 530. In which state are you most likely to be murdered? We can't say unless we take the two states' respective populations into consideration. To obtain a crime rate, we divide the number of reported crimes in a state by its population, and multiply the quotient by 100,000, as in the following comparison of California and Louisiana rates.

$$\text{Rate} = \frac{\text{CA Murders} = 1,972}{\text{CA Population} = 36,961,664} = .000053 \times 100,000 = 5.3$$

$$\text{Rate} = \frac{\text{LA Murders} = 530}{\text{LA Population} = 4,492,076} = .000138 \times 100,000 = 11.8$$

Thus, a person in Louisiana is at over twice the risk (11.8 versus 5.3) of being murdered compared to a person in California. This statement is based on the statewide rate; the actual risk will vary widely from person to person based on such factors as age, race, sex, socioeconomic status (SES), neighborhood, and urban versus rural residence.

The UCR separates crimes into two categories: **Part I offenses** (or **Index Crimes**), and **Part II offenses.** Part I offenses includes four violent (homicide, assault, forcible rape, and robbery) and four property offenses (larceny/theft, burglary, motor vehicle theft, and arson). Notice that these are all universally condemned mala in se offenses. Part I offenses correspond with what most people think of as "serious" crime. Part II offenses are treated as less serious offenses and are recorded based on arrests made rather than cases reported to the police. Part II offense figures understate the extent of criminal offending far more than is the case with Part I figures because only a very small proportion of these crimes result in arrest.

Table 2.1 is a page from the 2009 UCR listing all Part I and II crimes broken down by sex and percentage change in rates from 1999 to 2008. This provides us with an estimate of the number of times a given crime was reported to the police in 2009, and the male/female differences in arrests (as well as the increases/

Table 2.1　Ten-Year Arrest Trends for Part I and Part II Crimes by Sex						
	Male			**Female**		
Offense	**1999**	**2008**	**Change**	**1999**	**2008**	**Change**
Murder and nonnegligent manslaughter	6,636	6,292	−5.2	831	780	−6.1
Forcible rape	15,452	12,474	−19.3	179	148	−17.3
Robbery	54,658	64,844	+18.6	6,261	8,615	+37.6
Aggravated assault	228,525	202,645	−11.3	55,814	54,400	−2.5

(Continued)

Table 2.1 (Continued)

Offense	Male 1999	Male 2008	Change	Female 1999	Female 2008	Change
Burglary	149,875	157,341	+5.0	23,106	29,055	+25.7
Larceny-theft	461,632	431,212	−6.6	254,629	308,011	+21.0
Motor vehicle theft	60,540	43,801	−27.6	11,331	9,344	−17.5
Arson	8,317	7,116	−14.4	1,373	1,291	−6.0
Violent crime	305,271	286,255	−6.2	63,085	63,943	+1.4
Property crime	680,364	639,470	−6.0	290,439	347,701	+19.7
Other assaults	564,655	560,226	−0.8	169,665	196,577	+15.9
Forgery and counterfeiting	38,570	31,947	−17.2	24,253	19,631	−19.1
Fraud	114,020	80,973	−29.0	96,234	65,637	−31.8
Embezzlement	5,768	6,575	+14.0	5,634	7,039	+24.9
Stolen property; buying, receiving, possessing	56,110	53,172	−5.2	10,350	14,116	+36.4
Vandalism	135,146	137,165	+1.5	24,317	28,213	+16.0
Weapons; carrying, possessing, etc.	87,790	93,112	+6.1	7,492	7,480	−0.2
Prostitution and commercialized vice	19,762	12,133	−38.6	25,240	25,164	−0.3
Sex offenses (except forcible rape and prostitution)	48,800	40,876	−16.2	3,670	3,769	+2.7
Drug abuse violations	711,384	784,561	+10.3	155,256	185,201	+19.3
Gambling	4,481	2,227	−50.3	800	350	−56.3
Offenses against the family and children	65,172	51,268	−21.3	18,481	17,303	−6.4
Driving under the influence	714,457	667,017	−6.6	134,279	181,391	+35.1
Liquor laws	298,874	241,328	−19.3	84,444	89,999	+6.6
Drunkenness	370,924	347,399	−6.3	55,670	66,883	+20.1
Disorderly conduct	262,713	243,865	−7.2	82,332	89,530	+8.7
Vagrancy	13,302	12,037	−9.5	3,130	3,550	+13.4
All other offenses (except traffic)	1,687,374	1,724,690	+2.2	444,312	517,358	+16.4
Suspicion	3,563	790	−77.8	945	197	−79.2
Curfew and loitering law violations	56,843	40,275	−29.1	25,370	19,935	−21.4
Runaways	37,359	26,923	−27.9	54,048	34,363	−36.4

SOURCE: FBI (2010a).

decreases) provide interesting information regarding why these gender differences exist. Figure 2.1, the FBI's famous crime clock, further helps to put crime figures into perspective by indicating how often on an average day one of these crimes is committed. Remember that these are only rough estimates.

Cleared Offenses

If a person is arrested and charged for a Part I offense, the UCR records the crime as *cleared by arrest*. A crime may also be *cleared by exceptional means* when the police have identified a suspect and have enough evidence to support arrest, but the suspect could not be taken into custody immediately or at all. Such circumstances exist when the suspect dies or is in a location where the police cannot presently gain custody. For instance, he or she is in custody on other charges in another jurisdiction or is residing in a country with no extradition treaty with the United States.

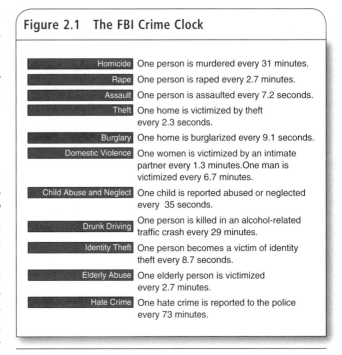

Figure 2.1   The FBI Crime Clock

Homicide	One person is murdered every 31 minutes.
Rape	One person is raped every 2.7 minutes.
Assault	One person is assaulted every 7.2 seconds.
Theft	One home is victimized by theft every 2.3 seconds.
Burglary	One home is burglarized every 9.1 seconds.
Domestic Violence	One women is victimized by an intimate partner every 1.3 minutes. One man is victimized every 6.7 minutes.
Child Abuse and Neglect	One child is reported abused or neglected every 35 seconds.
Drunk Driving	One person is killed in an alcohol-related traffic crash every 29 minutes.
Identity Theft	One person becomes a victim of identity theft every 8.7 seconds.
Elderly Abuse	One elderly person is victimized every 2.7 minutes.
Hate Crime	One hate crime is reported to the police every 73 minutes.

Source: FBI (2010a).

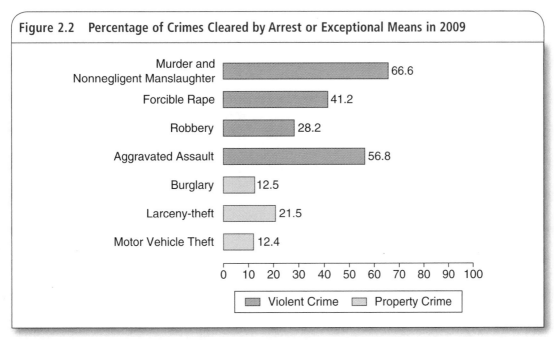

Figure 2.2   Percentage of Crimes Cleared by Arrest or Exceptional Means in 2009

- Murder and Nonnegligent Manslaughter: 66.6
- Forcible Rape: 41.2
- Robbery: 28.2
- Aggravated Assault: 56.8
- Burglary: 12.5
- Larceny-theft: 21.5
- Motor Vehicle Theft: 12.4

Violent Crime   Property Crime

Source: FBI (2010a).

As can be seen in Figure 2.2, which gives 2009 clearance rates, violent crimes are more likely to be cleared than property crimes because violent crime investigations are pursued more vigorously and because victims of such crimes may be able to identify the perpetrator(s).

Crime Trends

One thing about the UCR is that it is very useful for tracking crime trends. Table 2.2 shows trends from 1990 to 2009 (FBI, 2010a). Note that total crime dropped just over 40% (5,802.7−3,465.5 = 2337.2/5802.7 = .40278, or ≈ 40.3%) between those years. It is much easier to note that crime increased or decreased by some percentage over a specified time period than it is to explain why it did so, however. Despite the accumulation of tons of factual data, it is difficult to arrive at a sturdy conclusion that fits them together to everyone's satisfaction. Facts only describe events; they do not explain them. Any explanation for major fluctuations in crime rates requires that we have an understanding of the historical, social, political, economic, and demographic processes unfolding around the same time that increases or decreases in crime are recorded and how those processes interact. The effects of any particular process on crime may be immediate, such as a series of riots and general mayhem following some perceived injustice, or it may only be felt a decade or so down the road, such as an economic policy decision that later affects job creation. Whatever process or alleged cause we examine, you should keep in the forefront of your mind that just as there is no single cause of crime or criminality, there is no single cause that explains crime trends.

Examine the total UCR violent and property crime rates per 100,000 for 1963, 1993, and 2003, and ask yourself whether crime has gone up or down.

Year	Violent	Property
1963	168.2	2,021.1
1993	747.1	4,740.0
2003	475.8	3,591.2

If we compare 1993 with 2003, we conclude that crime dropped significantly, but if we take 1963 as our beginning year and compare it with 2003, we would conclude that crime has gone up significantly. Whether crime has "gone up" or "gone down" thus depends on where we choose to look. Interpretations of crime trends should be read with caution because the author may have chosen a beginning and ending year to support his or her favored explanation. So before we begin to congratulate or berate ourselves because the crime rate has gone up or down, it is wise to ask, "Compared to what year?"

Take also the murder rate trends from 1900 to 2006 presented in Figure 2.3. The graph looks like a rugged mountain range with peaks and troughs, indicating that at some points in American history murder rates were more than twice as high as they were at other points. The 1900 rate of 1.0 per 100,000 is highly suspect given the descriptions of life in such cities as New York and Boston at the turn of the century, as well as the still semi-civilized condition of much of the western United States. We should never take national statistics at face value unless we are very sure of their quality, and national reporting of crime statistics was in a terrible state in the early part of the 20th century.

With the advent of the UCR in 1930, national data became somewhat more reliable. The homicide rate started a steep climb after the Volstead Act prohibiting the production and sale of alcohol was passed in 1920 as gangs fought over the lucrative illegal alcohol market. The rate started to fall with the repeal of

Table 2.2 Crime Rates, 1990–2009

Year	Crime	Violent Crime	Property Crime	Murder	Rape	Robbery	Assault	Burglary	Larceny-Theft	Motor Vehicle Theft
1990	5,802.7	729.6	5,073.1	9.4	41.1	256.3	422.9	1,232.2	3,185.1	655.8
1991	5,898.4	758.2	5,140.2	9.8	42.3	272.7	433.4	1,252.1	3,229.1	659.0
1992	5,661.4	757.7	4,903.7	9.3	42.8	263.7	441.9	1,168.4	3,103.6	631.6
1993	5,487.1	747.1	4,740.0	9.5	41.1	256.0	440.5	1,099.7	3,033.9	606.3
1994	5,373.8	713.6	4,660.2	9.0	39.3	237.8	427.6	1,042.1	3,026.9	591.3
1995	5,275.0	684.5	4,590.5	8.2	37.1	220.9	418.3	987.0	3,043.2	560.3
1996	5,087.6	636.6	4,451.0	7.4	36.3	201.9	391.0	945.0	2,980.3	525.7
1997	4,927.3	611.0	4,316.3	6.8	35.9	186.2	382.1	918.8	2,891.8	505.7
1998	4,620.1	567.6	4,052.5	6.3	34.5	165.5	361.4	863.2	2,729.5	459.9
1999	4,266.6	523.0	3,743.6	5.7	32.8	150.1	334.3	770.4	2,550.7	422.5
2000	4,124.8	506.5	3,618.3	5.5	32.0	145.0	324.0	728.8	2,477.3	412.2
2001	4,162.6	504.5	3,658.1	5.6	31.8	148.5	318.6	741.8	2,485.7	430.5
2002	4,125.0	494.4	3,630.6	5.6	33.1	146.1	309.5	747.0	2,450.7	432.9
2003	4,067.0	475.8	3,591.2	5.7	32.3	142.5	295.4	741.0	2,416.5	433.7
2004	3,977.3	463.2	3,514.1	5.5	32.4	136.7	288.6	730.3	2,362.3	421.5
2005	3,900.5	469.0	3,431.5	5.6	31.8	140.8	290.8	726.9	2,287.8	416.8
2006	3,838.3	480.6	3,357.7	5.8	31.7	150.6	292.6	735.2	2,221.4	401.1
2007	3,748.8	472.0	3,276.8	5.7	30.5	148.4	287.4	726.0	2,186.3	364.6
2008	3,669.0	457.5	3,211.5	5.4	29.7	145.7	276.7	732.1	2,164.5	315.0
2009	3,465.5	429.4	3,036.1	5.0	28.7	133.0	262.8	716.3	2,060.9	258.8

SOURCE: FBI (2010a).

the Volstead Act in 1933, which effectively removed criminals from the alcohol business. It dropped even further during World War II when most young men were in uniform and overseas, showed a sharp rise when they returned, and then settled into a relatively peaceful period during the 1950s–early 1960s. Murder rates then started a precipitous rise beginning in the late 1960s. The late 1960s–early 1970s was a period of huge changes in American society. Opposition to the Vietnam War combined with the civil rights and feminist movements led to the widespread questioning of many of the fundamental values of American society. When values and norms are questioned, they become weaker in their ability to regulate behavior. Behavioral deregulation led to all kinds of experimentation with alternative lifestyles, including the use of drugs. The emergence of crack cocaine in the early 1980s led to a period of gang wars over territory, just like the gang wars over alcohol did in the 1920s. Crack cocaine is easier to make, conceal, and sell than barrels of beer or bottles of whiskey, so crack dealing is more of an "equal opportunity" enterprise than supplying illegal

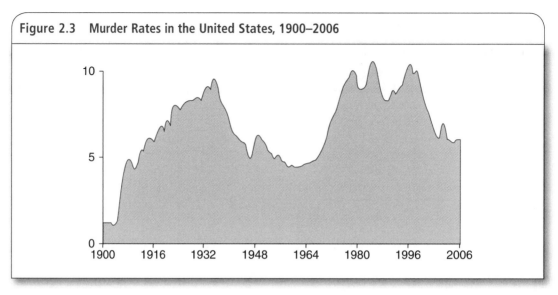

Figure 2.3 Murder Rates in the United States, 1900–2006

Source: Bureau of Justice Statistics, 2010. *Key facts at a glance.* Available at http://bjs.ojp.usdoj.gov/content/glance/hmrt.cfm

alcohol was. Numerous young "gangbangers" took advantage of the opportunity for easy money, sparking a decade-long street war with other like-minded individuals.

The decrease in the homicide rate in the early 1990s can be attributed to several factors including a large decrease in the crack market and in gang warfare as territories became consolidated by the strong pushing out the weak. Severe penalties for sale and possession of crack, and the danger from others trafficking in the same market, may have also driven out many dealers.

One of the biggest factors in the drop in homicide rates is medical and technological improvements. Cell phones for reporting incidents are everywhere, and emergency medical technicians are alerted and dispatched swiftly. Once hospitalized, victims have the benefit of all that medical experts have learned about treating violent traumas since the Vietnam War. It is estimated that we would be experiencing 5 times the murder rate today if medicine and technology were at the same level as in 1960 (Harris, Thomas, Fisher, & Hirsch, 2002), which is something to remember when comparing U.S. rates to those of less developed countries lacking America's medical and technological advantages. Thus, many other factors known and unknown have contributed to the fluctuations in the homicide rate we have observed over the course of the 20th century, which makes it always difficult to explain crime trends to everyone's satisfaction.

Problems With the UCR

UCR data have serious limitations that restrict their usefulness for criminological research, particularly research seeking to uncover causes of crime. Some of the more serious of these limitations are outlined below.

- The UCR data significantly underrepresents the actual number of criminal events in the United States each year. According to a nationwide victim survey, only 47% of victims of violent crime and 40% of victims of property crime indicated that they reported their victimization to the police

(Catalano, 2006). Victims are more likely to report violent crimes if injuries are serious, and are more likely to report property crimes when losses are high. Females (54.6%) are more likely than males (47%) to report violent victimization; males and females are about equally as likely (39%) to report property victimization.

- Federal crimes such as highly costly white-collar crimes such as stock market fraud, hazardous waste dumping, tax evasion, and false claims for professional services are not included in the UCR.
- Crimes committed in the jurisdictions of nonparticipating law enforcement agencies are not included in the data. Even with full voluntary compliance, all departments would not be equally efficient and thorough (or honest) in their record keeping.
- Crime data may be falsified by police departments for political reasons. The National Center for Policy Analysis (1998) reports that police departments in Philadelphia; New York; Atlanta; and Boca Raton, Florida, had underreported or downgraded crimes in their localities (and these are just the departments we know about).
- The UCR even underreports crimes that are known to the police because of the FBI's hierarchy rule. The **hierarchy rule** requires police to report only the highest (most serious) offense committed in a multiple-offense single incident to the FBI and to ignore the others. For instance, if a man robs five patrons in a bar, pistol-whips one patron who tried to resist, locks the victims in the beer cooler, and then rapes the female bartender, only the rape is reported to the FBI.

NIBRS: The "New and Improved" UCR

Efforts to improve the reliability and validity of official statistics are occurring all the time, with the most ambitious being the **National Incident-Based Reporting System (NIBRS).** NIBRS began in 1982 and is designed for the collection of more detailed and comprehensive crime statistics than the UCR (which it is supposed to replace). As opposed to the current UCR, which monitors only a relatively few crimes and gathers few details associated with them, NIBRS collects data on 46 "Group A" offenses and 11 "Group B" offenses. There is no hierarchy rule under NIBRS; it reports multiple victims, multiple offenders, and multiple crimes that may be part of the same incident. It also provides information about the circumstances of the offense and about victim and offender characteristics, such as offender–victim relationship, age, sex, and race of victims and perpetrators (if known). As of 2009, a total of 31 states covering 25% of the U.S. population were reporting crime incidents to NIBRS (Justice Research and Statistics Association, 2010). Unfortunately, many police departments lack the staffing and technical exper-

▲ **Photo 2.2** The use of technology by police has been credited in part for crime reduction in the 1990s.

tise to collect and process the wide and detailed range of information that is part of each crime incident their officers deal with, and administrators see little benefit to their department to justify the effort (Dunworth, 2001).

Because NIBRS data provide information about the offender and the victim (victims can identify physical characteristics of perpetrators), it can be used to try to resolve certain criminological issues. One issue is the disproportionately high rate of arrest for African Americans in the United States. The question for criminologists becomes this: Is the disproportion in arrests the result of disproportionately high black involvement in crime or the result of discriminatory arrest patterns of police?

This issue was explored by D'Alessio and Stolzenberg (2003) using NIBRS data from 17 states and 335,619 arrests for rape, robbery, and aggravated and simple assault. Their results indicate the odds of arrest for robbery, aggravated assault, and simple assault were significantly greater for white offenders than for black offenders, but there was no significant racial difference in the probability of arrest for rape. In other words, white offenders were more likely to be arrested for violent crimes other than rape than black offenders. The researchers concluded that the disproportionately high black arrest rate is likely attributable to their disproportionately higher involvement in crime. For instance, African Americans committed 5,278 robberies in those states for which only 21.4% were arrested; whites committed 2,620 robberies for which 30.8% were arrested. Similar results based on NIBRS data were found in Pope and Snyder's (2003) analysis of 102,905 violent incidents committed by juveniles; that is, white juveniles were significantly more likely to be arrested than black juveniles.

⬛ Crime Victimization Survey Data and Their Problems

Crime victimization surveys involve asking large numbers of people if they have been criminally victimized within some specified time frame regardless of whether they reported the incident to police. U.S. Census Bureau personnel interview a national representative sample of people age 12 or over on behalf of the Bureau of Justice Statistics (BJS) twice each year. This survey is known as the **National Crime Victimization Survey** (NCVS), and in 2008, a total of 77,852 people from 42,093 households were interviewed (Rand, 2009). The NCVS requests information on crimes committed against individuals and households, the circumstances of the offense, and personal information about victims (age, sex, race, income, and education level) and offenders (approximate age, sex, race, and victim–offender relationship). Figure 2.4 presents highlights from the 2009 NCVS report.

Victimization surveys have their own dark figures as well as other problems that make them almost as suspect as the UCR. Some of these problems include the following:

- Crimes such as drug dealing and all "victimless" crimes such as prostitution and gambling are not revealed in such surveys for obvious reasons. And because murder victims cannot be interviewed, this most serious of crimes is not included.
- Because NCVS only surveys households, crimes committed against commercial establishments such as stores, bars, and factories are not included. This exclusion results in a huge underestimate of crimes such as burglaries, robberies, theft, and vandalism.
- Victimization data do not have to meet any stringent legal or evidentiary standards in order to be reported as an offense; if the respondent says he or she was robbed, a robbery will be recorded. UCR data, on the other hand, passes through the legal sieve to determine whether the reported incident was indeed a robbery.
- Other problems involve memory lapses; providing answers the respondent thinks the interviewer wants to hear; forgetting an incident; embellishing an incident; and any number of other misunderstandings, ambiguities, and even downright lies that occur when one person is asking another about his or her life experiences.

Figure 2.4 Criminal Victimization, 2007–2008

Type of crime	Number of victimizations		Rates[a]		Percent change
	2007	2008	2007	2008	2007–2003[b]
All crimes	22,879,720	21,312,400	~	~	
Violent crimes[c]	5,177,130	4,856,510	20.7	19.3	−6.9%
Rape/sexual assault	248,280	203,830	1.0	0.8	−18.5
Robbery	597,320	551,830	2.4	2.2	−8.3
Assault	4,331,530	4,100,850	17.3	16.3	−6.0
Aggravated	858,940	839,940	3.4	3.3	−2.9
Simple	3,472,590	3,260,920	13.9	12.9	−6.8
Personal theft[d]	194,060	136,710	0.8	0.5	−30.1%
Property crimes	17,508,530	16,319,180	146.5	134.7	−8.1%*
Household burglary	3,215,090	3,188,620	26.9	26.3	−2.2
Motor vehicle theft	979,640	795,160	8.2	6.6	−19.9*
Theft	13,313,800	12,335,400	111.4	101.8	−8.6*

Source: Rand, M. (2009).

Note: Detail may not add to total because of rounding. Total population age 12 or older was 250,344,870 in 2007 and 252,242,520 in 2008. Total number of households was 119,503,530 in 2007 and 121,141,060 in 2008.

~Not applicable.

*Difference is significant at the 95%-confidence level.

- Consistent with the above, there are suggestions that just as underreporting plagues UCR data, over-reporting may plague NCVS data (O'Brien, 2001). Whatever the case may be, we find many anomalies when comparing the two sources of data. For instance, substantially more crimes appear in police records than NCVS victims claim to have reported to the police. The discrepancy is easily explained for burglary and motor vehicle theft because the NCVS does not include commercial establishments in their reports. It is more difficult to explain discrepancies in violent crime reports, however. One explanation is that the NCVS does not include victims who are under 12 years of age, whereas the UCR does, although it is difficult to believe that children under 12 account for 15 to 20% of the violent victimization known to the police.

NCVS researchers are aware of the many problems that arise when asking people to recall victimization and have initiated many interview improvements in their methodology, one of which is the *bounding interview*. This technique involves comparing reported incidents from the same household in the current interview with those reported 6 months prior. When a report appears to be a duplicate, the respondent is reminded of the earlier report and asked if the new report represents the incident previously mentioned or if it is different. Other techniques used to minimize some of the reported problems mention above are available on the NCVS website at www.icpsr.umich.edu/NACJD/NCVS/. Figure 2.5 provides an example of the kinds of questions asked by NCVS workers.

Figure 2.5 Examples of NCVS Victimization Questions

29.	**How were you attacked? Any other way?**	646
	Mark (X) all that apply.	1 ☐ Raped
		* 2 ☐ Tried to rape
		3 ☐ Sexual assault other than rape or attempted rape
	FIELD REPRESENTATIVE – *If raped, ASK –*	4 ☐ Shot
	Do you mean forced or coerced sexual intercourse?	5 ☐ Shot at (but missed)
	If No, ASK – **What do you mean?**	6 ☐ Hit with gun held in hand
		647 7 ☐ Stabbed/cut with knife/sharp weapon
		* 8 ☐ Attempted attack with knife/sharp weapon
	If tried to rape, ASK –	9 ☐ Hit by object (other than gun) held in hand
	Do you mean attempted forced or coerced sexual intercourse?	10 ☐ Hit by thrown object
		648 11 ☐ Attempted attack with weapon other than gun/knife/sharp weapon
	If No, ASK – **What do you mean?**	* 12 ☐ Hit, slapped, knocked down
		13 ☐ Grabbed, held, tripped, jumped, pushed, etc.
		14 ☐ Other – *Specify*

30.	**Did the offender THREATEN to hurt you before you were actually attacked?**	649
		1 ☐ Yes
		2 ☐ No
		3 ☐ Other – *Specify*

31.	**What were the injuries you suffered, if any? Anything else?**	655
	Mark (X) all that apply.	1 ☐ None – **SKIP** to 40
		* 2 ☐ Raped
		3 ☐ Attempted rape
	FIELD REPRESENTATIVE – *If raped and box 1 in item 29 is NOT marked, ASK –*	4 ☐ Sexual assault other than rape or attempted rape
		5 ☐ Knife or stab wounds
	Do you mean forced or coerced sexual intercourse?	6 ☐ Gun shot, bullet wounds
		656 7 ☐ Broken bones or teeth knocked out
	If No, ASK – **What do you mean?**	* 8 ☐ Internal injuries
		9 ☐ Knocked unconscious
	If attempted rape and box 2 in item 29 is NOT marked, ASK –	10 ☐ Bruises, black eye, cuts, scratches, swelling, chipped teeth
	Do you mean attempted forced or coerced sexual intercourse?	11 ☐ Other – *Specify*
	If No, ASK – **What do you mean?**	

32.	*ASK OR VERIFY –* **Were any of the injuries caused by a weapon other than a gun or knife?**	657
		1 ☐ Yes – Ask 33
		2 ☐ No – **SKIP** to 34

33.	**Which injuries were caused by a weapon OTHER than a gun or knife?**	658
	Enter code(s) from 31.	☐ ☐ ☐ Code Code Code

34.	**Were you injured to the extent that you received any medical care, including self treatment?**	659
		1 ☐ Yes – Ask 35
		2 ☐ No – **SKIP** to 40

35.	**Where did you receive this care? Anywhere else?**	660
	Mark (X) all that apply.	1 ☐ At the scene
		* 2 ☐ At home/neighbor's/friend's
		3 ☐ Health unit at work/school, first aid station at a stadium/park, etc.
		4 ☐ Doctor's office/health clinic
		5 ☐ Emergency room at hospital/emergency clinic
		6 ☐ Hospital (other than emergency room)
		7 ☐ Other – *Specify*

SOURCE: Catalano (2006).

◪ Areas of Agreement Between the UCR and NCVS

To the extent that two or more data sources tell us the same thing, our confidence in both is increased. The UCR and NCVS agree on the demographics of crime in that they both tell us that males, the young, the poor, and African Americans are more likely to be perpetrators and victims of crime than are females, older persons, wealthier persons, and persons of other races. Both sources also agree as to the geographic areas and times of the year and month when crimes are more likely to occur. Over a 3-year period, O'Brien (2001) found that NCVS victims reported that 91.5% of those who robbed them and 87.7% of their aggravated assault assailants were male, as were 91.2 and 84.3%, respectively, of those arrested for those offenses. Likewise, NCVS victims reported that 64.1% of those who robbed them and 40% of their aggravated assault assailants were African American. These percentages fit the UCR arrest statistics for race almost exactly: A total of 62.2% arrested for robbery were African American, as were 40% of those arrested for aggravated assault. Thus, the two data sets agree almost perfectly.

Comparisons of UCR and NCVS data have often proven very useful to resolve issues such as these. For instance, some feminist criminologists are of the opinion that women are becoming more "masculinized" as a result of assuming "male" roles in the workforce, and that this is reflected in the increased rates of female arrests for violent crimes. Darrell Steffensmeier and his colleagues (Steffensmeier, Zhong, Ackerman, Schwartz, & Agha, 2006) used a comparison of data trends reported in the UCR and the NCVS from 1980 to 2003 to explore the issue of whether the violent crime gap between males and females is closing. They found that both sources reported little or no changes in the gender ratio for violent crimes such as murder, rape, and robbery, but that the UCR reports indicated a sharp rise in assaults by females. Does this mean that women became more violent over the period examined, or does the increase reflect the behavior of the police more than the behavior of women? The authors conclude that net-widening policy shifts have escalated the arrest proneness of females for "criminal assault" (e.g., policing physical attacks/threats of marginal seriousness) rather than that women have become more violent. In other words, UCR increases in female arrests for assault are explained by changes in police policy, something that could not have been determined without examining both data sources. The addition of the NCVS and NIBRS to the nation's crime databases thus has great utility for settling some major quarrels among criminologists of different ideological persuasions, although not to the satisfaction of everyone.

Note from Table 2.1 that this trend was still in evidence when comparing UCR arrests for aggravated assault and simple (labeled "other" in the table) assault from 1999 to 2009. Female aggravated assault decreased by 2.5%, but female arrests for simple assault increased by 15.9%. The overall female increase in violent crime arrests of 1.4% is entirely a function of the huge increase in female robbery arrests over the period.

◪ Self-Report Crime Surveys and Their Problems

Self-report surveys of offending provide a way for criminologists to collect data without having to rely on government sources. Questionnaires used in these surveys typically provide a list of offenses and request subjects to check each offense they recall having committed and how often, and sometimes whether they have ever been arrested, and if so, how many times. Self-report surveys have relied primarily on college and high school students for subjects, although prison inmates and probationers/parolees have also been surveyed.

Several studies have addressed the issue of the accuracy and honesty of self-reported offenses in various ways, and the results have generally been encouraging, at least for uncovering the extent of minor offenses.

On average, known delinquents and criminals disclose almost 4 times as many offenses as the non-delinquents. Cartier, Farabee, and Prendergast's (2006) study of methamphetamine use and self-reported crime and recidivism among 614 California parolees, for instance, found that meth use was significantly related to self-reported crime and claim a high level of agreement between self-reported crime and actual crimes committed. Had differences such as these not been found, the validity of the self-report procedure would have been in doubt.

The greatest strength of self-report research is that researchers can correlate a variety of characteristics of respondents with their admitted offenses that go beyond the demographics of age, race, and gender. For instance, they can attempt to measure various constructs thought to be associated with offending, such as impulsiveness, empathy, and sensation seeking, as well as their peer associations and their attitudes. The evidence indicates that self-reported crime measures provide largely accurate information about some illegal act that occurred at some time in the respondent's life. However, there are a number of reasons why self-report crime surveys also provide a distorted picture of criminal involvement.

- The great majority of self-report studies survey "convenience" samples of high school and college students, populations in which we don't expect to find many seriously criminally involved individuals. Most self-report studies thus eliminate the very people we are most interested in gathering information about. One strength of the self-report method, however, is that is appears to capture the extent of illegal drug usage among high school and college students, something that neither the UCR nor the NCVS attempts to do.
- Self-report studies typically uncover only fairly trivial antisocial acts such as fighting, stealing items worth less than $5, smoking, and truancy. Almost everyone has committed one or more of these acts. These are hardly acts that help us to understand the nature of serious crime. A connected problem is that some researchers lump respondents who report one delinquent act together with adjudicated delinquents who break the law in many different ways and at many different times.
- Even though most people are forthright in revealing minor antisocial behaviors, most people do not have a serious criminal history, and those who do have a distinct tendency to underreport their crimes (Hindelang, Hirschi, & Weis, 1981). As the number of crimes people commit increases, so does the proportion of offenses they withhold reporting, with those arrested for the most serious offenses having the greatest probability of denial (Farrington, 1982).
- Males tend to report their antisocial activities less honestly than females and African Americans less honestly than other racial groups (Cernkovich, Giordano, & Rudolph, 2000; Kim, Fendrich, & Wislar, 2000). This evidence renders suspect any statements about gender or racial differences regarding antisocial behavior that are based on self-report data. When it comes to relying on self-report data to assess the nature and extent of serious crime, it does us well to remember the gambler's dictum: "Never trust an animal that talks."

We should not end on a pessimistic note about self-reports, however. Many of the major multi-million dollar longitudinal studies going on today have built-in safeguards against researchers naïvely taking subjects at their word. A number of studies verify self-report accounts with police records and other social agencies, a practice which further helps us to gain a grasp on the reliability of self-report studies. For instance, a large longitudinal (following the same people across the lifespan) cohort study (one that studies a set of individuals who share a common characteristic, such as being born in the same month in the same geographic area) showed that individuals from the lowest social class category reported 3.21 times more offenses than individuals from the highest social class category (Fergusson, Swain-Campbell, & Horwood, 2004). However, when researchers compared individuals from these two classes for official juvenile and adult convictions, the members of the lowest class had 25.82 times more officially recorded convictions than members of the highest class. Thus, the more actively involved delinquents/criminals do report more anti-social behavior than others, but they also tend to greatly underreport their antisocial actions.

White-Collar Crime: The FBI's Financial Crimes Report

The only "white collar" crimes, i.e., crimes committed by guile as opposed to force, listed in the UCR are embezzlement, forgery/counterfeiting, and fraud, which are mostly committed by individuals. There is, however, a separate accounting of major white-collar crimes committed by organized groups and corporations called the *Financial Crimes Report* (www.fbi.gov/stats-services/publications/financial-crimes-report-2009). This report is issued each year by the FBI and contains results of investigations carried out by the Financial Crimes Section (FCS) of the FBI. The role of the FCS is to oversee the investigation of financial fraud and to supervise the forfeiture of assets from individuals engaged in such crimes. The FCS is composed of the Asset Forfeiture/Money Laundering Unit (AF/MLU), the Economic Crimes Unit (ECU), the Health Care Fraud Unit (HCFU), the Forensic Accountant Unit (FAU), and the National Mortgage Fraud Team (NMFT).

In 2009, the FCS was investigating 1,510 cases of securities and commodities fraud and had recorded 412 indictments and 306 convictions. The FBI also conducted 350 investigations of money laundering, which resulted in 84 convictions. Corporate fraud is perhaps the major focus of the FBI today. Figure 2.6 shows

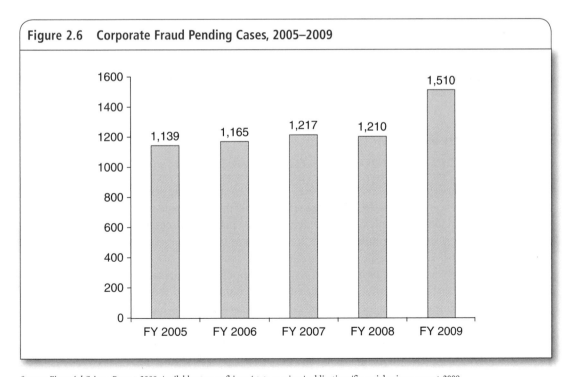

Figure 2.6   Corporate Fraud Pending Cases, 2005–2009

Source: *Financial Crimes Report, 2009.* Available at www.fbi.gov/stats-services/publications/financial-crimes-report-2009

the number of pending cases from 2005 to 2009. In 2009, investigations resulted in 156 convictions of corporate criminals, and numerous other cases are pending trials and plea agreements, although this is likely a drop in the bucket compared to the actual number of crimes committed by corporate criminals. During 2009, the FBI secured $6.1 billion in restitution orders and $5.4 million in fines from corporate criminals (www.fbi.gov/stats-services/publications/financial-crimes-report-2009).

⬚ The Dark Figure of Crime

The dark (or hidden) **figure of crime** is that portion of the total crimes committed each year that never comes to light. Figure 2.7 presents three diagrams that show the different dark figures for the three major measures of criminal behavior. (The dark figures are represented by the dark shading in each diagram.)

Each diagram shows the degree to which crimes of various levels of seriousness are most likely to be detected by each measure ("victimless" crimes excluded). In the top diagram displaying UCR data, you can see that very few trivial offenses are reported in official statistics, and most of those that are will be dismissed as unfounded by the police. For official statistics, then, the dark figures are highly concentrated at the nonserious end of the crime seriousness spectrum.

The middle diagram reveals that the dark figures for victimization data are primarily concentrated in the nonserious end of the spectrum also, although to a lesser degree than in the case of official data. The failure of victimization data to pick up these minor offenses is largely due to survey subjects not remembering all incidences of victimization.

In the bottom diagram, we see that most of the dark figures in the case of self-reports are concentrated in the upper end of the seriousness continuum rather than the lower end. This is partly due to the fact that (a) nearly all self-report surveys exclude the most persistent serious offenders from their subject pools, and (b) many of the most serious offenders who remain in self-report subject pools do not reveal the full extent of their criminal histories.

⬚ What Can We Conclude About the Three Main Measures of Crime in America?

All three main measures of crime in America are imperfect measures, and which one of them is "best" depends on what we want to know. UCR data are still probably the best single source of data for studying serious crimes, and indeed, the only one for studying murder rates and circumstances. For studying less serious but much more common crimes, either victimization or self-report survey data are best. If the interest is in drug offenses, self-reports are the preferable data source.

Because all three data sources converge on some very important points about crime, they enable us to proceed with at least some confidence in our endeavors to understand the whys of crime. The basic demographics of crime constitute the raw social facts that are the building blocks of our criminological theories. If street crime is concentrated among the lower socioeconomic classes and in the poorest neighborhoods, we can begin to ask such questions as, does poverty "cause" crime, or does some other variable(s) cause both? Is social disorganization in a neighborhood independent of the people living in it, or completely dependent on the people living in it? Why do females always and everywhere commit

Figure 2.7   Differing Proportions of Reported/Unreported Crimes for the Three Major Measures of Crime

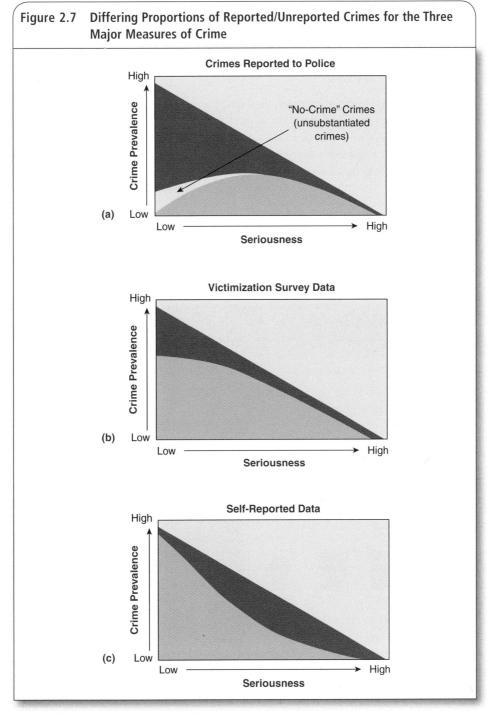

(a) **Crimes Reported to Police**

Crime Prevalence (High to Low) vs. Seriousness (Low to High)

"No-Crime" Crimes (unsubstantiated crimes)

(b) **Victimization Survey Data**

Crime Prevalence (High to Low) vs. Seriousness (Low to High)

(c) **Self-Reported Data**

Crime Prevalence (High to Low) vs. Seriousness (Low to High)

*Light shading = proportion of crimes reported. Dark shading = proportion not reported.

far fewer crimes (particularly the most serious crimes) than males? These and many dozens of other "why" questions can be asked once we have a firm grip on the raw facts supplied by the methods described in this chapter.

The FBI's Ten Most Wanted

One of the more interesting bits of criminological data provided by the government is the FBI's Ten Most Wanted list—a compilation of the worst of the worst criminals according to the FBI. The agency collects and distributes statistics on specific fugitive individuals wanted for particularly serious crimes. The "ten most wanted" individuals are listed every year and will change as fugitives are apprehended or die. Figure 2.8 provides the 2010 list, which includes a terrorist, a family murderer, major drug dealers, and organized crime figures.

Figure 2.8   The FBI's 10 Most Wanted Fugitives

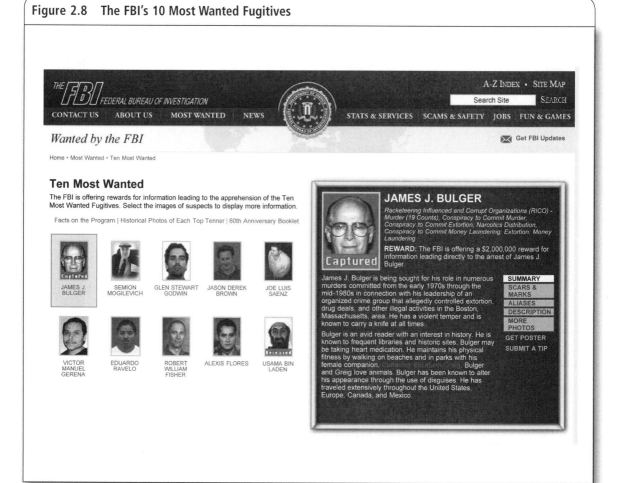

SUMMARY

- Crime and criminal behavior are measured in several ways in the United States. The oldest measure is the FBI's Uniform Crime Reports (UCR), which is a tabulation of all crimes reported to the police in most of the jurisdictions in the United States in the previous year. The UCR is divided into two parts: Part I records the eight Index crimes (murder, rape, robbery, aggravated assault, burglary, larceny/theft, and arson), and Part II records arrests made for all other crimes.
- UCR data seriously underestimate the extent of crime because the UCR only records reported crimes, ignores drug offenses, and only reports the most serious crime in a multiple-crime event. The problems with the UCR led to the implementation of the National Incident-Based Reporting System (NIBRS).
- The second major source of crime statistics is the National Crime Victimization Survey (NCVS). This survey consists of many thousands of interviews of householders throughout the United States, asking them about their crime victimization (if any) during the previous 6 months. The NCVS also has problems because it leaves out crimes against commercial establishments and relies exclusively on the memory and the word of interviewees.
- The third source of crime data is self-report data collected by criminologists themselves. The advantage of self-report data is that they are derived "from the horse's mouth," and typically the questionnaires used ask about "victimless" offenses not covered in either the UCR or NCVS. The major problems with self-report data is that it does not capture serious criminal behavior and is subject to dishonesty in the form of underreporting, especially underreporting by those most seriously involved in criminal activity.
- The UCR, NCVS, and self-report data come to different conclusions on a variety of points, but they agree about where, when, and among whom crime is most prevalent, and the fact that crime has fallen dramatically in the United States over the past decade. Taken together, then, we have a fairly reliable picture of the correlates of crime from which to develop our theories about explanatory mechanisms.

DISCUSSION QUESTIONS

1. Go to the website www.fbi.gov/wanted/topten/fugitives/fugitives.htm-1 for the FBI's ten most wanted fugitives, and research the background and crimes of one of the people listed there. Then write a one- to two-page summary and report to the class.

2. Do you think it wise to make "authoritative" statements or formulate theories of criminal behavior, especially serious criminal behavior, based on self-report data?

3. Can you think of other problems possibly associated with asking people about their delinquent or criminal behavior or their victimization other than those discussed in the chapter?

4. If you were the American "crime czar," what would you do to get the various law enforcement agencies to fully implement NIBRS—and no, you can't just order them to do so.

USEFUL WEBSITES

Bureau of Justice Statistics. www.ojp.usdoj.gov/bjs/

National Archive of Criminal Justice Data. www.icpsr.umich.edu/NACJD/

National Crime Victimization Survey Resource Guide. www.icpsr.umich.edu/NACJD/NCVS/

National Incident-Based Reporting System Resource Guide. www.icpsr.umich.edu/NACJD/NIBRS/

Uniform Crime Reports. www.fbi.gov/ucr/ucr.htm

CHAPTER TERMS

Cleared offenses

Crime rate

Dark figure of crime

Hierarchy rule

National Incident-Based
Reporting System (NIBRS)

National Crime Victimization
Survey (NCVS)

Part I offenses (or index crimes)

Part II offenses

Self-report surveys

Uniform Crime Reports (UCR)

CHAPTER

3

The Early Schools of Criminology and Modern Counterparts

"Lisa" is a 30-year-old mother of three children aged 8, 6, and 4. Her husband left her a year ago for another woman and his present whereabouts are unknown. Because Lisa only has a tenth-grade education and because she cannot afford child care costs, she was forced onto the welfare rolls. When Christmas came around, she had no money to buy her children any presents, so she took a temporary Christmas job at the local Wal-Mart store where she earned $1,200 over a 2-month period. Lisa did not report this income to the welfare authorities as required by law; a welfare audit uncovered her crime. The terrified and deeply ashamed Lisa pled guilty to grand theft, which carries a possible sentence of 2 years in prison, and was referred to the probation department for a presentence investigation report (PSI) and sentencing recommendation.

"Chris" is a 30-something male with a record of thefts and other crimes committed since he was 10 years old. Chris also pled guilty before the same judge on the same day and was likewise referred for a PSI. Chris had stolen money and parts totaling $1,200 from an auto parts store during one of his very brief periods of employment.

These two cases point to a perennial debate among criminal justice scholars, with one side favoring the so-called classical school position and the other favoring the positivist position. Both positions are ultimately about the role of punishment in deterring crime, but the classical position maintains that punishment should fit the crime and nothing else; i.e., all people

(Continued)

(Continued)

convicted of similar crimes should receive the same punishment regardless of any differences they may have. Both Lisa and Chris freely chose to commit the crime, and the fact that Chris has a record and Lisa does not is irrelevant. The positivist position is that punishment should fit the offender and be appropriate to rehabilitation. Lisa's and Chris's crimes were motivated by very different considerations, they are very different people morally, and blindly applying similar punishments to similar crimes without considering the possible consequences is pure folly. Think about these two cases as you read about classical and positivists thought regarding human nature, punishment, and deterrence.

Pre-Classical Notions of Crime and Criminals

Prior to the 18th century, explanations of a wide variety of phenomena, including criminal behavior, tended to be of a religious or spiritual nature. Good fortune and disaster alike were frequently attributed to good or evil supernatural forces. A simple extension of this worldview was to define crime as the result of demonic possession or the evil abuse of free will. Because of the legacy of the Christian idea of Original Sin, all human beings were considered born sinners, and so it made no sense to ask questions such as, "What causes crime?" The gift of the grace of God kept men and women on the straight and narrow, and if they deviated from this line, it was because God was no longer their guiding compass. That being so, it made sense to "beat the devil" out of them" by the most hideous and sadistic means to save their immortal souls.

Demonological explanations for criminal behavior began to wane in the 18th century with the beginning of a period historians call the **Enlightenment,** which was essentially a major shift in the way people began to view the world and their place in it. This new worldview questioned traditional religious and political values, such as absolute monarchy and demonic possession. In their place, they substituted humanism, rationalism, and a belief in the primacy of the natural over the supernatural world. Enlightenment thinkers believed in the dignity and worth of the individual, a view that would eventually find expression in the law and in the treatment of criminal offenders.

The Classical School

Modern criminology is the product of two main schools of thought: The classical school originating in the 18th century, and the positivist school originating in the 19th century. You may ask yourselves why a discussion of the "old masters" is necessary; after all, you don't see such discussions in physics, chemistry, or biology texts. The reason for this is that unlike those disciplines, modern criminology is still confronted by the same problems that confronted its pioneers, specifically the problem of explaining crime and criminality. Thus, their works are of more than passing interest to us.

The father of classical criminology is the Italian nobleman and professor of law, Cesare Bonesana, marchese di Beccaria. In 1764, Beccaria published a call for the reform of judicial and penal systems throughout Europe called *On Crimes and Punishment* (1764/1963). The book is a passionate plea to reform the criminal justice system, to humanize and rationalize the law, and to make punishment more just and humane. Beccaria placed emphasis on the idea that citizens give up certain rights in order to gain protection from the state (the so-called social contract). Although laws are supposed to be compacts between citizens and the state, they were often arbitrary and cruel. Beccaria believed that equals should be treated equally and unequals unequally according to relevant differences. By "equal" and "unequal," he was referring to the crimes offenders had committed, which was the only "relevant difference" to be considered, not their social standing, as was often considered at the time. Judges should not have the authority to interpret laws; they should simply apply the punishment for a given offense as it was statutorily defined by the legislature.

Beccaria (1764/1963) felt that the responsibility of determining the facts of a case should be placed in the hands of ordinary citizens, not simply given to a judge who had little idea what life was like for the ordinary person: "I consider an excellent law that which assigns popular jurors, taken by lot, to assist the chief judge . . . that each man ought to be judged by his peers" (p. 21). He was also very much against the practice of using torture to obtain confessions and other information from suspects, and was against the use of capital punishment (but he believed corporal punishment was appropriate for violent offenders). Capital punishment could not be an effective deterrent according to Beccaria because it was too quick; life imprisonment would be more effective.

In terms of the form of punishment, Beccaria (1764/1963) made three important observations regarding its application—certainty, swiftness, and severity—and makes clear that punishments must outweigh any benefits offenders get from crime if they are to be deterred from future crime.

> **Certainty:** "The certainty of punishment, even if it be moderate, will always make a stronger impression than the fear of another which is more terrible but combined with the hope of impunity" (p. 58).

> **Swiftness:** "The more promptly and the more closely punishment follows upon the commission of a crime, the more just and useful will it be" (p. 55).

> **Severity:** "For a punishment to attain its end, the evil which it inflicts has only to exceed the advantage derivable from the crime; in this excess of evil one should include the . . . loss of the good which the crime might have produced. All beyond this is superfluous and for that reason tyrannical" (p. 43).

Jeremy Bentham and Human Nature

Another prominent figure of the classical school was British lawyer and philosopher Jeremy Bentham. His major work, *Principles of Morals and Legislation* (1789/1948), is essentially a philosophy of social control based on the **principle of utility,** which posits that any human action

▲ **Photo 3.1**   Jeremy Bentham, often credited as the founder of University College London, insisted that his body be put upon display there after his death. You can visit a replica of it today.

should be judged moral or immoral by its effect on the happiness of the community. Thus, the proper function of the legislature is to pass laws aimed at maximizing the pleasure and minimizing the pain of the largest number in society—"the greatest good for the greatest number" (p. 151). If legislators are to legislate according to the principle of utility, they must understand human motivation, which for Bentham was easily summed up: "Nature has placed mankind under the governance of two sovereign masters, pain and pleasure. It is for them alone to point out what we ought to do, as well as to determine what we shall do" (p. 125).

The classical explanation of criminal behavior and how to prevent it can be derived from the Enlightenment assumption that human nature is hedonistic, rational, and endowed with free will. In many ways, the classical view of human nature can be likened to Thomas Sowell's (1987) constrained vision we discussed in Chapter 1. **Hedonism** is a doctrine whose central idea is that pleasure is the main goal of life. All other life goals are seen only as instrumentally desirable; that is, they are only desirable as means to the end of achieving pleasure or avoiding pain. Thus, hedonism is the greatest single motivator of human action.

Rational behavior is behavior that is consistent with logic. People are said to behave rationally when we observe a logical "fit" between the goals they strive for and the means they use to achieve them. The goal of human rationality is self-interest, and self-interest governs our behavior whether in conforming or deviant directions.

Hedonism and rationality are combined in the concept of the **hedonistic calculus,** a method by which individuals are assumed to logically weigh the anticipated benefits of a given course of action against its possible costs. If the balance of consequences of a contemplated action is thought to enhance pleasure or minimize pain, then individuals will pursue it; if not, they will not.

Free will enables human beings to purposely and deliberately choose to follow a calculated course of action. Therefore, if people seek to increase their pleasures illegally, they do so freely and with full knowledge of the wrongness of their acts, and thus society has a perfectly legitimate right to punish those who harm it.

It follows from these assumptions about human nature that if crime is to be deterred, then punishment (pain) must exceed the gain (pleasure) gained from it. Criminals will weigh the costs against the benefits of crime, and they will desist if, on balance, the costs exceed the benefits. Estimations of the value of pleasures and pains are to be considered with reference to four circumstances: intensity (severity), duration, certainty, and propinquity (how soon after the crime the pleasure or pain is forthcoming) (Bentham, 1764/1948, p. 151). In expressing these views, Bentham added to Beccaria's insight by showing how they cohered with human nature as classical scholars viewed it.

Bentham went somewhat further than Beccaria by devoting a great deal of energy (and his own money) to arguing for the development of prisons as punitive substitutes for torture, execution, or transportation to overseas penal colonies. He designed a prison called the *panopticon* ("all seeing" entity), which was to be a circular "inspection house" enabling guards to constantly see their charges, thus requiring fewer staff. Because prisoners could always be seen without seeing who was watching them or when they were being watched, the belief was that the perception of constant scrutiny would develop into self-monitoring. Bentham felt that prisoners could be put to useful work and thus pay for their own keep, with the hoped-for added benefit that they would acquire the habit of honest labor.

The Rise of Positivism

The explosion of technology and scientific knowledge in the 19th century led scholars to move away from classical assumptions and toward a more scientific view of human behavior. The increasingly popular view among criminologists of this period was that crime resulted from internal or external forces restricting individuals, biasing or even completely determining their behavior. This position became known as determinism. **Determinism** simply means that events have causes that preceded them. It does not mean that if X is present Y *will* occur; that is strong determinism, which has all but disappeared in science. Determinists only say that if X is present, then Y has a *certain probability* of occurring. Positivists were the main adherents of determinism.

Positivism is simply the application of the scientific method from which more *positive* knowledge can be obtained. Positivists insisted on divorcing science from metaphysics (talk of free will, and the like) and morals, and on looking only at what is, not what ought to be. The Enlightenment's flattering image of human beings gave way to the evolutionary view that we are different only in degree from other animals, and that science could explain human behavior just as well as it could explain events in the physical world.

Cartographic Criminology

Some of the earliest positivist attempts to leave the armchair and collect facts about crime in order to understand it were cartographers, i.e., scholars who employ maps and other geographic information in their research. Rather than asking why individuals commit crimes, cartographic criminologists were more interested in where and when criminal behavior is most prevalent. The two most important **cartographic criminologists** were André-Michel Guerry and Lambert Adolphe Jacques Quételet. Quételet compared crime rates in France across ages, sexes, and seasons and saw the same reflections in his data that we see today in the American crime data; that is, young males living in poor neighborhoods commit the most crime. He thought sociologically about crime before the discipline officially existed, writing that "society prepares the crime and the guilty is only the instrument by which it is accomplished" (quoted in Vold & Bernard, 1986, p. 132).

This cartographic method crossed the English Channel to influence British researchers Henry Mayhew and Joseph Fletcher. Using British crime data from the 1830s–1840s, both men independently mapped out the concentration of various kinds of criminal activity across England and Wales. They concluded that crime is concentrated in poor neighborhoods undergoing population changes. Many British cities experienced the same demographic changes in the early 1800s that American cities were to experience in the early 1900s. Rural people were flocking into the big cities to obtain work in the new factory system, and in the anonymity of these cities of strangers, social bonds were weakened as were informal social controls, morals declined, and crime flourished (Levin & Lindesmith, 1971).

Biological Positivism: Cesare Lombroso and the Born Criminal

Italian physician and anthropologist Cesare Lombroso published the first book devoted solely to the causes of criminality, which he titled *Criminal Man* (1876). His basic idea was that many criminals are evolutionary "throwbacks" to an earlier form of life. The term used to describe organisms resembling ancestral pre-human forms of life is **atavism.** Atavistic criminals could be identified by a number of measurable physical stigmata (physical markings), included protruding jaws, drooping eyes, large ears, twisted and flat noses, long arms relative to the lower limbs, sloping shoulders, and a coccyx that resembled "the stump of a tail" (Lombroso-Ferrero, 1911/1972, pp. 10–21). Lombroso was just one of many who sought to understand behavior with reference to the principles of evolution as they were understood at the time. If humans were at one end of the continuum of animal life, it made sense to many people that criminals—who acted "beastly" and lacked reasoned conscience—were biologically inferior beings belonging to an earlier evolutionary time.

* 1836   CESAR LOMBROSO   † 1909
Prof. de Clinique Psychiatrique et d'Antropologie
à l'Université Royale de Turin,·

▲ **Photo 3.2**   Cesare Lombroso (1836–1909)

Subsequent data did not support Lombroso's extreme views, so he modified them to include two other types in addition to his atavistic type: the **insane criminal**, and the **criminaloid.** Insane criminals bore some stigmata, but they were not born criminals; rather, they became criminal as a result of "an alteration of the brain, which completely upsets their moral nature" (Lombroso-Ferraro, 1911/1972, p. 74). Among the "insane" criminals were kleptomaniacs, nymphomaniacs, and child molesters. Criminaloids had none of the physical peculiarities of the born or insane criminal. They were further categorized as *habitual criminals,* who become so by contact with other criminals, the abuse of alcohol, or other "distressing circumstances"; *juridical criminals,* who fall afoul of the law by accident; and the *criminal by passion,* hot-headed and impulsive persons who commit violent acts when provoked.

Raffaele Garofalo: Natural Crime and Offender Peculiarities

Lombroso and two of his Italian contemporaries, Raffaele Garofalo and Enrico Ferri, founded what became known as the **Italian School** of criminology. Garofalo (1885/1968), who coined the term *criminology,* is perhaps best known for his efforts to formulate a "natural" definition of crime, wanting to anchor it in human nature. Garofalo believed that an act would be considered a crime if it was universally condemned, and it would be universally condemned if it offended the natural altruistic sentiments of probity (integrity, honesty) and pity (compassion, sympathy). Natural crimes are evil in themselves (mala in se), whereas other kinds of crimes (mala prohibita) are wrong only because they have been made wrong by the law.

Garofalo rejected the classical principle that punishment should fit the crime, arguing instead that it should fit the criminal, with the only question to be considered at sentencing being the danger the offender posed to society as judged by his or her peculiarities. *Peculiarities* are characteristics that place offenders at risk for further criminal behavior. There were four such categories, each meriting different forms of punishment: extreme, impulsive, professional, and endemic. Society could only be defended from *extreme criminals* by swiftly executing them, regardless of the crime for which they are being punished. *Impulsive criminals* (alcoholics and the insane) were to be imprisoned. *Professional criminals* are normal individuals who chose to commit their crimes and thus require "elimination," either by life imprisonment or transportation to a penal colony. Endemic criminals (those who committed crimes peculiar to a given region and mala prohibita crimes) could best be controlled by changes in the law.

Enrico Ferri and Social Defense

Enrico Ferri (1879/1917) believed that moral insensibility combined with low intelligence were the criminal's most marked characteristics: The criminal has "defective resistance to criminal tendencies and temptations, due to that ill-balanced impulsiveness which characterizes children and savages" (p. 11). Given this conception of criminals, his only rationale for punishment was **social defense.** This theory of punishment asserts that its purpose is not to deter or to rehabilitate but to defend society from criminal predation. Ferri reasoned that if criminals were not capable of basing their behavior on rational calculus, how could they be deterred? Rather, criminals must be locked up for as long as possible so that they no longer posed a threat to society.

◪ Neoclassicism: The Return of Choice and Deterrence

Rational Choice Theory

A combination of high crime rates, the failure of existing criminological theories to adequately account for them, and the emergence of a more conservative attitude in the 1980s saw a swing away from the ideals of the positivist school and back to the classical notion that offenders are free actors responsible for their own actions. Criminologists who embrace neoclassicism such as rational choice theorists are "soft" determinists because while they believe that criminal behavior is ultimately a choice made in the context of personal and situational constraints and opportunities. In other words, rational choice theorists substitute the extremes of the classical free will concept (our actions are free of any causal chains) for that of human agency. **Human agency** is a concept that maintains that humans have the capacity to make choices and the moral responsibility to make moral ones regardless of the internal or external constraints on one's ability to do so. According to rational choice theory, rationality is the quality of thinking and behaving in accordance with logic and reason such that one's reality is an ordered and intelligible system.

Rationality is both subjective and bounded, and unwanted outcomes can be produced by rational strategies. This is because we do not all make the same calculations or arrive at the same plan when pursuing the same goals, for we contemplate our anticipated actions with less-than-perfect knowledge, with different mind-sets, and with different reasoning abilities. In other words, we may think that we are behaving in a way that best serves our self-interest, i.e., rationally, but our behavior has brought us unwanted results because we were ignorant of some things and misinterpreted others.

Our social emotions (guilt, shame, embarrassment, etc.) also function to keep our temptations in check by "overriding" purely rational calculations of immediate gain (Mealey, 1995). We do the best we can to order our decisions relating to our self-interest with the knowledge and understanding we have about the possible outcomes of a particular course of action. All people have mental models of the world and behave rationally with respect to them, even if others might consider their behavior to be irrational. Criminals behave rationally from their private models of reality, but their rationality is constrained, as is everyone's, by ability, knowledge, and emotional input (Cornish & Clarke, 1986).

Rational choice theorists view crime in terms of Bentham's principle of maximizing pleasure and minimizing pain. People will choose crime if they perceive that its pleasures exceed the pains they might conceivably expect if discovered. The theory does not assume that we are all equally at risk to commit criminal acts, or that we do or do not commit crimes simply because we do or do not "want to." However, while the theory recognizes that factors such as temperament, intelligence, class, family structure, and neighborhood impact our choices (Clarke & Cornish, 1985), it largely ignores these factors in favor of concentrating on the conscious thought processes involved in making decisions to offend.

Rather than focusing on the nature and backgrounds of criminals, rational choice theorists simply assume criminally motivated offenders will always be with us and focus on the process of their choices to offend. This process is known as **choice structuring** and is defined as "the constellation of opportunities, costs, and benefits attaching to particular kinds of crime" (Cornish & Clarke, 1987, p. 993). Thus, criminal events require that motivated offenders meet situations that they perceive as an opportunity to acquire something they want, such as Willie Sutton viewing a bank. Each event is the result of a series of choice-structuring decisions to initiate the event, continue, or desist, and each particular kind of crime is the result of a series of different decisions that can only be explained on their own terms: The decision to rape is arrived at quite differently from the decision to burglarize.

Routine Activities Theory

Lawrence Cohen and Marcus Felson (1979) devised a neoclassical theory in the tradition of rational choice theory that may explain high crime rates in different societies and neighborhoods without invoking individual differences in criminal propensity. They do this by pointing to the routine activities in that society or neighborhood. Routine activities are defined as "recurrent and prevalent activities which provide for basic population and individual needs" (L. Cohen & Felson, 1979, p. 593). In other words, they are the day-to-day activities characterizing a particular community. In disorganized communities, the routine activities are such that they practically invite crime.

▲ **Photo 3.3**   A well protected security camera looms over this London apartment complex housing more than 3,000 people. Many residents are immigrants or asylum seekers and poverty is a huge challenge. There are 42 different languages spoken by children at the local schools, where 95% of the pupils are minorities. It is an area with a reputation for having a variety of social problems. Recently there have been efforts to establish a more coherent sense of community, but there are still substantial difficulties. Routine activities and social disorganization theories might best explain why there are so many problems in this complex.

According to Cohen and Felson (1979), crime is the result of (a) *motivated offenders* meeting (b) *suitable targets* that lack (c) *capable guardians*. If any one of these three elements is missing, crime is not likely to occur. Cohen and Felson take motivated offenders for granted and do not attempt to explain their existence. The theory is thus very much like rational choice theory in that it describes situations in which criminal victimization is likely to occur. In poor, disorganized communities, there is never a shortage of motivated offenders, and although the pickings are generally slim in such areas, victimization is more prevalent than in more affluent areas (Catalano, 2006). One of the obvious reasons for high victimization rates in poor, disorganized areas (besides the abundance of motivated offenders) is that they tend to lack capable guardians for either persons or property.

Routine activities theory looks at crime from the points of view both of the offender and of crime prevention efforts. A crime will only be committed when a motivated offender believes that he or she has found something worth stealing or someone to victimize that lacks a capable guardian. A capable guardian is a person or thing that discourages the motivated offender from committing the act. It can be the presence of a person, police patrols, strong security protection, neighborhood vigilance, or whatever. Because of disrupted families, transient neighbors, poverty, and all the other negative aspects of disorganized neighborhoods, except for police patrols, capable guardians are in short supply. Crime is a "situation," and crime rates can go up or down depending on how these situations (routine activities) change, without any changes at all in offender motivation. Recurring situations conducive to acquiring resources with minimal effort may also tempt more individuals to take advantage of them.

Deterrence and Choice: Pain Versus Gain

That people respond to incentives and are deterred by the threat of punishment is the philosophical foundation behind all systems of criminal law. Rational choice theory evolved out of deterrence theory, which can be encapsulated by the principle of operant psychology that states that *behavior is governed by its consequences*. That is, if a behavior is followed by some sort of reward, the behavior tends to be repeated; if it is followed by some sort of unpleasantness, it tends not to be repeated. A positive consequence of crime for criminals is that it affords them something they want for little effort; a negative consequence is the possible punishment attached to their crimes.

Deterrence is the prevention of criminal acts by the use or threat of punishment and may be either *specific* or *general*. **Specific deterrence** refers to the effect of punishment on the future behavior of the person who experiences the punishment. For specific deterrence to work, it is necessary that a previously punished person make a mental connection between an intended criminal act and the punitive consequences suffered as a result of similar acts committed in the past. Unfortunately, such connections, if made, rarely have the socially desired effect, either because memories of the previous consequences were insufficiently emotionally strong or because the offender discounted them.

Committing another crime after previously being punished for one is called **recidivism** ("falling back" into criminal behavior). Recidivism is a lot more common among ex-convicts than repentance and rehabilitation. Nationwide, about 33% of released inmates recidivate within the first 6 months, 44% within the first year, 54% by the second year, and 67.5% by the third year (M. Robinson, 2005, p. 222). These are just the ones who are caught, so we can safely say that there is very little specific deterrent effect.

The effect of punishment on future behavior also depends on the **contrast effect,** which is the distinction between the circumstances of punishment and the usual life experience of the person being punished. The prospect of incarceration is a nightmarish contrast for those who enjoy a loving family and a valued career. The mere prospect of experiencing the embarrassment of public disgrace threatening families and careers is a strong deterrent for people embedded in a prosocial lifestyle. For those lacking these things, punishment has little effect because the negative contrast between the punishment and their normal lives is minimal. Like many other things in life, the irony is that specific deterrence works best for those who need deterring the least, and works least for those who need deterring the most.

General deterrence is the preventive effect of the threat of punishment on the general population, that is, on *potential* offenders. The existence of a system of punishment for law violators deters a large but unknown number of individuals who might commit crimes if no such system existed. Reviews of deterrence research indicate that legal sanctions do have "substantial deterrent effect" (Nagin, 1998, p. 16), especially for *instrumental crimes* (crimes that bring material rewards) as opposed to *expressive crimes* (crimes that bring psychological rewards). This punishment effect is seen when it is applied swiftly and with certainty, but there is little evidence that increasing the severity (in the form of sentence length) has any effect (McCarthy, 2002). Overall, these findings support the classical notion that individuals do (subconsciously at least) calculate the ratio of expected pleasures to possible pains when thinking about their actions.

Figure 3.1 summarizes major differences between the classical and positivist schools.

Figure 3.1	Summary and Comparisons of the Classical and Positivist Schools	
	Classical	**Positivist**
Historical Period	18th-century Enlightenment, early period of Industrial Revolution	19th-century Age of Reason, mid–Industrial Revolution
Leading Figures	Cesare Beccaria, Jeremy Bentham	Cesare Lombroso, Raffaele Garofalo, Enrico Ferri
Purpose of School	To reform and humanize the legal and penal systems	To apply the scientific method to the study of crime and criminality
Image of Human Nature	Humans are hedonistic, rational, and have free will. Our behavior is motivated by maximizing pleasure and minimizing pain.	Human behavior is determined by psychological, biological, or social forces that constrain our rationality and free will.
Image of Criminals	Criminals are essentially the same as noncriminals. They commit crimes after calculating costs and benefits.	Criminals are different from noncriminals. They commit crimes because they are inferior in some way.
Definition of Crime	Strictly legal; crime is whatever the law says that it is.	Based on universal human abhorrence; crime should be limited to inherently evil (mala in se) acts.
Purpose of Punishment	To deter. Punishment is to be applied equally to all offenders committing the same crime. Judicial discretion is to be limited.	Social defense. Punishment is to be applied differently to different offenders based on relevant differences and should be rehabilitative.

Evaluation of Neoclassical Theories

Critics of neoclassical theories complain about the overemphasis on the rationality of human beings and the ignoring of social conditions that may make it rational for some to engage in crime (Curran & Renzetti, 2001). We do need to understand what turns some people into "motivated offenders"; i.e., what is it that makes some of us willing to expend one resource (our potential loss of freedom) to attain another (the fruits of crime)? Many of us don't spend our resources all that wisely because of a tendency to favor immediate gain over long-term consequences, and one wonders why some of us more strongly favor immediate gain than others. For instance, de Haan and Vos (2003) find the assumption of rational criminals to be mistaken. Based on interviews with a number of street robbers, they conclude that the affective (emotional) aspects of criminal behavior are more important than the rational aspects in motivating offending behavior. They explore the roles of impulsiveness, self-expression, moral ambiguity, and shame in the accounts of the crimes and lives of their interviewees to explain criminal behavior.

In response, neoclassical theorists might insist that they do not assume a model of "pure" rationality; rather, they assume a limited rationality constrained by ability, knowledge, and time (Cornish & Clarke, 1986, p. 1). They do insist, however, that deterrence has an impact on crime and claim that increased incarceration rates account for about 25% of the decline in violent crime over the last decade or so (Rosenfeld, 2000; Spelman, 2000). Unfortunately, we cannot determine from raw incarceration rates if we are witnessing a *deterrent* effect (Has violent crime declined because more people have perceived a greater punitive effect?) or an *incapacitation* effect (Has violent crime declined because more violent people are behind bars and thus not at liberty to commit violent crimes on the outside?). Furthermore, these theories do not claim to explore the role of outside forces in producing criminals, but rather they explore criminal events with the purpose of trying to prevent them. They seek to deny the motivated offender the opportunity to commit a crime by target hardening. In addition, the notion that individuals are responsible for their own actions meshes well with American values. If this assumption "grants society permission" (Williams & McShane, 2004, p. 242) to punish criminals who make purposeful decisions to flout the law, then so be it, for the act of punishment presupposes free human beings and thus dignifies them.

Policy and Prevention: Implications of Classical and Neoclassical Theories

The classical school was a school of philosophical jurisprudence bent on establishing a set of reformist moral values in criminal justice, not a school of empirical data collection and analysis attempting to build a theory of criminality. Regardless of their influence on criminological theory, the influence of the classical theorists on the legal and penal systems of Europe and North America was huge. Many European monarchs of the 18th century were moved to adopt their principles, and the American Constitution and the 1789 French *Declaration of the Rights of Man* were very much influenced by them. All criminal justice systems in the world assume the classical position that persons are free agents who deserve to be punished when they transgress the law. We may also recognize many of the ideas championed by Beccaria in such rights as freedom from cruel and unusual punishment, the right to a speedy trial, and the right to confront one's accusers, contained in the U.S. Bill of Rights and other documents at the heart of Western legal systems today. The emphasis on rationality, on free will, and on personal responsibility within the modern legal system reflects the once-radical image of human beings posited by the great Enlightenment thinkers.

What about crime prevention? If you were the kind of motivated rational criminal assumed by neoclassical theorists, what sort of questions would you ask yourself at the potential crime site before you made your decision to commit the crime or not? Among them would likely be, "Is there a quick way out of the area after the job is done?" "How vulnerable are the targets (Is the car unlocked? Is the door open? Is the girl alone?)?" "What are my chances of being seen by people in the area?" "If people in this area do see me, do they look likely to do something about it?" These policy implications boil down to trying to arrange things such that criminals will dissuade themselves from committing crimes by making their choice structuring as difficult as possible. This is what is meant by *target hardening*.

Neoclassical theorists would be especially likely to recommend that the police concentrate their efforts on so-called hot spots rather than spreading themselves around. *Hot spots* are places where not only serious crimes occur but also numerous minor antisocial acts such as public drunkenness and urination, fights, and vandalism. They would argue that such acts should not be ignored because they contribute to further deterioration of a neighborhood and invite worse crimes.

Rational choice and routine activities theories thus shift the policy focus from large and costly social programs such as antipoverty programs and toward target hardening. They shift attention away from policies designed to change offenders' attitudes and behavior toward making it more difficult and more costly for them to offend. Examples of target hardening include antitheft devices on automobiles, the use of vandal-resistant materials on public property, improved city lighting, surveillance cameras in stores and at public gathering places, check guarantee cards, banning the sale of alcohol at sporting events, neighborhood watches, and curfews for teenagers.

Environmental design is primarily concerned with **defensible space,** defined as "a model for residential environments which inhibit crime by creating the physical expression of a social fabric that defends itself" (Newman, 1972, p. 3). It endeavors to bring people together into a tribe-like sense of community by designing the physical environment so as to awaken the human sense of territoriality. The best possible physical environment for the growth of crime is the large barracks-like blocks of apartments with few entrances, few private spaces, and few demarcation barriers that say, "This space is mine." Families must be given back a sense of ownership, for if everything is "owned" in common (elevators, walkways and staircases, balconies, grass and shrubberies), then no one takes care of it and it deteriorates rapidly. Streets must be blocked off, both to generate a sense of belonging to "my special little neighborhood" and so that criminals cannot easily access or escape them.

The answer to crime for classical deterrence theorists is punishment. "Get tough" messages are the kinds of simple, easily implemented solutions that policy makers love—build prisons and fill 'em up. But getting tough is expensive, as many legislative bodies have found out. Many states started putting more and more offenders behind bars for longer periods and implemented mandatory sentencing laws in the 1980s, but soon found their prisons so overcrowded that the courts intervened. This resulted in the repeal of some states' mandatory sentencing laws and the institution of early release programs. Thus, releasing offenders to the streets became the solution to a current problem, but that solution *was* the problem a few years earlier (Gilsinian, 1991).

This goes to show how remarkably complicated and even perverse policy decisions can be, and why we should not judge a criminological theory based on its impact (or lack of) on public policy.

What About Deterrence: Is the United States Hard or Soft on Crime?

The United States is perceived to be soft on crime by many laypersons, but if we define hardness or softness in terms of incarceration rates, the numbers do not support the perception. Figure 3.2 shows incarceration rates per 100,000 for selected countries in 2004. Using incarceration rates per 100,000 *citizens* is not the same as the rate per 100,000 *criminals,* however. The greater incarceration rate in the United States is justified if it has more criminals than these other countries. Of course, no one knows how many criminals any country has, but we can get a rough estimate from a country's crime rates. For instance, the U.S. homicide rate is about 5 times that of England and Wales, which roughly matches the United States' 5 times greater incarceration rate. However, when it comes to property crimes, Americans are about in the middle of the pack of nations in terms of the probability of being victimized (less than in England and Wales, incidentally). This fact notwithstanding, burglars serve an average of 16.2 months in prison in the United States, compared with 6.8 months in Britain and 5.3 months in Canada (Mauer, 2005), which makes the United States harder on crime than its closest cultural relatives and suggests that we may be overusing incarceration to address our crime problem. (Alternatively, from a crime control perspective, these other nations can be seen as underutilizing incarceration at the expense of raising crime rates.)

If we define hardness/softness in terms of alternative punishments or the conditions of confinement, then the United States is "soft" on crime, although a better term would be *humane.* For instance, although China is listed by Mauer (2005) as having an incarceration rate more than 5 times lower than the United States, it is the world's leader in the proportion of its criminals it executes each year. Also, punishment in

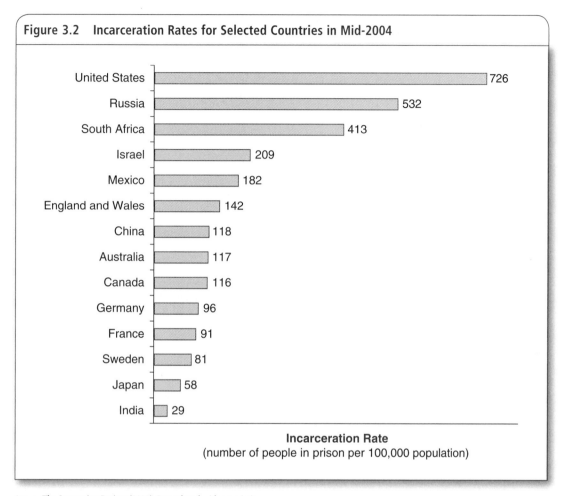

Figure 3.2   Incarceration Rates for Selected Countries in Mid-2004

Incarceration Rate
(number of people in prison per 100,000 population)

Source: The Sentencing Project (2005). Reproduced with permission.

some fundamentalist Islamic countries such as Saudi Arabia and Afghanistan under the Taliban often includes barbaric corporal punishments for offenses considered relatively minor in the West. Drinking alcohol can get the drinker 60 lashes, robbers may have alternate hands and feet amputated, and women accused of "wifely disobedience" may be subjected to corporal punishment such as a "spanking" with a bamboo cane (Walsh & Hemmens, 2011).

So, is the United States softer or harder on crime than other countries? The answer obviously depends on how we conceptualize and measure the hardness and softness and with which countries we compare ourselves. Compared with countries that share our democratic ideals, we are tough (because of our retention of the death penalty, some would even say barbaric) on crime; compared with countries most distant from Anglo/American ideals, we are extremely soft, and for that we should be grateful. But regardless of hardness or softness, we have to remember that the effects of deterrence depend far more on the certainty and swiftness of punishment, not its severity, and most assuredly on the contrast effect.

SUMMARY

- The classical school of criminology began during the Enlightenment with the work of Cesare Beccaria, whose aim was to reform an arbitrary and cruel system of criminal justice.
- Jeremy Bentham, best known for his concept of the hedonistic calculus, was another leading figure. The hedonistic calculus summarized the classical notion of human nature as hedonistic, rational, and possessed of free will.
- The positivist school aimed at substituting the methods of science for the armchair philosophizing of the classicists; i.e., they sought measurable causes of behavior.
- The cartographic criminologists such as Guerry, Quételet, Mayhew, and Fletcher were among the first positivists. These scholars studied maps and statistics to pinpoint where and when crime was most likely to occur.
- Cesare Lombroso is widely considered the father of criminology. His work was much influenced by evolutionary thought as he understood it. Lombroso saw criminals as atavistic "throwbacks" to an earlier evolutionary period, who could be identified by a number of bodily stigmata.
- Other early positivists included Raffaele Garofalo and Enrico Ferri. Garofalo was interested in developing a "natural" definition of crime, and in generating categories of criminals for the purpose of determining what should be done with them. Ferri was instrumental in formulating the concept of social defense as the only justification for punishment.
- Neoclassical theories reemerged in the form of rational choice and routine activities theories in the 1980s. These theories assume that humans are rational and self-seeking, although rationality is bounded by knowledge levels and thinking abilities. They downplay personal and background factors that influence choices in favor of analyzing the processes leading to offenders' choices to offend.
- Routine activities theory looks at a criminal event as a motivated offender meeting a suitable target and lacking a capable guardian. These ideas show how crime rates can go up or down without a change in the prevalence of motivated offenders by increasing/decreasing suitable targets and capable guardians.

DISCUSSION QUESTIONS

1. If humans are primarily motivated by the hedonistic calculus, is simple deterrence the answer to the crime problem?

2. What advantages (or disadvantages) does positivism offer us over classicism?

3. Is Ferri's social defense rationale for punishment preferable to one emphasizing rehabilitation of offenders?

4. Use any search engine and type in "Beccaria preventing crime." How do Beccaria's ideas compare with those of the positivists on preventing crime? What is Beccaria's idea of "real crime," and how does it compare with Garofalo's?

USEFUL WEBSITES

Biological Positivism. www.woodfin.org/classes/criminology/biological%28T%29.html

Classical and Positivist Schools of Criminology. http://en.wikipedia.org/wiki/Criminology

Enrico Ferri. www.crimetheory.com/Theories/Positivist.htm

Rational Choice Theory. http://privatewww.essex.ac.uk/~scottj/socscot7.htm

Routine Activities Theory. http://en.wikipedia.org/wiki/Routine_activity_theory

CHAPTER TERMS

Atavism	Determinism	Insane criminal
Cartographic criminologists	Deterrence	Positivism
Choice structuring	Free will	Rational
Classical School	General deterrence	Rational choice theory
Contrast effect	Hedonism	Routine activities theory
Criminaloid	Hedonistic calculus	Social defense
Defensible space	Human agency	Specific deterrence

4

Social Structural Theories

On June 15, 1975, 12-year-old Kody Scott graduated from elementary school in Los Angeles. During the ceremony, his thoughts were on "the hood" and his one ambition in life, which was to join the Eight Tray Crips; become a "ghetto star"; and major in murder, robbery, and general mayhem. He went straight from the graduation to his initiation into the gang, which involved taking part in the gunning down of 15 members of a rival faction of L.A.'s other notorious gang, the Bloods. Two years later, during a robbery in which the victim tried to run, Kody beat and stomped the man into a coma. A police officer at the scene said that "whoever did this is a monster," a name Kody proudly took as his street moniker. Monster did time in juvenile detention and then served several prison terms. During one of these terms, he converted to Afrocentric Islam and changed his name to Sanyika Shakur. He also wrote *Monster: The Autobiography of an L.A. Gang Member,* which provides a frightening portrayal of the violence of ghetto life. Shakur was paroled in 1995; returned to prison on parole violations in 1996, 1997, and 1998; and again was incarcerated for a shooting in 2000. Paroled again sometime later, he was rearrested in 2004 for "battery with great bodily harm" and again sent to prison.

Shakur was allegedly the illegitimate son of an ex–football player named Dick Bass. His mother subsequently married another man and had four more children. She divorced their father when Shakur was 6 years old and had to raise the children alone. Shakur was mistreated by his stepfather and never included in family outings. He spent almost all his childhood in the wild and chaotic streets, which he says was the only thing that really interested him. As you read this chapter about disorganized neighborhoods, blocked opportunities to legitimate success, and lower-class values, try to imagine Shakur at the center of it all and how these things may have shaped his life.

▧ The Social Structural Tradition

Almost all sociological theories of crime touch on **social structure** to some degree. Social structure is the framework of social institutions—the family, educational, religious, economic, and political institutions that operate to structure the patterns of relationships members of a society have with one another. Structural theorists maintain that wholes (societies, institutions, groups) are greater than the sums of their parts, that these wholes are real and enjoy an existence of their own, separate from their individual members. Although groups cannot exist apart from the individuals who comprise them, once formed they take on an existence independent of any one individual. Structural theorists work from assumptions made from general models of society and deduce everyday experiences of individuals from those models. Their philosophy is summed up in the proposition that society prepares the crime and individuals are only the instruments that give it life. They are thus interested in seeking the social structural causes of group crime rates rather than the answer to why particular individuals commit crimes.

Structural criminologists tend to assume that human nature is socially constructed, an unconstrained vision position that avers human traits and characteristics are specific to local cultures. This is in opposition to the constrained vision that maintains that there is a human nature common to all cultures, although cultures determine its expression (Mallon, 2007). Given the socially constructed assumption, the task of structural criminologists is to discover why social animals commit antisocial acts. If human nature is socially constructed, the presence of antisocial individuals reflects defective social practices such as competitiveness, poverty, racism, inequality, and discrimination rather than defective human materials.

Structural theorists follow one of two general models of society—the consensus perspective or the conflict perspective. This chapter examines the **consensus perspective** (sometimes known as the **functionalist perspective**), which views society as a system of mutually sustaining parts and characterized by broad normative consensus. Society is viewed analogously to the body integration of diverse organs (heart, liver, kidneys, etc.). When any one of these parts (the organs of the body or the institutions of society) malfunctions, all the other parts are adversely affected.

▧ The Chicago School of Ecology/Social Disorganization

The first criminological theory to be developed in the United States was the Chicago School of human ecology developed at the University of Chicago in the 1920s and 1930s primarily through the works of Clifford Shaw and Henry McKay (1972). If you look back to Table 1.1 in Chapter 1, you will see that social disorganization theory is ranked number 4 in popularity by contemporary criminologists.

Early social ecologists viewed the city as a kind of super organism with "natural areas" that were adaptive for different ethnic groups (Little Italy, Chinatown, etc.). When natural areas are eroded by "alien" ethnic groups, large increases in deviant behavior ensue. In their analysis of Cook County Juvenile Court records spanning the years from 1900 to 1933, Shaw and McKay noted that the majority of delinquents always came from the same neighborhoods. This suggested the existence of natural areas that facilitated crime and delinquency independent of other factors. Their findings also increased confidence among sociologists in their assertion that the environment was more important than ethnic group or individual differences in explaining criminal behavior. It was not claimed that residential areas "caused" crime, but rather that it was heavily concentrated in certain neighborhoods regardless of the ethnic identities of their residents.

Previous research in social ecology characterized the spatial patterns of U.S. cities as radiating outwards from central business and industrial areas in a series of concentric circles, or *ecological zones*, such as in the city of Chicago. As shown in Figure 4.1, Zone I was the Loop area of Chicago; Zone II was the factory zone around which earlier Chicago residents had built their homes, but was then inhabited by the poorest residents. This was the so-called **transition zone** where social changes leading to delinquency mostly occurred. This was the zone inhabited by newly arrived individuals looking for work because the cheapest rents were there. In response to increasing factory expansion, some Zone II residents invaded Zone III, the zone of working-class homes, making this zone less desirable, and those who could afford to do so moved out. Successive waves of poor foreign and native immigrants moved into these old inner-city neighborhoods, just as they had in the British cities studied by Mayhew and Fletcher in the 1800s, discussed in Chapter 3. This process had a rippling effect, like a stone dropped in a lake. Successive waves of newcomers

---

Figure 4.1   Zone Map of Male Delinquents in Chicago, 1925–1933

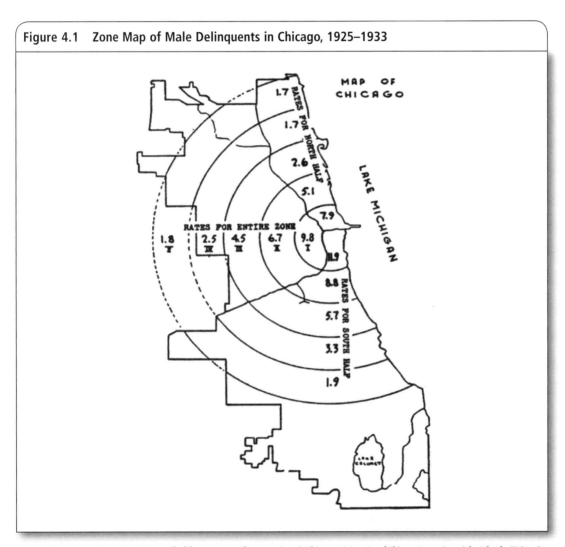

to the poorest neighborhoods precipitate constant movement from zone to zone as more established groups seek to escape the intrusion of the newcomers. Urban decay and crime are consequences of this population movement. Note that delinquency rates decreased linearly from 9.8 incidents per 100 juveniles in the poorest zone to 1.8 per 100 in the most affluent zone.

The decline in crime and delinquency that occurred from the inner city outwards was not in itself theoretically significant or useful. What Shaw and McKay (1972) needed was a mechanism that explained it. The mechanism they came up with was **social disorganization,** by which they meant the breakdown of the power of informal community rules to regulate conduct. Social disorganization is created by the continuous redistribution of neighborhood populations, bringing with them a wide variety of cultural traditions sometimes at odds with traditional American middle-class norms of behavior. A neighborhood invaded by members of alien racial or ethnic groups becomes rife with conflicting values and conduct norms and loses its sense of community.

Social disorganization impacts crime and delinquency in two ways. First, the lack of informal social controls within them facilitates crime by failing to inhibit it. Second, in the absence of prosocial values, a set of values supporting antisocial behavior is likely to develop to fill the vacuum. Slum youths thus have both negative and positive inducements to crime and delinquency, represented by the absence of social controls and the presence of delinquent values, respectively. These conditions are transmitted across generations until they become intrinsic properties of the neighborhood. Figure 4.2 is a diagrammatic presentation of the theory.

Shaw and McKay (1972) worked under the assumption that effective neighborhoods were characterized by warm emotional bonds based on shared ethnicity and values, and that social control was born from this shared intimacy. Although rooted in the social disorganization approach, Robert Sampson (2004) updated the notion of neighborhood control of crime without reference to the narrow focus of traditional ethnicity-based emotional ties with his concept of collective efficacy. **Collective efficacy** is the shared power of a group of connected and engaged individuals to influence the maintenance of public order. Modern neighborhoods exercise social control based on shared rational goals utilizing shared expectations that others can be counted on to take action to prevent crime (neighborhood watches, voluntary associations, demand for police services, and so on). As we would expect from individuals bonded by shared interests rather than shared personal ties, neighborhood collective efficacy is dynamic and task-specific rather than static and generalized. A study of Chicago neighborhoods found that crime rates were low when collective efficacy was high. We should note that the same things that predict low collective efficacy—concentrated poverty, lack of home ownership, rundown buildings, family disruptions, and so on—are the things that predict high social disorganization; that is, collective efficacy and social disorganization are negatively related with respect to causal function (Sampson, Raudenbush, & Earls, 1997).

Figure 4.2   Diagrammatic Presentation of Social Disorganization Theory

| Influx of immigrants into cities looking for work and congregating in poorest areas | → | Value conflicts and decrease in formal and informal social controls lead to **Social Disorganization** | → | Deterioration of neighborhood and development of delinquent values | → | **Delinquency and Crime** |

Evaluation of Social Ecology/Social Disorganization Theory

Shaw and McKay's (1972) theory points out that crime is concentrated in socially disorganized areas inhabited by economically deprived people. But causal direction has always been a problem: Are neighborhoods run-down and criminogenic because people with personal characteristics conducive to both crime and poverty populate them, or do neighborhoods somehow "cause" crime independent of the characteristics of people living there? After all, when formerly blighted areas become "gentrified" and middle-class people move in, the neighborhood is no longer "criminogenic." Some theorists argue that ecological factors have no independent effect on crime once the human composition of areas is taken into consideration. Others argue the opposite, while still others argue that people and places are equally important in explaining crime.

The "people versus places" argument was best stated by Ruth Kornhauser's (1978) question: "How do we know that area differences in delinquency rates result from the aggregated characteristics of communities rather than the characteristics of individuals selectively aggregated into communities" (p. 104). In other words, do neighborhoods make people or do people make neighborhoods? Obviously, choices people make are often constrained by factors beyond their power to control, but surely people bear some responsibility for the state of their environments. The people vs. places issue has been well researched with the overall conclusion being that neighborhoods do have effects independent of individual differences, although individual differences matter (Webster, MacDonald, & Simpson, 2006). Related to this issue, Osgood and Chambers' (2003) large study of social disorganization theory in a rural setting also traced social disorganization to high population turnover and ethnic diversity, but the authors identified high rates of female-headed households as the most important factor in explaining crime rates.

Ecological theory cannot account for why the majority of people in disorganized neighborhoods do not commit serious crimes, or why among those who do, a very small minority commits the majority of them. The concept of **ecological fallacy** states that we cannot make inferences about individuals and groups on the basis of information derived from the larger population of which they are a part. Even Shaw and McKay (1972) showed that Asians and Jews living in high-crime areas had very low crime rates. To find high crime rates in a neighborhood with a large Asian population living there and then to assume that Asians commit crimes at a rate matching the neighborhood rate is an example of the ecological fallacy. Low Asian crime rates in high-crime neighborhoods also suggests that the rejection of group and individual differences as explanations for crime and delinquency was premature.

⬚ The Anomie/Strain Tradition

French sociologist Émile Durkheim provided criminology with one of its most revered and enduring concepts: anomie. **Anomie** is a term meaning "lacking in rules" or "normlessness." Durkheim was greatly concerned with social solidarity and the threat posed to it by social change. He distinguished between mechanical and organic solidarity. **Mechanical solidarity** exists in small, isolated, pre-state societies in which individuals share common experiences and circumstances, and thus common values and strong emotional ties to the collectivity. This is basically the same idea at the societal level that Shaw and McKay emphasized at the neighborhood level. Under these circumstances, informal social controls are strong and antisocial behavior is minimal. **Organic solidarity** is characteristic of modern societies with high degrees of occupational specialization and diversity of experiences and circumstances. This diversity weakens common values and social bonds and antisocial behavior grows. Durkheim argued that because crime is found

▲ **Photo 4.1** Émile Durkheim (1858–1917)

at all times and in all societies, it is a normal and inevitable social phenomenon, and even socially useful (functional) in that it serves to identify the limits of acceptable behavior. Too much repression of deviant behavior would lead to a pathological conformity that would stifle creativity, progress, and personal freedom. Crime is one of the prices we pay for personal freedom and for social progress. Durkheim (1982) even asserted that when crime drops significantly below average levels (such as in wartime), it is sign that something is wrong (p. 102).

Although Durkheim (1982) believed that crime was "a normal phenomenon of normal sociology," he did not imply that the criminal "is an individual normally constituted from the biological and psychological points of view" (pp. 106–107). He also asserted that in any society there are always individuals who "diverge to some extent from the collective type," and among them is the "criminal character" (p. 101). Durkheim thus left room for psychologists and biologists to get into the criminology game.

Durkheim (1951) also set the stage for later extensions of anomie theory by addressing "strains" when he wrote that although human beings are similar in their "essential qualities, [there is] one sort of heredity will always exist, that of natural talent" (p. 251). The crux of the crime problem at the individual level is that "human activity naturally aspires beyond assignable limits and sets itself unattainable goals" (p. 247). When people get less than they expect, they are ripe for criminal behavior.

✉ **Robert Merton's Extension of Anomie Theory**

Robert Merton (1938) expanded anomie theory to develop an explanation of crime that has come to be known as **strain theory.** The central feature of Merton's theory is that American culture defines monetary success (the "American Dream") as the predominant cultural goal to which all its citizens should aspire, while at the same time American social structure restricts access to legitimate means of attaining this goal for certain segments of the population. The disjunction between cultural goals and the structural impediments to achieving them is the anomic gap in which crime is bred.

Rather than anomie being an occasional condition arising in periods of rapid social change as Durkheim saw it, for Merton anomie is a permanent condition of society caused by this disjunction. According to Merton, being unable to attain culturally defined wants legitimately invites frustration (strain), and may result in efforts to obtain them illegitimately. Merton disagreed with Durkheim's belief that acquisitiveness is intrinsic to human nature, viewing it instead as a culturally generated characteristic. Merton claims that American culture and social structure actually exert pressure on some people to engage in *nonconforming* (criminal) behavior rather than conforming behavior. Thus, society is the cause of anomie, not the victim of it as in Durkheim's view.

Merton identified five **modes of adaptation** that people adopt in response to this societal pressure, all of which, with the exception of conformity, are deviant.

1. *Conformity* is the most common mode of adaptation because most people have the means of legally attaining cultural goals at their disposal. Conformists accept the success goals of American society, and the prescribed means of attaining them (hard work, education, persistence, dedication).

2. *Ritualism* is the adaptation of the 9-to-5 slugger who has given up on ever achieving material success, but who nevertheless continues to work within legitimate boundaries because he or she accepts the legitimacy of the opportunity structure.

3. *Innovation* is the mode of adaptation most associated with crime. For Merton, crime is an innovative avenue to success—a method by which deprived people (and the not-so-deprived) get what they have been taught by their culture to want.

4. *Retreatism* is adopted by those who reject both the cultural goals and the institutionalized means of attaining them. Retreatists drop out of society and often take refuge in drugs, alcohol, and transience and are frequently in trouble with the law.

5. *Rebellion* is the adaptation of those who reject both the goals and the means of American society but wish to substitute alternative legitimate goals and alternative legitimate means. Rebels may be committed to some alternative sociopolitical ideal, such as socialism.

Merton never systematically explored why certain individuals took on one adaptation rather than another. In a sense, his theory of crime is about the envy and resentment that his innovators feel about being left out of the American Dream. The power of relative deprivation (the discontent felt when comparing what one has with what others have and discovering that one has less than one believes one should have) is often accompanied by negative self-feelings, which in turn may motivate adoption of deviant coping patterns (Stiles, Liu, & Kaplan, 2000). Whatever the route to one adaptation or the other may be, Merton's anomie strain theory has generated a great deal of interest and theoretical extension over the decades, as we will see below.

Institutional Anomie Theory

Institutional anomie theory (IAT) extends anomie theory and claims, "High crime rates are intrinsic to American society: in short, at all social levels, *America is organized for crime*" (Messner & Rosenfeld, 2001, p. 5). Messner and Rosenfeld show that inequality in the United States is not an aberration of the American Dream but an expression of it: "A competitive allocation of monetary rewards [that] requires both winners and losers, and winning and losing have meaning only when rewards are distributed unequally" (p. 9). From this position, a meritocracy is both criminogenic and unfair. Sawhill and Morton (2007) also question the fairness of the American meritocracy when they write that "people are born with different genetic endowments and are raised in different families over which they have no control, raising fundamental questions about the fairness of even a perfectly functioning meritocracy" (p. 4).

IAT posits that the root of the problem is the subjugation of all other social institutions to the economy in the United States. American culture devalues the non-economic function and roles of other social institutions and obliges them to accommodate themselves to economic requirements. Messner and Rosenfeld (2001) claim that a great deal of the focus of the family, religion, education, law, and government is brought to bear on instilling in Americans the beliefs and values of the marketplace to the detriment of the institution-specific beliefs and values they are supposed to inculcate. The dominance of the economy thus disrupts and devalues the prosocial functioning of the other institutions and substitutes an overweening concern for the pursuit of monetary rewards, which IAT sees as profoundly criminogenic.

Messner and Rosenfeld's (2001) plan to reduce crime is one of **decommodification.** *Commodification* means the transformation of social relationships formerly untainted by economic considerations into commodities; thus, *de*commodification refers to policies intended to free social relationships from economic considerations by freeing the other social institutions from the domination of the economy. For instance, few people complain about finding time for their jobs as opposed to finding time for their family. And how many students go to college "to get a job" rather than for the love of learning? Freeing people from economic domination would allow individuals to construct their lives unconstrained by market considerations. According to IAT, this would reduce what its adherents see as cut-throat competition, which in return would reduce crime.

Robert Agnew's General Strain Theory

Robert Agnew has made several attempts to fine-tune and reformulate strain theory, culminating in his **general strain theory (GST).** Agnew (2002) identifies several other sources of strain besides the disjunction between expectations and actual achievements. Strain can also result from the removal of positively valued stimuli, such as the loss of a boyfriend/girlfriend or having to move to a new neighborhood. These problems may induce delinquency or crime via efforts to prevent or regain the loss via illegal means or to

▲ **Photo 4.2**   Strain theories presume that lower-class citizens envy the rich but, lacking the means to become productive, often turn to crime as an alternative pathway. These brand-new condos were built by tearing down some of the neighboring high-rise slum buildings in Cabrini Green. Police patrol the streets that separate the new housing, unaffordable to the ghetto residents, making sure that the new condos are safe from burglaries or robberies.

gain revenge on those deemed responsible for the loss. Strain also arises from the presentation of negative stimuli such as child abuse/neglect and negative school experiences. These can lead to delinquency and crime via efforts to escape negative stimuli by running away from home and truancy.

We all experience multiple strains throughout our lives, but the impact of strain differs according to its *magnitude, recency, duration,* and *clustering* (Miseries that cluster together produce a whole greater than the sum of its parts and may overwhelm coping resources.). Agnew (2002) tells us that strain can result in crime and delinquency through the development of a generally negative attitude about other people: "Repeated or chronic strain may lead to a hostile attitude—a general dislike and suspicion of others and an associated tendency to respond in an aggressive manner" (p. 119).

For Agnew, strain is primarily the result of *negative emotions* that arise from negative relationships with others and not only from Merton's blocked opportunities to financial success. Although Agnew added much to the understanding of strain, his greatest contribution is to remind us that the most important factor is not strain per se but *how one copes with it*. Although none of us is happy when we are strained and may curse and throw things, few of us cope with it by committing crimes. How we cope with strain depends on things such as the level of social support we enjoy; the number, frequency, duration, and intensity of the strain-inducing circumstances we face; and what kind of person we are. According to Agnew (2002), the individual traits that differentiate between people who cope poorly with strain and others who cope well include "temperament, intelligence, creativity, problem-solving skills, interpersonal skills, self-efficacy, and self-esteem" (p. 123). He goes on to say, "These traits affect the selection of coping strategies by influencing the individual's sensitivity to objective strains and the ability to engage in cognitive, emotional, and behavioral coping" (p. 123). Of course, these may also be the traits that move people into one or the other of Merton's adaptations.

⊠ Subcultural Theories

Subcultural theories evolved primarily from the anomie/strain tradition via Merton's many students. Subcultures emerge when a significant number of people feel alienated or are segregated from the larger culture and forge a lifestyle that is different from the mainstream culture. Albert Cohen (1955) proposed a subcultural theory explaining how lower-class youths adapt to the limited avenues of success open to them. Cohen maintained that most criminal behavior in lower-class neighborhoods is not a rational method of acquiring assets, but is rather an expression of **short-run hedonism,** meaning that the actor is seeking immediate gratification of his or her desires without regard for any long-term consequences. Much delinquent behavior is also non-utilitarian, malicious, and negative in the sense that it turns middle-class norms of behavior upside down (e.g., destroying rather than creating).

Because lower-class boys cannot adjust to what Albert Cohen (1955) called **middle-class measuring rods,** they experience a status frustration and spawn an oppositional culture with behavioral norms that are consciously contrary to those of the middle class. Cohen saw criminal subcultures as a kind of mass reaction formation to the problem of blocked opportunities, although he saw **status frustration,** not blocked opportunity, as the real problem. Lower-class youth desire approval and status like everyone else, but seek it via alternate means. To gain status and respect, members of criminal subcultures establish "new norms, new criteria of status which defines as meritorious the characteristics they *do* possess, the kinds of conduct of which they *are* capable" (p. 66). These status criteria are most often physical in nature, such as being ready and able to respond violently to challenges to one's manhood, or gaining a reputation as a "stud."

Another influential extension of strain theory was Cloward and Ohlin's (1960) **opportunity structure theory.** Cloward and Ohlin accepted that delinquents and criminals want middle-class financial success, but have little interest in the usual indicators of middle-class economic success, preferring "big cars," "flashy clothes," and "swell dames" (p. 96). Their biggest contribution, however, was to point out that just as there are barriers to achieving legitimate success, there are also barriers to achieving illegitimate (criminal) success—it takes more than talent and motivation to make it within either the legitimate or illegitimate opportunity structures. To obtain illegitimate opportunities, would-be crooks need a friend, relative, or acquaintance who can show them "the ropes." Youths born into an established and organized delinquent subculture—the illegitimate "opportunity structure"—have a career advantage over "wannabe" outsiders. Individuals within an illegitimate opportunity structure join *criminal gangs*. The best example of this type of gang is organized crime gangs such as the Mafia, which has a pool of aspiring "sponsored" recruits.

Cloward and Ohlin (1960) identified two other gang types that develop from the frustration in lower-class culture: conflict gangs and retreatist gangs. *Conflict gangs* are generated in slum areas with a high degree of transience and instability as opposed to stable areas with an established illegitimate opportunity structure. Members of these loose-knit gangs commit senseless acts of violence and vandalism, and their efforts to make a living from criminal activity tend to be "individualistic, unorganized, petty, poorly paid and unprotected" (p. 73). *Retreatist gangs* are more "escapist" in their attitudes than conflict gangs in that almost all of their members abuse drugs and/or alcohol. In both conflict and retreatist gangs, the concern is not with remote goals but rather with the immediate gratification of present wants.

⬚ Walter Miller's Theory of Focal Concerns

Walter Miller's (1958) theory was based on a very large study sponsored by the National Institute of Health and conducted by Miller and seven trained social workers who maintained daily contact with subjects "for a total time period of about thirteen worker years" (p. 6). Miller took issue with the idea that gangs are formed as a *reaction* to status deprivation. Criminals may resent the middle class, but it is not a matter of "If you can't join 'em, lick 'em," because their resentment is born out of envy for what middle-class people have, not for what they are. Middle-class traits such as hard work, stable habits, and responsibility are not appealing to them. Miller asserted that lower-class values must be viewed on their own terms and not as simple negations of those of the middle class. He also identified six **focal concerns** that are part of a value system and lifestyle that has emerged from the realities of life on the bottom rung of society. These interrelated focal concerns are as follows:

1. *Trouble* is something to stay out of most of the time, but life is trouble and it is something that confers status if it is the right kind (being able to handle one's self).

2. *Toughness* is very important to the status of lower-class males—being strong, brave, macho, sexually aggressive, unsentimental, and "not taking any shit."

3. *Smartness* refers to street smarts, the ability to survive on the streets using one's wits.

4. *Excitement* is the search for fun, often defined in terms of fighting, sexual adventurism, gambling, and getting drunk or stoned.

5. *Fate* is a belief that the locus of control is external to oneself.

6. *Autonomy* means personal freedom, being outside the control of authority figures such as teachers, employers, and the police, and thus being able to "do my own thing."

The hard-core lower-class lifestyle typified by these focal concerns catches those engaged in it in a web of situations that virtually guarantee delinquent and criminal activities. The search for *excitement* leads to sexual adventures in which little preventative care is taken (*fate*), and the desire for personal freedom (*autonomy*) is likely to preclude marriage if pregnancy results. Miller was concerned about the fact that many lower-class males thus grow up in homes lacking a father or any other significant male role model. This leaves them with little supervision and leads them to seek their male identities in what Miller (1958) called "one-sex peer units" (male gangs) (p. 14).

Miller's ideas are given strong support by Elijah Anderson's (1999) ethnographic work in African American neighborhoods in Philadelphia. The concentration of disadvantages in such neighborhoods has spawned a hostile oppositional culture spurning most things valued by middle-class America, as in "rap music that encourages its young listeners to kill cops, to rape, and the like" (p. 107). Anderson points out that although there are many "decent" families in these neighborhoods, the cultural ambiance is set by "street" families, which often makes it necessary for decent people to "code-switch" (adopt street values) to survive. Striving for education and upward mobility is viewed as "dissing" the neighborhood, and street people do what they can to prevent their "decent counterparts from . . . acting white" (p. 65). The street code is primarily a campaign for respect ("juice") achieved by exaggerated displays of manhood, defined in terms of toughness and sexuality.

Evaluation of the Anomie/Strain and Subcultural Tradition

Despite its impact on theory generation in criminology, the anomie/strain tradition does not seem to be strongly in favor among contemporary criminologists. Looking back at Table 1.1 in Chapter 1, Merton's theory was checked by only 2 of the 379 criminologists, Messner and Rosenfeld's IAT was checked by 6, and Agnew's GST by 7.

Because of the emphasis on monetary success in the United States, anomie/strain theory should best explain rational crimes netting perpetrators monetary gain. For instance, Bartol and Bartol (1989) claim that anomie theory's strength lies in its "ability to explain why utilitarian crime rates are so high in one society (e.g., the United States) and so low in another (e.g., England)" (p. 110). Ever since the mid-1980s, however, England has had higher rates of utilitarian crimes (e.g., burglary, auto theft) than the United States (Kesteren, Mayhew, & Nieuwbeerta, 2000). It is non-utilitarian crimes such as murder, rape, and assault that are more common in the United States than in England, which supports Albert Cohen's contention that much of American lower-class crime is non-utilitarian and malicious rather than rationally instrumental.

Agnew (1997) points out a potentially fatal flaw in Merton's anomie/strain theory. A strict interpretation of it should lead us to predict a sharp increase in criminal behavior in late adolescence/early adulthood when many such individuals begin to seriously enter the job market. If there is a disjunction between cultural goals and structural impediments to achieving them, a number of young adults entering the job market will feel its bite for the first time and respond with one of Merton's deviant adaptations. However, just when the alleged cause of criminal behavior becomes most salient for young people, we observe a significant decrease in antisocial behavior among them rather than an increase.

Another flaw is that Merton never attempts to explain what it is that sorts individuals into his modes of adaptation beyond citing class-based socialization practices. In other words, he is basically saying that social class causes social class, and social class determines mode of adaptation. Agnew's general strain theory has attempted to account for the mode adopted by noting that how we deal with strain is more important than its existence because it is something that we all experience in our lives. Dealing with strain productively leads to positive outcomes; dealing with it negatively leads to destructive outcomes.

Subcultural theories look at patterned ways of life in areas whose inhabitants set themselves apart and pride themselves in their distinctiveness. It is this patterned way of life that sustains delinquent values and goals.

However, a number of theorists have cast doubt as to whether there are distinct lower-class subcultures in this sense. It is difficult to imagine that lower-class subculture arose by deliberately taking steps to turn middle-class norms on their head, as Albert Cohen's reaction formation hypothesis supposes. Nevertheless, there do seem to be areas in our cities in which middle-class values are disdained, not because they are defined as middle class, but rather because they demand self-control, delayed gratification, and the disciplined application of effort.

Walter Miller's work was criticized by Tittle (1983) who asked if the middle and upper classes embrace values that are opposite of those of the lower class: Do they "value weakness, stupidity, boredom, and dependency?" (p. 341). This is a strange criticism because Miller (1958) explicitly states that these concerns are not confined to the lower classes; it is their meaning and the ways they are expressed "that differs both in rank order and weighting" (p. 6). Presumably we all understand that toughness, excitement, smartness, and autonomy mean vastly different things to middle- and lower-class individuals.

As we have seen, Elijah Anderson supports Albert Cohen and Walter Miller in his contention that the street code is primarily a campaign for respect ("juice") achieved by exaggerated displays of manhood, defined in terms of toughness and sexuality. These displays contribute greatly to the high rates of violence and out-of-wedlock births in the kinds of neighborhoods that Anderson (and Cohen and Miller before him) describes. If these theorists are correct about the role of fate in lower-class life, then the whole anomie/strain argument about blocked opportunities may be well off base. If lower-class individuals perceive their opportunities in a fatalistic "live-for-the-moment" way, or spurn them as antithetical to the street code, it is their visions of reality and the values imparted by their subculture that are blocking their legitimate success.

The anomie/strain tradition does not link poverty straightforwardly to poverty; rather, it sees it as having criminogenic effects only when coupled with a competitive culture that ties self-worth to monetary success. This leads us to a major issue in criminology: Does poverty cause crime or does crime cause poverty? Prominent criminologist Robert Sampson (2000) notes, "Everyone believes that 'poverty causes crime' it seems; in fact, I have heard many a senior sociologist express frustration as to why criminologists would waste time with theories outside the poverty paradigm. The reason we do . . . is that the facts demand it" (p. 711). Frank Schmalleger (2004) also notes that the underlying assumption of all structural theories is that the "root causes" of crime are poverty and various social injustices. But as he also points out, "Some now argue the inverse of the 'root causes' argument, saying that poverty and what appear to be social injustices are produced by crime, rather than the other way around" (p. 223). This is an extremely important issue, the answer to which points to opposite policy implications. Of course, neither may cause the other because some third set of variables may cause both crime and poverty.

✉ Gangs Today

Figures from the 2009 National Youth Gang Survey (NYGS) estimated that there were 27,900 gangs and 774,000 gang members in 2007 and that approximately half of all homicides in Chicago and Los Angeles in that year were gang related (National Gang Center, 2009). Gang membership has increased dramatically since theorists such as Albert Cohen were writing in the 1950s and 1960s. The increase has been largely attributed to the loss of millions of manufacturing jobs in the United States (J. Moore & Hagedorn, 2001), a loss that has hit our most vulnerable citizens, the young and the uneducated, hardest. Deindustrialization has set in motion a chain of events that has created a large segment of the population that is economically marginalized and socially isolated from mainstream culture.

Marginalized and isolated people (mainly African Americans and Hispanics) have become known as the "underclass" and the "truly disadvantaged" (J. Wilson, 1987, p. 8), and the neighborhoods where they live are fertile soil for the growth of gangs. The 2009 NYGS survey of the ethnic composition of gangs found that 49.5% were Hispanic, 35.2% were African American/black, 8.5% were white, and 6.8% were "other" (National Gang Center, 2009). It is estimated that over 25% of black males between the ages of 15 and 24 in Los Angeles County are members of one of the nation's two most notorious youth gangs, the Crips or the Bloods (Shelden, Tracy, & Brown, 2001, p. 28).

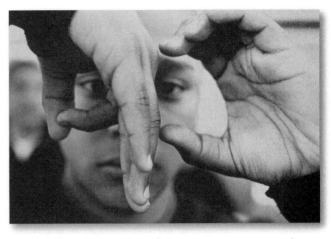

▲ **Photo 4.3**   Young member of the Cypress gang in Los Angeles flashing a gang sign. Gangs claim to replace the family cohesiveness the youth may not have at home, while symbols such as colors and signs clearly demarcate members from outsiders or rival gangs.

Why Do Young People Join Gangs?

Irving Spergel (1995) writes that "Youths join gangs for many reasons: status, security, money, power, excitement, and/or new experiences" (pp. 108–109). Joining a gang has become a survival imperative in some areas where unaffiliated youths are likely to be victimized. Having "homies" watching your back makes you feel safe and secure. Gang membership also provides a means of satisfying "belongingness" needs—having a place in the world among people who care. Gang members often display their belonging through initiation rites, secret gang signals, special clothing, "colors," and tattoos, all of which shout out loud, "I belong! I'm valued!" A youth camp counselor describes this function of gangs well: "The gang serves emotional needs. You feel wanted. You feel welcome. You feel important. And there is discipline and there are rules" (Bing, 1991, p. 12).

Social institutions (especially the family and the economy) satisfy most of the needs for most of us, but in the virtual absence of the influence of these institutions in the lives of those most affected by the economic and demographic transitions of the past few decades, the gang offers an attractive substitute means of achieving their needs. Thus, the gang functions for many of its members as (1) a family, (2) a friendship group, (3) a play group, (4) a protective agency, (5) an educational institution, and (6) an employer.

However, Martin Sanchez-Jankowski (2003) would like us to focus less on the psychological rewards of gang membership and more on the relationship between the structure of society and the structure of gangs. He argues in the tradition of IAT, stating that the competitive nature of American society requires winners and losers, and gangs are composed of society's losers. Sanchez-Jankowski also sees gang members in Mertonian fashion as accepting the economic principles of America and seeking economic success within the confines of the opportunities available to them.

An ongoing concern in all cities struggling to contain gangs is gang violence. The factors most often leading to gang violence according to the agencies responding to the NYGS is provided in Figure 4.3. For instance, 73.5% of the agencies mentioned drugs as a major factor influencing gang violence. Note also that 43.1% of the agencies report the emergence of new gangs in their areas, and 21.8% report migration from outside the United States. This indicates that gangs will continue to be a big problem for a long time to come.

Table 4.1 provides a useful summary of the key concepts, strengths, and weaknesses of the theories presented in this chapter.

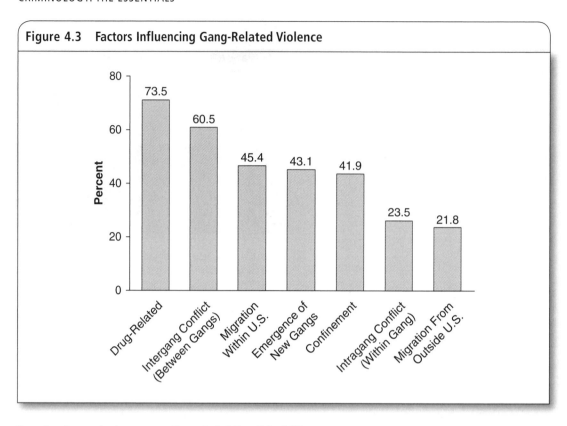

Figure 4.3 Factors Influencing Gang-Related Violence

Policy and Prevention: Implications of Social Structural Theories

If socially disorganized slum neighborhoods are the "root cause" of crime, what feasible policy strategies might be recommended to public policy makers? One of the first things you might want to suggest would be the strengthening of community life, but how do we go about it? Clifford Shaw (1972) began by securing funds for the **Chicago Area Project (CAP),** which consisted of a number of programs aimed at generating or strengthening a sense of community in neighborhoods with the help and cooperation of schools, churches, recreational clubs, trade unions, and businesses. Athletic leagues, various kinds of clubs, summer camps, and many other activities were formed to busy the idle hands of the young. "Street corner" counselors were hired to offer advice to youths and to mediate with the police on the youths' behalf when they got into trouble. Neighborhood residents were encouraged to form committees to resolve neighborhood problems.

Despite the money and energy invested in the CAP between 1932 and 1957, its effects were never evaluated in any systematic way. Similar programs in other cities had a number of positive outcomes, but their impact on crime and delinquency rates was negligible. Writing about the overall impact of CAP-type

Table 4.1	Summarizing Social Structural Theories		
Theory	**Key Concepts**	**Strengths**	**Weaknesses**
Social Disorganization	Poverty concentrates people of different cultural backgrounds and generates cultural conflict. The breakdown of informal social controls leads to social disorganization, and peer group gangs replace social institutions as socializers.	Explains high crime rates in certain areas. Accounts for intergenerational transmission of deviant values and predicts crime rates from neighborhood characteristics.	Cannot account for individuals and groups in the same neighborhood who are crime-free or why a few individuals commit a highly disproportionate share of crime.
Anomie (Durkheim)	Rapid social change leads to social deregulation and the weakening of restraining social norms. This unleashes "insatiable appetites," which some seek to satisfy through criminal activity.	Emphasizes the power of norms and social solidarity to restrain crime and points to situations that weaken them.	Concentrates on whole societies and ignores differences in areas that are differentially affected by social deregulation.
Anomie/Strain (Merton)	All members of American society are socialized to want to attain monetary success, but some are denied access to legitimate means of attaining it. These people may then resort to crime to achieve what they have been taught to want.	Explains high crime rates among the disadvantaged and how cultural norms create conflict and crime. Explains various means of adapting to strain.	Does not explain why individuals similarly affected by strain do not react (adapt) similarly.
Institutional Anomie	America is literally organized for crime due to its overweening emphasis on the economy and material success. All other institutions are devalued and must accommodate themselves to the requirements of the economy.	Explains why crime rates are higher in America than in other capitalist societies. Points to decommodification as crime-reduction strategy.	Concentrates on single cause of crime. Should predict high rates of property crime in America rather than violent crime, but the opposite is true.
General Strain	There are multiple sources of strain, and strain differs along numerous dimensions. Strain is the result of negative emotions that arise from negative relationships with other as well as from sociocultural forces. Individual characteristics help determine whether we cope poorly or well with strain.	Reminds us that strain is multifaceted and that how we cope with it is more important than its existence. Adds individual characteristics to theory.	Criticized by structural theorists as reductionist because it fails to explore structural origins of strain.
Subcultural	Much delinquency is short-run hedonism rather than utilitarian. Lower-class youths cannot live up to middle-class measuring rods and thus develop status frustration. They seek status in ways peculiar to the subculture. Subcultural youths do not have equal illegitimate opportunities for attaining success. Those who do join criminal gangs; those who don't join retreatist and conflict gangs and engage in mindless violence and vandalism.	Extends the scope of anomie theory and integrates social disorganization theory. Focuses on processes by which lower-class youths adapt to their disadvantages and shows that illegitimate opportunities are also denied to some. Explains the patterned way of life that sustains delinquent values and goals.	Explains subcultural crime and delinquency only. There is some question as to whether a distinct lower-class culture exists in the sense that it is supported by proscriptive values that require antisocial behavior.

programs, Rosenbaum, Lurigio, and Davis (1998) concluded that there were "few positive program effects. The local programs did not affect official crime rates and in some cases were associated with adverse change in survey-based victimization rates" (p. 214).

The ideas of anomie/strain theory had tremendous impact on public policy via President Lyndon Johnson's War on Poverty. The reasoning was that since crime is perceived as an activity engaged in mostly by the poor, fewer poor people would mean fewer crimes. The poverty rate for families in 1963 was 15.9 and in 1993 was 12.3 (U.S. Census Bureau, 2004), but the expected bonus of crime reduction did not materialize. As billions of tax dollars were spent on antipoverty programs during the three decades separating 1963 and 1993, crime rates soared. Specifically, as the poverty rate in the United States *decreased* by about 23 percent, we saw an overall *increase* in the crime rate of approximately 350 percent (Walsh & Ellis, 2007).

If crime is caused by a disjunction between cultural values emphasizing success for all and social structure denying access to legitimate means of achieving it to some, rather than poverty per se, then the cure for crime is to increase opportunities or dampen aspirations. The latter option is not acceptable to policy makers of either the right or the left—or to the general public either—so we are left with the task of trying to increase opportunities. Following in the footsteps of CAP, Richard Cloward and Lloyd Ohlin (1960) developed a delinquency prevention project known as *Mobilization for Youth (MFY),* which concentrated on expanding legitimate opportunities for disadvantaged youths via a number of educational, training, and job placement programs. MFY programs received generous private, state, and federal funds and served as models for such federal programs as Head Start, the job corps, the Comprehensive Employment and Training Act (CETA), affirmative action, and many others (LaFree, Drass, & O'Day, 1992).

Some unknown number of people was diverted from a life of crime because of the opportunities presented to them by such programs, but unfortunately their heyday occurred at the same time that the United States was undergoing a huge jump in crime, from 1965 to 1980. This unfortunate convergence provided conservatives and neoclassical criminologists with arguments against the use of social welfare policies to combat crime and in favor of the kinds of crime control mechanisms addressed in Chapter 3.

The policy recommendation of institutional anomie theory would be to tame the power of the market via decommodification. For instance, the decision to have children could be freed from economic considerations by granting government-guaranteed maternity leave benefits and family allowances/income support, and higher education could be accessible to all people with talent without regard for the financial ability to pay, in other words, policies that ensure an adequate level of material well-being that is not so completely dependent on an individual's performance in the marketplace.

If this sounds too socialistic, you may be surprised to learn that almost every item on the 1928 economic platform of the American Socialist Party has been adopted in the United States. These items include a 40-hour workweek, unemployment benefits, social security, public works, legal trade unions, child labor laws, and government unemployment offices. Other reforms predating 1928 such as a graduated income tax, free education for all children, and the abolition of child labor have been so integrated into American life that few today would call them "socialist" or "un-American," although they all have their origins in left-wing thought (Walsh, 2009a, p. 255).

Any policy recommendations derived from subcultural theories would not differ from those derived from ecological or anomie/strain theories. Changing a subculture is extremely difficult. Insofar as a subculture is a patterned way of life, we cannot attack the problem in parts and expect to change the whole. One possible strategy would be to disperse "problem families" throughout a city rather than concentrating them in block-type projects as is typically done. But even if this was politically feasible, rather than breaking up the subculture and its values, it may result in its displaced carriers "infecting" areas previously insulated from deviant values.

The gang problem suggests obvious policy recommendations in theory, such as increasing low-skill work opportunities by preventing American companies from moving them overseas and strengthening the other social institutions for which gangs are a substitute. Gangs will always be a problem while legitimate social institutions in our poorest areas are too weak to provide young people with their basic needs.

SUMMARY

- Social structural theories focus on social forces that influence people to commit criminal acts. Ecological theory emphasizes that "deviant places" can cause delinquent and criminal behavior regardless of the personal characteristics of individuals residing there. Such areas are characterized by social disorganization, which results from diverse cultural traditions within slum areas.
- One of the most interesting early findings of this perspective is that the same slum areas continued to have the highest crime rates in a city regardless of the ethnic or racial composition of its inhabitants. More recent ecological studies find that neighborhoods do have effects independent of the people who live in them, but most effects are mediated by individual differences.
- Collective efficacy is the opposite of social disorganization, but does not focus on emotional bonds tied to ethnicity. The concept is about a neighborhood's ability to mobilize its residents as an effective force to fight problems, including crime.
- Anomie/strain theories focus on the strain generated by society's emphasis on success goals coupled with its denial of access to legitimate opportunities to achieve it to some. Merton's strain theory focuses on the ways people adapt to this situation via conformity, ritualism, retreatism, rebellion, and innovation (the modes of adaptation). Although the latter four modes are "deviant," they are not all criminal. The innovator and the retreatist modes are considered the most criminal.
- Institutional anomie theory argues that the United States is literally organized for crime because the institutional balance of power strongly favors the economy. All other American institutions are subordinate to our highly competitive economy, and the competition would be meaningless if there were not both winners and losers.
- General strain theory argues that there are many other sources of strain other than Merton's disjunction between goals and means. These strains result in negative emotions that adversely affect relationships with others and may lead to crime. The important thing is not strain, however, but how people cope with it. Among the many attributes Agnew lists as coping resources are temperament, intelligence, and self-esteem.
- Subcultural strain theories have slightly different emphases. Albert Cohen noted that lower-class boys, knowing that they cannot live up to the middle-class measuring rod, form oppositional gangs that perpetuate an oppositional subculture. These gangs usually reject both the goals and the means of middle-class society, as gauged by the malicious and non-utilitarian nature of many of their crimes.
- Walter Miller augments Cohen's assertion that lower-class culture is oppositional to middle-class culture with his theory of focal concerns. Focal concerns—trouble, toughness, smartness, excitement, fate, and autonomy—are behavioral norms of lower-class culture that command strong emotional attention. Miller's thesis is supported by later work done by Elijah Anderson.
- Cloward and Ohlin emphasize that people have differential access to illegitimate, as well as legitimate, means to success and that sociological and psychological factors limit a person's access to both.
- Youth gangs have been noted throughout recorded history. The prevalence of gangs in the United States is greater than ever before and has been attributable to the deindustrialization of America. Deindustrialization

has affected minorities the most and has tended to leave a sizable number of them marginalized from mainstream society and living in disorganized neighborhoods. The gang becomes an attractive option to many of these youths because it offers them some of the things that the ineffective social institutions in those neighborhoods do not, such as a sense of belonging.

DISCUSSION QUESTIONS

1. What is your position on the "kinds of people vs. kinds of places" argument in ecological theory? Do places matter, independent of the people living in them?

2. Is the American stress on material success a good or bad one overall? Is greed "good"? Does it drive the economy? Would we be better off psychologically with less?

3. Are lower-class delinquents reacting against middle-class values as Albert Cohen contends, or is there a lower-class culture with its own set of values and attitudes to which delinquents are conforming, as Walter Miller contends?

4. Debate whether poverty causes crime, crime causes poverty, or something else causes both.

USEFUL WEBSITES

Chicago Area Project. www.chicagoareaproject.org/

Émile Durkheim. www.emile-durkheim.com/

General Strain Theory. www.criminology.fsu.edu/crimtheory/agnew.htm

The Chicago School of Ecology. http://userpages.umbc.edu/~lutters/pubs/1996_SWLNote96-1_Lutters, Ackerman.pdf

Youth Gangs. www.ncjrs.gov/pdffiles/167249.pdf

CHAPTER TERMS

Anomie

Conformity

Consensus or functionalist perspective

Decommodification

Focal concerns

General strain theory

Institutional anomie theory

Mechanical solidarity

Middle-class measuring rods

Modes of adaptation

Opportunity structure theory

Organic solidarity

Short-run hedonism

Social disorganization

Social structure

Status frustration

Strain theory

Transition zone

Social Process Theories

The social structural theories discussed in the previous chapter only explain part of the possible reason that Kody Scott chose the path in life he did. Not all who experience the same conditions turn out the same way; indeed, only one of Kody's brothers ran afoul of the law. The social process theories discussed in this chapter take us a step further in understanding Kody's choices. Two of the theories in this chapter tell us that criminal behavior is learned in association with peers and that we choose to repeat behaviors that are rewarding to us. After he shot and killed the Bloods gang members in his initiation, Kody tells us that he lay in bed that night feeling guilty and ashamed of his actions and that he knew they were wrong. Nevertheless, when the time came to do the same thing again, he chose to do what his peers told him to do because he valued their praise and approval more than anything else in life. His fellow Crips also provided him with rationales and justifications for his actions that neutralized his guilt.

Another theory in this chapter tells us that labels have the power to make us live up to them; we have seen how Kody proudly accepted the label of "Monster" and how he did his best to live up to it. Other theories stress the importance of attachment to social institutions, but he tells us that his "homeboys" were his only family and that his only commitment and involvement were to and with them and their activities.

On a personal level, he plainly lacked self-control; he was impulsive, hedonistic, and angry. Theorists in this chapter tell us that self-control is developed by consistent parental monitoring, supervision, and discipline, but his weary single mother lacked the time, resources, or incentive to provide Kody with proper parenting. From the youngest age, he came and went as he pleased. His autobiography makes it plain that he was something of a "feral child," big enough, mean enough, and guiltless enough to be free to satisfy any and all urges as they arose. This chapter details many of the social processes by which Kody came to be the monster he claims himself to be.

⌧ The Social Process Tradition

Social process criminologists operate from a sociological perspective known as **symbolic interactionism**. Symbolic interactionists focus on how people interpret and define their social reality and the meanings they attach to it in the process of interacting with one another via language (symbols). Social process theorists believe that if we wish to understand social behavior, we have to understand how individuals subjectively perceive their social reality, and how they interact with others to create, sustain, and change it. The processes most stressed are socialization and cultural conflict; i.e., they seek to describe criminal and delinquent socialization (how antisocial attitudes and behavior are learned) and how social conflict "pressures" individuals into committing antisocial acts. Some process theories focus on the reverse process of learning prosocial attitudes and behavior in the face of temptations to do otherwise. All social process theories represent the joining of sociology and psychology to varying extents.

⌧ Differential Association Theory

Differential association theory (DAT) is the brainchild of Edwin Sutherland, whose ambition was to devise a theory that could explain both individual criminality and aggregate crime rates by identifying conditions that must be present for crime to occur and that are absent when crime is absent. His theory attempted to provide a mechanism by which factors such as social disorganization led to

▲ **Photo 5.1**   Youthful racist skinheads in London give the fascist salute. Differential association theory would argue that if the people you spend most of your time with espouse deviant values, you are likely to adopt these as well.

crime. Like Walter Miller, Sutherland saw lower-class culture has having its own integrity and he disdained the phrase "social disorganization" as insulting, substituting "differential social organization" instead. Although Sutherland explicitly denied the role of psychology in crime and delinquency, his theory is implicitly psychological in that it focuses on the process of becoming delinquent via subjective social definitions of reality and attitude formation.

DAT takes the form of nine propositions outlining the process by which individuals come to acquire attitudes favorable to criminal behavior, which may be summarized simply by saying that criminal behavior is learned in intimate social groups. By emphasizing social learning, DAT guides criminologists away from the notion that criminal behavior is the result of biological or psychological abnormalities, or invented anew by each criminal. Criminality is not the result of individual traits, nor learned from impersonal communication from movies or magazines and the like. The learning of criminal behavior involves the same mechanisms involved in any other learning, and includes specific skills and techniques for committing crimes, as well as the motives, rationalizations, justifications, and attitudes of criminals.

The theory asserts that humans take on the hues and colors of their environments, blending in and conforming with natural ease. Most Americans probably like baseball, hot dogs, apple pie, and Chevrolets, as a Chevrolet commercial used to remind us. But do we prefer these things over, say, soccer, bratwurst, strudel, and Volkswagens because the former are demonstrably superior to the latter, or simply because we are Americans and not Germans? We view the world differentially according to the attitudes, beliefs, and expectations of the groups around which our lives revolve; it could hardly be otherwise, particularly in our formative years. Sutherland's basic premise is that delinquent behavior is learned just as readily as we learn to play the games, enjoy the food, and drive the cars that are integral parts of our cultural lives.

The key proposition in DAT is this: "A person becomes delinquent because of an excess of definitions favorable to violations of law over definitions unfavorable to violations of law" (Sutherland & Cressey, 1974, p. 75). Learning criminal conduct is a process of modeling the self after and identifying with individuals we respect and value. **Definitions** refer to the meanings our experiences have for us, how we see things—our attitudes, values, and habitual ways of viewing the world.

Definitions become favorable to law violation according to the *frequency, duration, priority, and intensity* of exposure to them. That is, the earlier we are exposed to criminal definitions, the more often we are exposed to them, the longer they last, and the more strongly we are attached to those who supply us with them, the more likely we are to commit criminal acts when opportunities to do so arise. As we have already seen, antisocial definitions are more likely to be learned in lower-class neighborhoods. In such neighborhoods, children are surrounded by antisocial definitions (the code of the streets, focal concerns) and cannot help being influenced by them, regardless of their individual characteristics. Figure 5.1 illustrates the process of crime described in differential association theory, from its origins in lower-class areas, to the process of learning definitions favorable to law violation, and finally to crime and delinquency.

Evaluation of Differential Association Theory

DAT is in the unconstrained vision camp in that it assumes that antisocial behavior is learned, not something that comes naturally in the absence of prosocial training. As one early critic (no doubt a constrained visionary) of DAT asked, "What is there to be learned about simple lying, taking things that belong to another, fighting and sex play?" (Glueck, 1956, p. 94). Individuals certainly learn to get better at doing these things in their associations with other like-minded individuals, but do they have to be taught them, or do they have to be taught how to curb them, what constitutes moral behavior, and how to consider the rights and feelings of others?

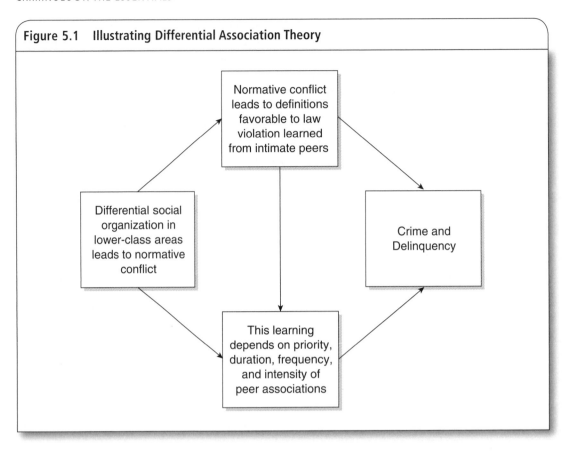

Figure 5.1   Illustrating Differential Association Theory

DAT is also criticized for ignoring individual differences in the propensity to associate with antisocial peers. Individual traits do sort people into different relationship patterns—as numerous studies of relationship patterns attest (Rodkin, Farmer, Pearl, & Van Acker, 2000). Differential association may thus be a case of birds of a feather flocking together than of innocents joining a flock and then changing their feathers, and their flocking facilitates and accentuates their activities, but does not "cause" them. Reviews typically find that delinquent behavior *precedes* gang membership, and that association with other delinquents simply speeds up and enhances delinquency among the predisposed rather than being a stimulator of uncharacteristic behavior among the innocent. As Gottfredson (2006) summarized a number of studies addressing this issue, "The evidence is consistent with the proposition that much of the variance in peer effects on delinquency is attributable to the selection effect of like individuals associating together" (p. 92).

Despite his later rejection of individual differences, Sutherland recognized that they affect relationship patterns in his 1939 statement that "Individual differences among people in respect to personal characteristics or social situations cause crime only as they affect differential association or frequency and consistency of contacts with criminal patterns" (p. 8). Sutherland was specifying a path in which differential association clearly *intervenes* between individual differences and crime as follows:

Individual differences→ Contact with criminal patterns→Crime

In defense of DAT, the concept of differential social organization accounts for the associations people have. Children associate, play, and become friendly with individuals in the neighborhoods their parents provide. In certain neighborhoods, delinquent peers may indeed "cause" delinquency among youths who are otherwise insulated from it, as well as facilitate and accelerate it among others attracted to it. The causal order criticism may be valid for children growing up in better neighborhoods with equal access to both pro- and antisocial peers, but not for kids growing up in the urban slums where prosocial peers are rare. Ronald Akers (1999) responded to the "birds of a feather" adage with the equally pithy reply, "If you lie down with dogs you get up with fleas" (p. 480).

Mark Warr (2000) criticizes DAT for having a singular vision of peer influence. Warr makes a distinction between two approaches to the influence of delinquent peers—compliance and private acceptance. *Compliance* is "going through the motions" of delinquent activity without privately accepting the appropriateness of what one is doing. *Private acceptance* refers to both the public and private acceptance of the attitudes, values, and behavior of the delinquent group. Warr says that DAT was formulated only with the idea of private acceptance in mind while totally ignoring the idea of compliance.

Ronald Akers' Social Learning Theory

As we saw in Table 1.1 in Chapter 1, social learning theory, associated primarily with Ronald Akers, was the theory most frequently chosen by criminologists surveyed in 2007 as the one that best explains variance in criminal behavior (50 out of 379, or 13.2% chose it). Akers' **social learning theory (SLT)** goes beyond looking solely at learned "definitions favorable" to getting involved in delinquency to look at mechanisms that lead individuals to either continue or desist from it. Akers and his colleague, Robert Burgess (1966), applied the powerful concepts of operant psychology to this issue, and claimed that it was the differential reinforcement of behavior that either amplified or extinguished criminal behavior. **Differential reinforcement** is defined as "the balance of anticipated or actual rewards and punishments that follow or are consequences of behavior" (Akers, 2009, p. 67). While psychological principles are central to SLT, Akers insists that it is in the same sociological tradition of DAT and that it retains all the processes found in Sutherland's theory, albeit modified and clarified.

Operant psychology is a theory of learning that asserts that behavior is governed by its consequences. When we behave, we receive feedback from others that we interpret in terms of the positive or negative consequences they have for us. Behavior has two general consequences: It is reinforced or it is punished. Behavior that has positive consequences for the actor is said to reinforce that behavior, making it more likely that the behavior will be repeated in similar situations. Behavior that is punished is less likely to be repeated and may even be extinguished (see Figure 5.2).

Figure 5.2   Illustrating Types of Reinforcement and Punishment

Reinforcement Increases Behavior	Punishment Decreases Behavior
Positive Reinforcement	**Positive Punishment**
(something rewarding received)	(something punishing applied)
Negative Reinforcement	**Negative Punishment**
(something punishing avoided)	(something rewarding lost)

Reinforcement is either positive or negative. The loot from a robbery and status achieved by facing down rivals are examples of *positive reinforcement. Negative reinforcement* occurs when some aversive condition is avoided or removed, such as the removal of a street reputation as a "punk" following some act of bravado. Both types of reinforcement strengthen criminal behavior and thus result in its amplification.

Punishment, which leads to the weakening or eliminating of the behavior preceding it, can also be positive or negative. *Positive punishment* is the application of something undesirable, such as a prison term. *Negative punishment* is the removal of a pleasant stimulus, such as the loss of status in a street gang. The acquisition of Sutherland's "definitions favorable" to antisocial behavior (or prosocial behavior, for that matter) thus depends on each individual's history of reinforcement and punishment.

In any peer group, each member has reciprocal effects on every other member via his or her participation in the reinforcement/punishment process. Of course, what is reinforcing for some may be punishing for others. For teens who value the approval of their parents and teachers, an arrest is punishment; for teens who value the approval of antisocial peers, such an outcome is a reinforcer since it marks them officially as a "badass." The social context is thus an extremely important component of SLT because most learning takes place in the presence of others who provide both the social context and the available reinforcers or punishments.

Discrimination is another important component of SLT. Whereas reinforcements or punishments *follow* behavior, discriminative stimuli are present *before* the behavior occurs. Discriminative stimuli are clues that signal whether a particular behavior is likely to be followed by reward or punishment. In other words, discrimination involves learning to distinguish between stimuli that have been reinforced or punished in the past and similar stimuli you expect will result in the same response in the future. For instance, an unlocked car with the keys in it is a discriminative stimulus that signals "immediate reward" for the criminal, but for the average person it probably signals nothing other than how foolish the owner is because he or she has never previously been rewarded for stealing a car.

Evaluation of Social Learning Theory

Many of the same criticisms applicable to DAT also apply to SLT, so they won't be repeated here. SLT adds some meat to DAT by specifying how definitions favorable to law violation are learned by operant conditioning, although it neglects the role of individual differences in the ease or difficulty with which persons learn. Some people find general hell-raising more exciting (and thus more reinforcing) than others. Some people are more susceptible to short-term rewards because they are especially impulsive, and some are better able to appreciate the long-term rewards of behaving well. According to Cao (2004), in common with DAT, SLT assumes "a passive and unintentional actor who lacks individuality" and is "better at explaining the transmission of criminal behavior than its origins." Moreover, because of their "limited conception of human nature, learning theories generally also ignore the differential receptivity of individuals to criminal messages." In other words, some individuals are more ready to engage in aggressive behavior than others because of the nature of their personalities and will find such behavior is reinforced in delinquent areas. As criminologist Gwynn Nettler (1984) so aptly put it, "Constitutions affect the impact of environment. What we learn and how well we learn it depends on constitution . . . The fire that melts the butter hardens the egg" (p. 295).

In response to criticism that SLT neglects individual differences, Akers (1999) replies that it does not: "An individual in a low-crime group or category who is nevertheless more exposed to criminal associations, models, definitions, and reinforcement than someone in a high-crime group or category will have a higher probability of committing criminal or deviant acts" (p. 482). But this is an explanation in terms of different *environments,* not in terms of different *individuals.* As if to vindicate Cao's (2004) observation, it assumes automaton-like individuals entering different environments at one end and emerging out the other as criminal or not criminal based solely on their exposure to different environments.

◪ **Social Control Theory**

To ensure a peaceful and predictable social existence, all societies have created mechanisms designed to minimize nonconformity and deviance that we may collectively call social control, which may be defined as any action on the part of others, deliberate or not, that facilitates conformity to social rules. In many senses, both Durkheim's anomie theory and Shaw and McKay's ecological theory are control theories because they point to situations or circumstances (anomie or social disorganization) that lessen social control of individuals' behavior. Social control may be direct, formal, and coercive, as exemplified by uniformed symbols of the state. But indirect and informal social control is preferable because it produces prosocial behavior regardless of the presence or absence of external coercion. Obeying society's rules of proper conduct because we believe that the rules are right and just, not simply because we fear formal sanctions, means that we have our own internalized police officer and judge in the form of something called a conscience.

Travis Hirschi's Social Bond Theory

There are a variety of control theories, but the most popular is Travis Hirschi's (1969) **social bond theory.** Previous theories we have examined assume that crime is something learned by basically good people living in bad environments, and they ask, "What causes crime?" Control theorists believe that this question reveals a faulty understanding of human nature, and that the real question is not why some people behave badly, but why most of us behave well most of the time. They tell us that we behave well if our ties to prosocial others are strong, but we may revert to predatory self-interest if they are not. After all, children who are not properly socialized hit, kick, bite, steal, and scream, whenever the mood strikes them. They have to be taught not to do these things, which in the absence of training "come naturally." In this tradition, it is society that is "good" and human beings, in the absence of the proper training, who are "bad." Gwynn Nettler (1984) said it most colorfully: "If we grow up 'naturally,' without cultivation, like weeds, we grow up like weeds—rank" (p. 313). Indeed, virtually all developmental experts agree that "weediness" will be the natural outcome for children who are not subjected to controls. For example, a longitudinal observational study of children from 2 to 12 years of age found that the frequency of hitting, biting, and kicking peaked at 27 months and declined by about 66% by age 12 (Tibbetts & Hemmens, 2010). From this perspective, which is one with a constrained vision of human nature, criminals are simply children grown strong.

Social control theory is thus about the role of social relationships that bind people to the social order and prevent antisocial behavior. Antisocial behavior will emerge automatically if controls are lacking; it needs no special learning or motivating factors since human beings are assumed to be naturally self-centered. The classical assumption of self-interested persons anxious to experience pleasure and avoid pain is still there, but the theory tries to account for why some people pursue their self-interest in legitimate ways and others do not, with primary emphasis on socialization practices that do or do not produce individuals capable of reining in their natural instincts.

The Four Elements of the Social Bond

Hirschi formulated his theory with some foundational facts about the "typical" criminal, and found him to be a young male who grew up in a fatherless home in an urban slum, with a history of difficulty in

▲ **Photo 5.2**   Attachment to parents provides the foundation for attachment to social institutions and to commitment to a prosocial career, involvement in prosocial activities, and belief in prosocial values.

school, and who is unemployed. From this he deduced that those most likely to commit crimes lack the four elements of the social bond that form the foundation of prosocial behavior—*attachment, commitment, involvement,* and *belief.*

Attachment is the emotional component of conformity and refers to the emotional bonds existing between the individual and key social institutions such as the family and the school. Attachment to conventional others is the foundation for all other social bonds because it leads us to feel valued, respected, and admired, and to value the favorable judgments of those to whom we are attached. Much of our behavior can be seen as attempts to gain favorable judgments from people and groups we care about. Lack of attachment to parents and lack of respect for their wishes easily spills over into a lack of attachment and respect for the broader social groupings of which the child is a part. Much of the controlling power of others outside the family lies in the threat of reporting misbehavior to parents. If the child has little respect for parental sanctions, the control exercised by others will have little effect because parental control has little effect.

Commitment is the rational component of conformity and refers to a lifestyle in which one has invested considerable time and energy in the pursuit of a lawful career. People who invest heavily in a lawful career have a valuable stake in conformity, and are not likely to risk it by engaging in crime. School dropouts and the unemployed do not have a strong investment in conventional behavior and therefore risk less by committing crime. Acquiring a stake in conformity requires disciplined application to tasks that children do not relish but which they complete in order to gain approval from parents. Attachment is thus the essential foundation for commitment to a prosocial lifestyle.

Involvement is a direct consequence of commitment; it is part of an overall conventional pattern of existence. Involvement is a matter of time and energy constrictions placed on us by the demands of our lawful activities that reduce exposure to illegal opportunities. Conversely, noninvolvement in conventional activities increases the possibility of exposure to illegal activities.

Belief refers to the acceptance of the social norms regulating conduct. Persons lacking attachment, commitment, and involvement do not believe in conventional morality. A belief system empty of conventional morality is concerned only with narrow self-interest. It is important to realize that crime is not motivated by the absence of any of the social bonds; their absence merely represents social deficiencies that result in a reduction of the potential costs of committing it.

Gottfredson and Hirschi's Low Self-Control Theory

With colleague Michael Gottfredson, Hirschi has moved away from explaining crime and delinquency in terms of social control and toward explaining it in terms of self-control. **Self-control** is defined as the "extent to which [different people] are vulnerable to the temptations of the moment" (Gottfredson & Hirschi, 1990, p. 87).

Self-control theory accepts the classical idea that crimes are the result of unrestrained natural human impulses to enhance pleasure and avoid pain. These impulses often lead to crimes, which Gottfredson and Hirschi define as "acts of force or fraud undertaken in pursuit of self-interest" (p. 15). Most crimes, they assert, are spontaneous acts requiring little skill and earn the criminal very little short-term satisfaction. People with low self-control possess the following personal traits that put them at risk for criminal offending:

- They are oriented to the present rather than to the future, and crime affords them immediate rather than delayed gratification.
- They are risk-taking and physical as opposed to cautious and cognitive, and crime provides them with exciting and risky adventures.
- They lack patience, persistence, and diligence, and crime provides them with quick and easy ways to obtain money, sex, revenge, and so forth.
- They are self-centered and insensitive, so they can commit crimes without experiencing pangs of guilt for causing the suffering of others (pp. 89–90).

▲ **Photo 5.3**   Jail inmate yearning for the outside. Does his situation reflect low self-control, labeling, or something else?

Low self-control is established early in childhood, tends to persist throughout life, and is the result of incompetent parenting. Children do not learn low self-control; low self-control is the default outcome that occurs in the absence of adequate socialization. Parental warmth, nurturance, vigilance, and the willingness to practice "tough love" are necessary to forge self-control in their offspring. Other factors that may result in low self-control include parental criminality (criminals are not very successful in socializing their children); family size (the larger the family, the more difficult it is to monitor behavior); a single-parent family (two parents are generally better than one); and working mothers, which negatively impacts the development of children's self-control if no substitute monitor is provided (Gottfredson & Hirschi, 1990, pp. 100–105).

Gottfredson and Hirschi (1990) argue that children acquire or fail to acquire self-control in the first decade of life, after which the attained level of control remains stable across the life course. Subsequent experiences, situations, and circumstances have little independent effect on the probability of offending. Because low self-control is a stable component of a criminal personality, most criminals typically fail in anything that requires long-term commitment, such as school, employment, and marriage, because such commitments get in the way of immediate satisfaction of their desires.

Low self-control is a necessary but not sufficient cause of criminal offending. What accounts for variation in criminal offending are the different opportunities criminals encounter that are conducive to committing crimes. A criminal **opportunity** is a situation that presents itself to someone with low self-control by which he or she can immediately satisfy needs with minimal mental or physical effort (Gottfredson & Hirschi, 1990, pp. 12–13). Crime is thus the result of people with low self-control meeting a criminal opportunity, and by virtue of differential placement in the social structure, some individuals are exposed to more criminal opportunities than others.

⬚ Integrating Social Control and Self-Control Theories

Given the emphasis on parental guidance in the development of self-control, many criminologists wondered for a long time why the elements of the *social bond* are absent in self-control theory. After all, both self-control and social control (or social bond) theories assume a natural inclination to pursue selfish interests with little effort and without regard for others, and both maintain that what distinguishes law-abiding people from criminals are the controls that prevent the former from acting on their natural impulses. Hirschi (2004) has addressed this concern by assuming that, like self-control, "differences in social control [or social bonds] are stable, that social control and self-control are the same thing" (p. 543). He welds the two theories together with two simple sentences: "Self-control is the set of inhibitions one carries with one wherever one goes. Their character may be initially described by going to the elements of the bond identified by social control theory: Attachment, commitment, involvement, and belief" (pp. 543–544). Hirschi is now saying that self-control and the social control (social bonds) exercised by others in the family, school, and workplace mutually affect one another constantly across the life course and that self-control and social bond theories may now be considered integrated. This integration is illustrated in Figure 5.3.

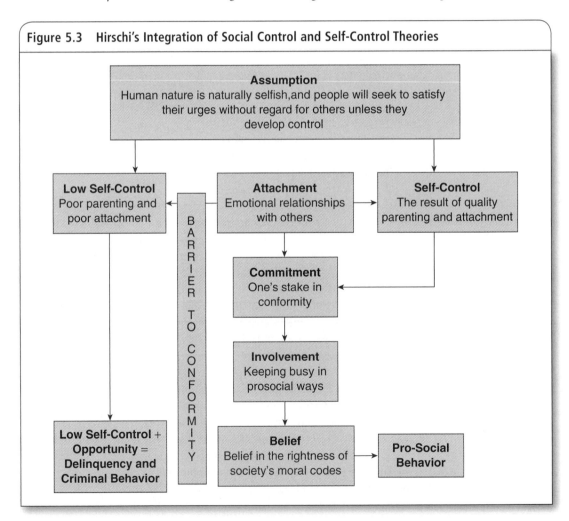

Figure 5.3   Hirschi's Integration of Social Control and Self-Control Theories

Evaluation of Social Control and Self-Control Theories

Criminologists ranked social control and self-control theories as the most empirically supported theories for decades, but they were only ranked 3 and 5, respectively, in the 2007 survey discussed in Chapter 1. However, if we combine both theories and treat them as one, as Hirschi now does, then viewed as a single theory they get the most "votes" at 66. Both versions of control theory agree that the family is central to the development of the mechanisms that affect criminal behavior, and because of this they have been criticized for neglecting social structure (Grasmick, Tittle, Bursik, & Arneklev, 1993). Critics feel that if the family is so important, the social, economic, and political factors that impede stable and nurturing families should be addressed. Control theorists might respond that whatever those things may be, they are not within the purview of control theory, which attempts to explain the *consequences* of weak and disrupted families, not the *reasons* they are disrupted.

A major criticism of self-control theory arises from the claim that it is a *general* theory meant to explain *all* crime. Although most common street crimes are impulsive, spontaneous acts, many others are not. White-collar criminals, serial killers, and terrorists, among others, typically plan their crimes extensively. It is too simplistic to claim that crime can be explained by the single tendency of self-control.

Self-control theory has also been criticized for attributing variation in people's level of self-control solely to variation in parental behavior and ignoring child effects. The child development literature makes it clear that socialization is a two-way street in which parental behavior is shaped by the evocative behavior of the child just as much as the child's behavior is shaped by its parents (Harris, 1998). Low self-control may be something children bring with them to the socialization process rather than a product of the failure of that process. This is supported by a number of studies that have found a strong genetic component to low self-control (Wright & Beaver, 2005; Wright, Beaver, Delisi, & Vaughn, 2008). As Lilly et al. (2007) point out, "research suggests that parents may affect levels of self-control less by their parenting styles and more by genetic transmission" (p. 110). This does not mean that parenting doesn't matter; it only means that if we are to understand self-control at a more sophisticated level, we must use genetically informed samples and assess the influences of children's behavior on parenting styles.

Labeling Theory: The Irony of Social Reaction

Labeling theory (LT) takes seriously the power of bad labels to stigmatize and by doing so, to evoke the very behavior the label signifies. The labeling perspective is interesting and provocative; unlike other theories, it does not ask why crime rates vary, or why individuals differ in their propensity to commit crime. Rather, it asks why some behaviors are labeled criminal and not others, and thus shifts the focus from the *actor* (the criminal) to the *reactor* (the criminal justice system).

LT is often traced to Frank Tannenbaum's *Crime and Community* (1938), which emphasized that a major part in the making of a criminal is the process of identifying and labeling a person as such—the "dramatizing of evil" (p. 20). Tannenbaum viewed the labeling of a person as a "criminal" as a self-fulfilling prophesy (a definition of something which becomes true if it is believed and acted upon), which means that processing law violators through the criminal justice system, rather than deterring them from future criminal behavior, may embed them further in a criminal lifestyle.

For labeling theorists, crime and other forms of deviance have no objective reality and are defined into existence rather than discovered. There is no crime independent of cultural values and norms, which are embodied in the judgments and reactions of others. To put it simply, no act is by its "nature" criminal, because acts do not have natures until they are witnessed, judged good or bad, and reacted to as such by others.

LT distinguishes between primary deviance and secondary deviance (Lemert, 1974). **Primary deviance** is the initial nonconforming act that comes to the attention of the authorities. Primary deviance can arise for a wide variety of reasons, but these are of little interest because they have only marginal effects on the offender's self-concept as a criminal or noncriminal, and it is the individual's self-concept that is crucial in labeling theory. Primary deviance is of interest to labeling theorists only insofar as it is detected and reacted to by individuals with power to pin a stigmatizing label on the rule-breaker. Being caught in an act of primary deviance is either the result of police bias or sheer bad luck; the real criminogenic experience comes *after* a person is caught and labeled. The central concern of LT is thus to explain the consequences of being labeled.

Secondary deviance results from society's reaction to primary deviance. The stigma of a criminal label may result in the person becoming more criminal than he or she would have been had the person not been caught. This may occur in two ways. First, labeled persons may alter their self-concepts in conformity with the label ("Yes, I am a criminal, and will act more like one in future."). Second, the label may exclude them from conventional employment and lead to the loss of conventional friends. This may lead them to seek illegitimate opportunities to fulfill their financial needs, and to seek other criminals to fulfill their friendship needs, which further strengthens their growing conception of themselves as "really" criminal. The criminal label becomes a self-fulfilling prophesy because it is a more powerful label than other social labels that such offenders may claim.

▧ Sykes and Matza's Neutralization Theory

Sykes and Matza's (2002) **neutralization theory** (NT) is a learning theory that attacks differential association's failure to explain why some people drift in and out of crime rather than being consistently criminal. It also runs counter to the assumption of DAT and the subcultural theories that give the impression that criminal behavior is endowed with positive value and condoned as morally right. It is difficult to believe that criminals do not know "deep down" that their behavior is wrong: "if there existed in fact a delinquent subculture such that the delinquent viewed his behavior as morally correct," he or she would show no shame when caught, but would instead show "indignation or a sense of martyrdom" (Sykes & Matza, 2002, p. 145). NT suggests that delinquents know their behavior is wrong, but they justify it on a number of grounds. In other words, they "neutralize" any sense of shame or guilt for having committed some wrongful act, which means that they are at least minimally attached to conventional norms. NT also runs counter to labeling theory because it shows how delinquents *resist* labeling rather than passively accepting it. Sykes and Matza's five **techniques of neutralization** are listed below.

1. *Denial of responsibility* shifts the blame for a deviant act away from the actor: "I know she's only 6, but it was she who seduced me."

2. *Denial of injury* is an offender's claim that no "real" offense occurred because no one was harmed: "He got his car back, and his insurance covers the damage, doesn't it?"

3. *Denial of victim* implies that the victim got what he or she deserved: "I guess I did beat her up, but she kept nagging; hell, she asked for it!"

4. *Condemnation of the condemners* involves attempts by the offender to share guilt with the condemners (parents, police, probation officers) by asserting that their behavior is just bad as his or hers is: "You drink booze; I smoke grass. What's the difference?"

5. *Appeal to higher loyalties* elevates the offender's moral integrity by claiming altruistic motives: "I have to cover my homies' backs, don't I?"

The motive behind the employing of these techniques is assumed to be the maintenance of a non-criminal self-image by individuals who have committed a criminal act and have been asked to explain why. Such individuals "define the situation" in a way that mitigates the seriousness of their acts and simultaneously protects the image they have of themselves as noncriminals. A less benign interpretation of the use of these techniques is that rather than trying to protect their self-images, they are seeking to mitigate their punishment, or at least to "share" it with some convenient other. Conversely, intensive interviews with hardcore criminals indicate that they strive to maintain an image consistent with inner-city street codes, not with conventional ones; i.e., "they neutralize being good rather than being bad" (Topalli, 2005, p. 798).

⊠ Evaluation of Labeling and Neutralization Theories

After a period of great popularity in the 1970s, labeling theory is practically out of the criminological picture today (see Table 1.1 in Chapter 1). Labeling theory has been criticized for coming close to claiming that the original "causes" of deviant behavior (primary deviance) do not matter. If the causes of primary deviance do not matter, then efforts to control crime via various structural changes would be abandoned in favor of reliance on labeling theory's "non-interventionism" ("Leave the kids alone. They'll grow out of it."). This advice may be prudent for teenage pot smokers or runaways, but hardly wise for teenage robbers and rapists. LT advises that such delinquents should be "treated" rather than "punished." But since its proponents insist that there is nothing intrinsically bad about any action, what is the point of treatment? What is there to be treated?

One of the positive elements of neutralization theory is that it eliminates much of the over-determined image of subcultural values implied in subcultural theories. Many delinquents are no more completely committed to antisocial values than they are to prosocial values. Neutralization techniques are not viewed as "causes" of antisocial behavior; rather, they are a set of justifications that loosen moral constraints and allow offenders to drift in and out of antisocial behavior because they are able to "neutralize" these constraints.

One of the major problems with the theory is that it says nothing about the origins of the antisocial behavior the actors seek to neutralize. To be a causal theory of criminal behavior rather than an explanation of the post hoc process of rationalization, it would have to show that individuals *first* neutralize their moral beliefs and *then* engage in antisocial acts. Some studies have found that neutralization techniques were able to explain future deviance, but this is no surprise, since persons in a position where they have to explain their offending behavior are more likely than those not in such a position to offend in the future—past is prologue, regardless of our explanations of it.

⊠ Policy and Prevention: Implications of Social Process Theories

Where we see the cause(s) of crime is where we assume we will find the solution. However, although they deal with different units of analysis, very few policy recommendations not discussed in ecological and strain theories can be gleaned from DAT or SLT. The bottom line for all subcultural theories is that lower-class

Table 5.1	Summarizing Social Process Theories		
Theory	**Key Concepts**	**Strengths**	**Weaknesses**
Differential Association	Crime is learned in association with peers holding definitions favorable to law violation. Most likely to occur in differentially organized (lower-class) neighborhoods.	Explains the onset of offending and the power of peer pressure.	Neglects possibility of like seeking like (birds of a feather). Does not make distinction between private accepters and temporary compliers.
Social Learning	Definitions favorable to law violation depend on history of reinforcement and punishment. Excess rewards for criminal behavior perpetuate it.	Adds powerful concepts of operant psychology to explain how people learn criminal behavior. Links sociology to psychology.	Neglects individual differences affecting what is reinforcing to whom and the ease or difficulty with which one learns.
Social Bond	Bonds to social institutions prevent crime, which otherwise comes naturally. The bonds are attachment, commitment, involvement, and belief.	The most popular and empirically supported theory. Emphasizes importance of the family and provides workable policy recommendations.	Neglects structural variables contributing to family instability and to loss of occupational opportunities. Neglects differences in the ease with which attachment is achieved.
Self-Control	Low self-control explains all crime and analogous acts. Low self-control occurs in the absence of proper parenting. Exposure to criminal opportunities explains differences in criminal behavior among low self-control individuals.	Identifies a single measurable trait as responsible for many antisocial behaviors. Accords well with the impulsive nature of most criminal behavior. Links sociology to psychology.	Claims too much for a single trait. Neglects child influences on parenting behavior and the effects of genes on low self-control.
Labeling	Crime has no independent reality. Original primary deviance is unimportant; what is important is the labeling process, which leads to secondary (continuing) deviance. Labeling people criminal leads them to organize their self-concepts around that label.	Explains consequences of labeling with a "master status." Identifies the social construction of crime and points to the power of some (the powerful) to criminalize the acts of others (the powerless).	The neglect of causes of primary deviance. Advice that criminals should be treated, not punished, contradicts the theory that says there is nothing intrinsically bad about crime and therefore there is nothing to "treat."
Neutralization	Delinquents and criminals learn to neutralize moral constraints and thus their guilt for committing crimes. They drift in and out of crime.	Emphasizes that criminals are no more fully committed to antisocial attitudes than they are to prosocial attitudes. Shows how criminals handle feelings of guilt.	Says nothing about the origins of behavior being neutralized. More a theory of antisocial rationalization than of crime.

neighborhoods harbor values and attitudes conducive to criminal behavior. Thus, if learning crime and delinquency within a particular culture is the problem, then changing relative aspects of that culture is the answer. However, we have already seen that attempts to do that have met with only meager success at best.

Because DAT concerns itself with the influence of role models in intimate peer groups, the provision of prosocial role models to replace antisocial ones is an obvious thought. Probation and parole authorities have long recognized the importance of keeping convicted felons away from each other, making it a revocable offense to "associate with known felons." As every probation and parole officer knows, however, this is easier said than done. Programs that bring youths together for prosocial purposes, such as sports leagues and community projects, might be high on the agenda of any policy maker using differential association as a guide. But the lure of "the streets" and of the friends they have grown up with remains a powerful force retarding rehabilitation. The good news is that most delinquents will desist as they mature, and the breakup of the friendship group by the incarceration, migration, death, or marriage of some of its members will break the grip of antisocial behavior for many of the remaining members (Sampson & Laub, 1999).

The policy implications derivable from social control and self-control theories have to do with the family. Given the importance of nurturance and attachment, both versions of control theory support the idea of early family intervention designed to cultivate these things. Families with children in almost all advanced nations receive support via family allowances and receive paid maternal leave, but such programs do not exist in the United States, which shows that politics and ideology dictate the direction of criminal justice policy more than criminological theory.

Other attempts to increase bonding to social institutions would concentrate on increasing children's involvement in a variety of prosocial activities and programs centered in and around the school. These programs provide prosocial models, teach moral beliefs such as personal responsibility, and keep youths busy in meaningful and challenging ways. Social control theory might recommend more vocationally oriented classes to keep less academically inclined students bonded to school.

Neither version of control theory would advise increased employment opportunities as a way to control crime. The assumption of control theory is that people who are attached and who possess self-control will do fine in the job market as it is, and increasing job opportunities for those lacking attachment and self-control will have minimal effect. Because low self-control is the result of the absence of inhibiting forces typically experienced in early childhood, Gottfredson and Hirschi (1997) are pessimistic about the ability of less powerful inhibiting forces (such as the threat of punishment) present in later life to deter crime. They also see little use in satisfying the wants and needs other theories view as important in reducing crime (reducing poverty, improving neighborhoods, etc.) because crime's appeal is its provision of immediate gains and minimal cost. In short, "society" is neither the cause of nor the solution to crime.

Gottfredson and Hirschi advocate some of the same policies (e.g., target hardening) supported by rational choice and routine activities to reduce criminal opportunities. However, the most important policy recommendation is to strengthen families and improve parenting skills, especially skills relevant to teaching self-control. It is only by working with and through families that society can do anything about crime in the long run. Gottfredson and Hirschi's (1997) most important recommendation is stated as follows: "Delaying pregnancy among unmarried girls would probably do more to affect the long-term crime rates than all the criminal justice programs combined" (p. 33).

Labeling theory has had an effect on criminal justice policy far in excess of what its empirical support warrants. If it is correct that official societal reaction to primary deviance amplifies and promotes more of the same, the logical policy recommendation is that we should ignore primary deviance for the sake of alleviating secondary deviance. Labeling theory recommends that we allow offenders to protect their self-images as noncriminals by not challenging their "techniques of neutralization." Juveniles must be particularly protected from labeling.

The only policy implication of neutralization theory is the exact opposite of that of labeling theory; i.e., criminal justice agents charged with managing offenders (probation/parole officers, etc.) should strongly challenge their excuses. If offenders come to believe their own rationalizations, rehabilitative efforts will become more difficult. Thus, offenders must be shown that their thinking patterns have negative long-term consequences for them.

SUMMARY

- Social process theories emphasize how people perceive their reality and how these perceptions structure their behavior. DAT is a learning theory that emphasizes the power of peer associations and the definitions favorable to law violation found within them to be the cause of crime and delinquency.

- SLT adds to DAT by stressing the mechanisms by which "definitions favorable" are learned. Behavior is either reinforced (rewarded) or punished. Behavior that is rewarded tends to be repeated; behavior that is punished tends not to be. Discriminative stimuli provide signals for the kinds of behaviors that are likely to be rewarding or punishing, and are based on what we have learned about those stimuli in the past.

- Control theories are in many ways the opposite of DAT and SLT because they don't ask why people commit crimes, but why most of us do not. Crime comes naturally to those who are not either socially or self-controlled. Hirschi speaks of the social bonds (attachment, commitment, involvement, and belief) that keep us on the straight and narrow. These are not causes of crime; rather, they are bonds, the absence of which allows our natural impulses to emerge.

- Gottfredson and Hirschi's self-control theory moves the focus from social control to self-control, although our experiences within the family are still vital to learning self-control. Low self-control must be paired with a criminal opportunity for crime to occur. Hirschi has now integrated social control and self-control theories, stating that they are "the same thing."

- Labeling theory is not interested in why some people commit crimes (primary deviance), believing that the only thing that differentiates criminals from the rest of us is that they have been caught and labeled. The real problem is the affixing of a deviant label because it changes the person's self-concept, and he or she then engages in secondary deviance in conformity with the label.

- Sykes and Matza's techniques of neutralization theory is contrary to labeling theory because it focuses on individuals' attempts to resist being labeled criminal by offering justifications or excuses for their behavior.

DISCUSSION QUESTIONS

1. Without indicating a particular theory, does the social structural or social process approach to explaining crime and criminality make sense to you? If so, why?

2. Compare differential association theory with control theory in terms of their respective assumptions about human nature. Which assumption makes more sense to you?

3. Is a delinquent or criminal label applied to someone sufficient in most cases to change a person's self-concept enough to lead him or her to continue offending?

4. Gottfredson and Hirschi claim that parents are to blame for an individual's lack of self-control. Are there some children who are simply more difficult to socialize than others, and if so, are they rather than their parents at fault for their lack of self-control?

5. Why is attachment the most important of the four social bonds?

USEFUL WEBSITES

Control Theories of Crime. http://sitemason.vanderbilt.edu/files/l/l3Bguk/soccon.pdf

Differential Association Theory. www.criminology.fsu.edu/crimtheory/sutherland.html

Learning Theories of Crime. http://en.wikipedia.org/wiki/Social_learning_theory

Social Control Theories. http://en.wikipedia.org/wiki/Social_control_theory

CHAPTER TERMS

Attachment	Involvement	Secondary deviance
Belief	Labeling theory	Social bond theory
Commitment	Neutralization theory	Social control
Definitions	Operant psychology	Social learning theory
Differential association theory	Primary deviance	Symbolic interactionism
Differential reinforcement	Punishment	Techniques of neutralization
Discrimination	Reinforcement	

CHAPTER

6

Critical Theories: Marxist, Conflict, and Feminist

At the heart of the theories in this chapter is social stratification by class and power, and they are the most "politicized" of all criminological theories. Sanyika Shakur, aka Kody Scott, came to embrace this critical and politicized view of society as he grew older and converted to Afrocentric Islam. Shakur was very much a member of the class Karl Marx called the "lumpenproletariat" (a German word meaning "rag proletariat"), which is the very bottom of the class hierarchy. Many critical theorists would view Shakur's criminality as justifiable rebellion against class and racial exploitation. Shakur wanted all the material rewards of American capitalism, but he perceived that the only way he could get them was through crime. He was a total egoist, but many Marxists would excuse this as a trait that is nourished by capitalism, the "root cause" of crime. From his earliest days, he was on the fringes of a society he clearly disdained. He frequently referred to whites as "Americans" to emphasize his distance from them, and he referred to black cops as "Negroes" to distinguish them from the "New African Man." He called himself a "student of revolutionary science" and "rebellion," and advocated a separate black nation in America.

Conflict concepts dominated Shakur's life as he battled the Bloods as well as other Crips "subsets" whose interests were at odds with his set. It is easy to imagine his violent acts as the outlets of a desperate man struggling against feelings of class and race inferiority. Perhaps he was only able to achieve a sense of power when he held the fate of another human being in his hands. His fragile narcissism often exploded into violent fury whenever he felt himself being "dissed." How much of Shakur's behavior and the behavior of youth gangs in general are explained by the concepts of critical theories? Is violent conflict a justifiable response to class and race inequality in a democratic society, or are there more productive ways to resolve such conflicts?

⬚ The Conflict Perspective of Society

Although all sociological theories of crime contain elements of social conflict, consensus theories tend to judge alternative normative systems from the point of view of mainstream values, and they do not call for major restructuring of society. Theories presented in this chapter do just that, and concentrate on power relationships as explanatory variables to the exclusion of almost everything else. They view criminal behavior, the law, and the penalties imposed for breaking it, as originating in the deep inequalities of power and resources existing in society. For conflict theorists, the law is not a neutral system of dispute settlement designed to protect everyone, but rather the tool of the privileged who criminalize acts that are contrary to their interests.

You don't have to be a radical or even a liberal to acknowledge that great inequalities of wealth and power exist in every society and that the wealthy classes have the upper hand in all things. History is full of examples: Plutarch wrote of the conflicts generated by disparity in wealth in Athens in 594 B.C. (Durant & Durant, 1968, p. 55), and U.S. President John Adams (1778/1971) wrote that American society in the late 18th century was divided into "a small group of rich men and a great mass of poor engaged in a constant class struggle" (p. 221).

⬚ Karl Marx and Revolution

Karl Marx is the father of critical criminology. The core of Marxism is the concept of **class struggle:** "Freeman and slave, patrician and plebian, lord and serf, guildmaster and journeyman, in a word, oppressor and oppressed, stood in constant opposition to one another" (Marx & Engels, 1948, p. 9). The oppressors in Marx's time were the owners of the means of production (the **bourgeoisie**), and the oppressed were the workers (the **proletariat**). The bourgeoisie strives to keep the cost of labor at a minimum, and the proletariat strives to sell its labor at the highest possible price. These opposing goals are the major source of conflict in a capitalist society. The bourgeoisie enjoys the upper hand because capitalist societies have large armies of unemployed workers eager to secure work at any price, thus driving down the cost of labor. According to Marx, these economic and social arrangements—the material conditions of people's lives—determine what they will know, believe, and value, and how they will behave.

Marx and his collaborator Friedrich Engels (1948) were disdainful of criminals, describing them in terms that would make a New York cop proud: "The dangerous class, the social scum, that rotting mass thrown off by the lowest layers of the old society" (p. 22). These folks came from a third class in society—the **lumpenproletariat**—who would play no decisive role in the expected revolution. For Marx and Engels (1965) crime was simply the product of unjust and alienating social conditions—"the struggle of the isolated individual against the prevailing conditions" (p. 367). This became known as the **primitive rebellion hypothesis,** one of the best modern statements of which is Bohm's (2001): "Crime in capitalist societies is often a rational response to the circumstances in which people find themselves" (p. 115).

▲ **Photo 6.1**   For Karl Marx (1818–1883), the resolution of social problems such as crime would be achieved through the creation of a socialist society characterized by communal ownership of the means of production and an equal distribution of the fruits of these labors.

Another concept that is central to critical criminology is alienation (Smith & Bohm, 2008). **Alienation** is a condition that describes the distancing of individuals from something. For Marx, most individuals in capitalist societies were alienated from work (which they believed should be creative and enjoyable), which led to alienation from themselves and from others. Work is central to Marx's thought because he believed that while nonhuman animals instinctively act on the environment *as given* to satisfy their immediate needs, humans distinguish themselves from them by consciously *creating* their environment instead of just submitting to it. Alienation is the result of this discord between one's *species being* and one's behavior (e.g., mindlessly noncreative work as opposed to creative work). Marx thought that wage-labor dehumanizes human beings by taking from them their creative advantage over other animals—robbing them of their species being (in effect, their human nature) and reducing them to the level of animals.

When individuals become alienated from themselves, they become alienated from others and from their society in general. Alienated individuals may then treat others as mere objects to be exploited and victimized as they themselves are supposedly exploited and victimized by the capitalist system. Since the great majority of workers do not experience their work as creative activity, they are all dehumanized ritualists or conformists (to borrow from Merton's modes of adaptation). If we accept this notion, then perhaps one can view criminals as heroic rebels struggling to rehumanize themselves, as some Marxist criminologists have done.

Willem Bonger: The First Marxist Criminologist

Dutch criminologist Willem Bonger's *Criminality and Economic Conditions* (1905/1969) is the first work devoted to a Marxist analysis of crime. For Bonger, the roots of crime lay in the exploitative and alienating conditions of capitalism, although some individuals are at greater risk for crime than others because people vary in their "innate **social sentiments**"—*altruism* (an active concern for the well-being of others) and its opposite, *egoism*—(a concern only for one's own selfish interests). Bonger believed that capitalism generates egoism and blunts altruism because it relies on competition for valuable resources, setting person against person and group against group, leaving the losers to their miserable fates. Thus, all individuals in capitalist societies are infected by egoism, and all are therefore prone to crime—the poor out of economic necessity, the rich and the middle classes from pure greed. Poverty was a major cause of crime for Bonger, but it worked by way of its effects on family structure (broken homes) and poor parental supervision of their children. Because of his emphasis on family structure and what he saw as the moral deficits of the poor, Bonger has been criticized by other Marxists, but he firmly believed that only by transforming society from capitalism to socialism would it be possible to regain the altruistic sentiment and reduce crime.

Modern Marxist Criminology

Contrary to Marx, modern Marxist criminologists tend to excuse criminals. William Chambliss (1976) views some criminal behavior as "no more than the 'rightful' behavior of persons exploited by the extant economic relationships" (p. 6), and Ian Taylor (1999) sees the convict as "an additional victim of the routine operations of a capitalist system—a victim, that is of 'processes of reproduction' of social and racial inequality" (p. 151). David Greenberg (1981) even elevated Marx's despised lumpenproletariat to the status of revolutionary leaders: "[C]riminals, rather than the working class, might be the vanguard of the revolution" (p. 28). Marxist criminologists also appear to view the class struggle as the *only* source of *all* crime and to view "real" crime as violations of human rights, such as racism, sexism, imperialism, and capitalism, and accuse other criminologists of being parties to class oppression. Tony Platt even wrote that "it is not too farfetched to characterize many criminologists as domestic war criminals" (quoted in Siegel, 1986, p. 276).

In the 1980s, Marxists calling themselves **left realists** began to acknowledge that predatory street crime is a *real* source of concern among the working class, who are the primary victims of it. Left realists understood that they have to translate their concern for the poor into practical, *realistic* social policies. This theoretical shift signals a move away from the former singular emphasis on the political economy to embrace the interrelatedness of the offender, the victim, the community, and the state in the causes of crime. It also signals a return to a more orthodox Marxist view of criminals as people whose activities are against the interests of the working class as well as those of the ruling class. Although unashamedly socialist in orientation, left realists have been criticized by more traditional Marxists who see their advocacy of solutions to the crime problem within the context of capitalism as a sellout (Bohm, 2001).

▧ Conflict Theory: Max Weber, Power and Conflict

In common with Marx, Max Weber (1864–1920) saw societal relationships as best characterized by conflict. They differed on three key points, however: First, while Marx saw cultural ideas as molded by the economic system, Weber saw a culture's economic system being molded by its ideas. Second, whereas Marx emphasized economic conflict between only two social classes, Weber saw conflict arising from multiple sources, with economic conflict often being subordinate to other conflicts. Third, Marx envisioned the end of conflict with the destruction of capitalism, while Weber contended that it will always exist, regardless of the social, economic, or political nature of society, and that it was functional because of its role in bringing disputes into the open for public debate.

Even though individuals and groups enjoying great wealth, prestige, and power have the resources necessary to impose their values on others with fewer resources, Weber viewed the various class divisions in society as normal, inevitable, and acceptable, as do many contemporary conflict theorists (Curran & Renzetti, 2001). As opposed to Marx's concentration on two great classes (the bourgeoisie and the proletariat) based only on economic interests, Weber focused on three types of social group that form and dissolve as their interests change—class, party, and status. A *class* group shares only common economic interests, and *party* refers to political groups. *Status* groups were the only truly social groups because members hold common values, live common lifestyles, and share a sense of belonging. For Weber, the law is a resource by which the powerful are able to impose their will on others by criminalizing acts that are contrary to their class interests. Because of this, wrote Weber, "criminality exists in all societies and is the result of the political struggle among different groups attempting to promote or enhance their life chances" (quoted in Bartollas, 2005, p. 179).

George Vold produced a version of conflict theory that moved conflict away from an emphasis on value and normative conflicts (as in the Chicago ecological tradition) to include conflicts of interest. Vold saw social life as a continual struggle to maintain or improve one's own group's interests—workers against management, race against race, ecologists against land developers, and the young against adult authority—with new interest groups continually forming and disbanding as conflicts arise and are resolved. Conflicts between youth gangs and adult authorities were of particular concern to Vold, who saw gangs in conflict with the values and interests of just about every other interest group, including those of other gangs (as in the Crips versus the Bloods, for example). Gangs are examples of **minority power groups,** or groups whose interests are sufficiently on the margins of mainstream society that just about all their activities are criminalized. Minority power groups are excellent examples of Weber's status groups in which status depends almost solely on adherence to a particular lifestyle: "Status honour is normally expressed by the fact that above all else a specific *style of life* is expected from all those who wish to belong to the circle" (Weber, 1978, p. 1028).

We have already discussed this kind of status group in terms of young males in so-called honor subcultures who literally risk life and limb in the pursuit of status as it is defined in those subcultures. Vold's theory concentrates entirely on the clash of individuals loyally upholding their differing group interests, and is not concerned with crimes unrelated to group conflict.

Like Weber, Vold viewed conflict as normal and socially desirable. Conflict is a way of assuring social change, and in the long run, a way of assuring social stability. A society that stifles conflict in the name of order stagnates and has no mechanisms for change short of revolution. Since social change is inevitable, it is preferable that it occur peacefully and incrementally (evolutionary) rather than violently (revolutionary). Even the 19th-century arch conservative British philosopher Edmund Burke saw that conflict is functional in this regard, writing that "A state without the means of some change is without means of its conservation" (quoted in Walsh & Hemmens, 2000, p. 214).

Conflict criminology differs from Marxist criminology in that it concentrates on the *processes* of value conflict and lawmaking rather than on the social structural elements underlying those things. It is also relatively silent about how the powerful got to be powerful and makes no value judgments about crime (Is it the activities of "social scum" or of "revolutionaries"?); conflict theorists simply analyze the power relationships underlying the act of criminalization.

Because Marxist and conflict theories are frequently confused with one another, Table 6.1 summarizes the differences between them on key concepts.

Table 6.1 **Comparing Marxist and Conflict Theory on Major Concepts**

Concept	Marxist	Conflict
Origin of conflict	It stems from the powerful oppressing the powerless (e.g., the bourgeoisie oppressing the proletariat under capitalism).	It is generated by many factors, regardless of the political and economic system.
Nature of conflict	It is socially bad and must and will be eliminated in a socialist system.	It is socially useful and necessary and cannot be eliminated.
Major participants in conflict	The owners of the means of production and the workers are engaged in the only conflict that matters.	Conflict takes place everywhere, between all sorts of interest groups.
Social class	Only two classes are defined by their relationship to the means of production, the bourgeoisie and proletariat. The aristocracy and the lumpenproletariat are parasite classes that will be eliminated.	There are a number of different classes in society defined by their relative wealth, status, and power.
Concept of the law	It is the tool of the ruling class that criminalizes the activities of the workers harmful to its interests and ignores its own socially harmful behavior.	The law favors the powerful, but not any one particular group. The greater the wealth, power, and prestige a group has, the more likely the law will favor it.

(Continued)

Table 6.1	(Continued)	
Concept	Marxist	Conflict
Concept of crime	Some view crime as the revolutionary actions of the downtrodden, others view it as the socially harmful acts of "class traitors," and others see it as violations of human rights.	Conflict theorists refuse to pass moral judgment because they view criminal conduct as morally neutral with no intrinsic properties that distinguish it from conforming behavior. Crime doesn't exist until a powerful interest group is able to criminalize the activities of another less powerful group.
Cause of crime	The dehumanizing conditions of capitalism. Capitalism generates egoism and alienates people from themselves and from others.	The distribution of political power that leads to some interest groups being able to criminalize the acts of other interest groups.
Cure for crime	With the overthrow of the capitalist mode of production, the natural goodness of humanity will emerge, and there will be no more criminal behavior.	As long as people have different interests and as long as some groups have more power than others, crime will exist. Since interest and power differentials are part of the human condition, crime will always be with us.

⬚ Peacemaking Criminology

Peacemaking criminology is a fairly recent addition to the growing number of theories in criminology and has drawn a number of former Marxists into its fold. It is situated squarely in the postmodernist tradition (a tradition that rejects the notion that the scientific view is better than any other view, and which disparages the claim that any method of understanding can be objective). In its peacemaking endeavors, it relies heavily on "appreciative relativism," a position that holds that all points of view, including those of criminals, are relative, and all should be appreciated. It is a compassionate and spiritual criminology that has much of its philosophical roots in humanistic religion.

▲ **Photo 6.2**   The friendly presence of police at a large ethnic festival demonstrates the peacekeeping approach to crime prevention.

Peacemaking criminology's basic philosophy is similar to the 1960s hippie adage, "Make love, not war," without the sexual overtones. It shudders at the current "war on crime" metaphor and wants to substitute "peace on crime." The idea of making peace on crime is perhaps best captured by Kay Harris (1991) when she writes that we

need to reject the idea that those who cause injury or harm to others should suffer severance of the common bonds of respect and concern that bind members of a community. We should relinquish the notion that it is acceptable to try to "get rid of" another person whether through execution, banishment, or caging away people about whom we do not care. (p. 93)

While recognizing that many criminals should be incarcerated, peacemaking criminologists aver that an overemphasis on punishing criminals escalates violence. Richard Quinney (1975) has called the American criminal justice system the moral equivalent of war and notes that war naturally invites resistance by those it is waged against. He further adds that when society resists criminal victimization, it "must be in compassion and love, not in terms of the violence that is being resisted" (quoted in Vold, Bernard, & Snipes, 1998, p. 274).

In place of imprisoning offenders, peacemaking criminologists advocate **restorative justice,** which is basically a system of mediation and conflict resolution. Restorative justice is primarily oriented toward justice by repairing the harm caused by the crime and typically involves face-to-face confrontations between victim and perpetrator to arrive at a mutually agreeable solution to "restore" the situation as much as possible to what it was before the crime (Champion, 2005). Restorative justice has been applauded because it humanizes justice by bringing victim and offender together to try to correct the wrong done, usually in the form of written apologizes and payment of restitution. Although developed for juveniles and primarily confined to them, restorative justice has also been applied to nonviolent adult offenders in a number of countries in addition to the United States. The belief behind restorative justice is that, to the extent that both victim and victimizer come to see that justice is attained when a violation of one person by another is made right by the violator, the violator will have taken a step toward reformation and the community will be a safer place in which to live.

Evaluation of Critical Theories

It is often said that Marxist theory has very little that is unique to add to criminology theory: "When Marxist theorists offer explanations of crime that go beyond simply attributing the causes of all crime to capitalism, they rely on concepts taken from the same 'traditional' criminological theories of which they have been so critical" (Akers, 1994, p. 167). Marxists tend to ignore empirical studies, preferring historical, descriptive, and illustrative research. The tendency to romanticize criminals as revolutionaries has long been a major criticism of Marxist criminologists, although because of the influence of left realists they are less likely to do this today.

Can Marxists claim support for their argument that capitalism causes crime and socialism "cures" it? It may be true that capitalist countries in general have higher crime rates than socialist countries, but the question is whether the Marxist interpretation is correct. Lower crime rates in socialist societies may have more to do with repressive law enforcement than with any altruistic qualities intrinsic to socialism.

Marxist criminology also seems to assume that the conditions prevailing in Marx's time still exist today in advanced capitalist societies. People from all over the world have risked everything to get into capitalist countries because those countries are where human rights are most respected and human needs most readily accessible. Left realism realizes this and is more the reform-minded "practical" wing of Marxism than a theory of crime that has anything special to offer criminology. Indeed, "working within the system" has produced numerous changes in American society that used to be considered socialist, such as those mentioned under the policy implications of institutional anomie theory in Chapter 4.

Conflict theory is challenging and refreshing because its efforts to identify power relationships in society have applications that go beyond criminology. But there are problems with it as a theory of criminal behavior. It has even been said that "[c]onflict theory does not attempt to explain crime; it simply identifies social conflict as a basic fact of life and a source of discriminatory treatment" (Adler, Mueller, & Laufer, 2001, p. 223).

Conflict theory's assumption that crime is just a "social construct" without any intrinsic properties diminishes the suffering of those who have been assaulted, raped, robbed, and otherwise victimized. These acts *are* intrinsically bad (mala in se) and are not arbitrarily criminalized because they threaten the privileged world of the powerful few. Humans worldwide react with anger, grief, and a desire for justice when they or their loved ones are victimized by a mala in se crime. There is wide agreement among people of various classes in the United States and around the world about what crimes are—laws exist to protect everyone, not just "the elite" (Walsh & Ellis, 2007).

Peacemaking criminology urges us to make peace on crime, but what does such advice actually mean? As a number of commentators have pointed out, "being nice" is not enough to stop others from hurting us (Lanier & Henry, 2010). It is undoubtedly true that the reduction of human suffering and achieving a truly just world will decrease crime, as advocates of this position contend, but they offer us no notion of how this can be achieved beyond counseling that we should appreciate criminals' points of view and not be so punitive.

◪ Policy and Prevention: Implications of Critical Theories

The policy implications of Marxist theory are straightforward: Substitute socialism for capitalism and crime will be reduced. Modern Marxists realize that this is unrealistic, a fact underlined for them by the collapse of Marxist states across Eastern Europe. They also realize that the emphasis on a single cause of crime (the class struggle) and romanticizing criminals is equally unrealistic. Rather than throw out their entire ideological agenda, left realists now temper their views while still maintaining their critical stance toward the "system." Policy recommendations made by left realists have many things in common with those made by ecological, anomie, and routine activities theorists. Community activities, neighborhood watches, community policing, dispute resolution centers, and target hardening are among the policies suggested.

Because crime is viewed as the result of conflict between interest groups with power and wealth differences, and since conflict theorists view conflict and the existence of social classes as normal, it is difficult to recommend policies *specifically* derived from conflict theory. We might logically conclude from this view of class and conflict that if these things are normal and perhaps beneficial, then so is crime in some sense. If we want to reduce crime, we should equalize the distribution of power, wealth, and status, thus reducing the ability of any one group to dictate what is criminalized. Generally speaking, conflict theorists favor programs such as minimum wage laws, sharply progressive taxation, a government-controlled comprehensive health care system, paid maternal leave, and a national policy of family support as a way of reducing crime (Currie, 1989).

◪ Feminist Criminology

The Concepts and Concerns of Feminist Criminology

Feminist criminology sits firmly in the critical/conflict camp of criminology. Feminists see women as oppressed both by gender inequality (their social position in a sexist culture) and by class inequality (their economic position in a capitalist society). But there is no one feminist positions on crime or on anything else. Some feminists believe the answer to women's oppression is the overthrow of the two-headed monster—capitalism and patriarchy; others simply seek reform. In the meantime, they all want to be able to interpret female crime from a feminist perspective.

The core concept of most feminist theorizing is patriarchy. **Patriarchy** literally means "rule of the father," and is a term used to describe any social system that is male dominated at all levels, from the family to the highest reaches of government, and supported by the belief of male superiority. A patriarchal society is one in which "masculine" traits such as competitiveness, aggressiveness, autonomy, and individualism are valued, and "feminine" traits such as intimacy, connection, cooperation, nurturance, while appreciated, are downplayed (Grana, 2002). Sociologist Joan Huber (2008) looks at the origins of patriarchy in the different reproductive roles of the sexes, with our ancestral mothers caught in a continuous cycle of gestation and lactation that has barred women from public life, first by virtue of their reproductive role, and later also by customs and laws that justified their exclusion. Huber maintains that to understand gender differences in almost any behavior, we must understand its evolutionary logic.

Feminist criminologists wrestle with two major concerns, the first being, "Do traditional male-centered theories of crime apply to women?" This is known as the **generalizability problem,** which has been defined as "the quest to find theories that account equally for male and female offending" (Irwin & Chesney-Lind, 2008, p. 839).

Many feminist criminologists have concluded that male-centered theories have limited applicability to females because they focus on male frustration in their efforts to obtain success goals (status, resources) and ignore female relationship goals (marriage, family) (Leonard, 1995). Some feminist scholars believe that no feminist-specific general theory is possible and that they must be content to focus on crime-specific "mini-theories" (K. Daly & Chesney-Lind, 2002).

An example is Meda Chesney-Lind's (1995) concept of *criminalizing girls' survival* in which she describes a sequence of events related to efforts of parents and social control agents to closely supervise girls, and notes that girls are more likely than boys to be reported for status offenses. She also notes that girls are more likely to be sexually abused than boys, that their assailants are more likely to be family members, and that a likely response is for girls to run away from home. When girls run away from such homes, they are returned by paternalistic juvenile authorities who feel it is their duty to "protect" them, which reinforces the girl's feeling that "nobody cares" and strengthens her resolve not to get caught again. When a girl is on the streets, she has to do something to survive: steal money, food, or clothing; use and sell drugs; and/ or engage in prostitution, which may then become lifetime patterns of behavior. Chesney-Lind's point is that girls' victimization and their response to it are shaped by their status in a patriarchal society in which males dominate the family and define their daughters as property. Thus, patriarchy combines with paternalism to force girls to live "lives of escaped convicts."

Others wonder why a special feminist theory is needed, since most female offenders tend to be found in the same places as their male counterparts—i.e., among single-parent families located in poor, socially disorganized neighborhoods. Male and female crime rates increase or decrease together across different

▲ **Photo 6.3**   Girls who run away from home often find themselves on the streets where they are preyed upon and often forced into prostitution, shoplifting, and drug trafficking.

nations and communities, indicating that females are responsive to the same environmental conditions as males (Campbell, 2009). Individual-level correlates of male offending such as low self-control, low IQ, attention-deficit/hyperactivity disorder (ADHD), and so forth, are also correlated with female offending (Moffitt et al., 2001).

Thus, "Males and females are not raised apart and exposed to an entirely different set of developmental conditions" (Bennett, Farrington, & Huesmann, 2005, p. 280). Males and females may be affected to different degrees by the same risk factors, but criminogenic risk factors are still risk factors for both males and females (Steffensmeier & Haynie, 2000).

Given this evidence, K. Daly and Chesney-Lind (2002) ask, "why do such similar processes produce a distinctive, gender-based [male] structure to crime and delinquency?" (p. 270)—a question that leads us to the gender ratio problem, feminist criminologists' second major concern.

The **gender ratio problem** is this: "What explains the universal fact that women are far less likely than men to involve themselves in criminal activity?" (K. Daly & Chesney-Lind, 2002, p. 270). Eileen Leonard (1995) contends that the fact of huge gender differences in criminal behavior is not in dispute by feminists or anyone else: "Women have had lower rates of crime in *all nations,* in *all communities* within nations, for *all age groups,* for *all periods in recorded history,* and for practically *all crimes*" (p. 55). Why this is so has been called the "single most important fact that criminology theories must be able to explain" (Bernard, Snipes, & Gerould, 2010a, p. 299). Figure 6.1 shows percentages of males and females arrested for seven of the eight FBI index crimes (FBI, 2009) in 2008. There are about 9 males for every 1 female arrested for murder and robbery, but less than 2 males for every female arrested for larceny-theft. The more violent the offense, the more males dominate in its commission.

Women's Liberation and Crime

Two early attempts to address the gender ratio problem were Freda Adler's (1975) **masculinization hypothesis** and Rita Simon's (1975) **emancipation hypothesis,** both of which looked at the effect of the women's liberation movement (now simply called the women's movement) on female offending. In Adler's view, as females increasingly adopt "male" roles, they will increasingly masculinize their attitudes and behavior, and will thus become as crime-prone as men. Simon's view was that increased participation in the workforce affords women greater opportunities to commit job-related crime, and that there was no reason for them to first undergo Adler's masculinization. Neither hypothesis proved useful in explaining the gender crime ratio because male/female arrest rates have not varied by more than 5 percentage points over the past 40 years (Campbell, 2009).

Another interpretation linking women's liberation to female crime is the **economic marginalization hypothesis.** This perspective argues that both Adler and Simon neglected to pay sufficient attention to patriarchy and the extent to which males control female labor and sexuality. Research has suggested that much of female crime is related to economic need, and that women's poverty and crime rates have risen together (Hunnicutt & Broidy, 2004). According to this hypothesis, both the increasing crime and poverty rates are indirectly related to the women's movement. Specifically, the woman's movement has generated efforts by women to free themselves from the power of men, but by doing so they have freed men from their traditional roles as providers. According to this hypothesis, the decline in male respect for women has led to a large increase in out-of-wedlock births and divorce. These things have led to female-headed households and the "feminization of poverty," which has led many women to engage in economically related crimes such as prostitution, drug sales, and shoplifting to support themselves (Reckdenwald & Parker, 2008).

Power-Control Theory

Despite doubts that a general feminist theory of criminal behavior is possible, there have been some attempts to formulate one, including John Hagan's (1989) power-control theory. **Power-control theory** views gender differences in antisocial behavior as a function of power differentials in the family, and states that these arise from the positions the spouses occupy in the workforce. Where fathers are the sole bread-winner and mothers are housewives and/or have menial jobs, a *patriarchal family* structure results, especially if the father is in a position of authority at work. The patriarchal family is one in which the workplace experiences are reproduced, and it is said to be "unbalanced" in favor of the father. Patriarchal families are viewed as granting greater freedom to boys to prepare them for traditional male roles, while daughters are socialized to be feminine, conforming, and domesticated.

The *egalitarian family* develops in the absence of large differences between the work roles of parents and is one in which the responsibility for child rearing is shared. Power relations in such families are said to be "balanced," and parents socialize male and female children similarly. Similarity of treatment will tend to lead to sons and daughters developing similar traits, attitudes, and behaviors, which implies that girls from such families will have increased rates of delinquent involvement, similar to the rates for their male counterparts. Hagan (1989) claims that while there will be large gender differences in delinquency among children from patriarchal families, egalitarian families will show smaller gender differences. According to Siegel (1992), it is not only middle-class girls who will increase their offending: "Power-control theory, then, implies that middle-class youth of *both sexes* will have higher crime rates than their lower-class peers" (p. 270).

Structured Action Theory: "Doing Gender"

Central to James Messerschmidt's (1993) **structured action theory** is the concept of hegemonic masculinity, which is the cultural ideal of masculinity that men are expected to live up to. **Hegemonic masculinity** is "defined through work in the paid-labor market, the subordination of women, heterosexism, and the driven uncontrollable sexuality of men" (p. 82). Although it is about living up to masculine ideals and distancing the masculine self from femininity, hegemonic masculinity is also a way to maintain patriarchy (Connell & Messerschmidt, 2005).

For Messerschmidt (1993), gender is something males and females demonstrate and accomplish rather than something they automatically are by virtue of biological sex. "Doing gender" is an ongoing dynamic process by which males express their masculinity to audiences of both sexes so that it may be socially validated. This expression can take many forms according to culture and social context, which jointly informs males of the appropriate norms of masculine behavior. To project a positive masculine image to the world, a man must learn the relevant cultural definitions of masculinity. Traditional middle-class ways of doing masculinity (proving one's manhood) include being successful in a career, having and providing for a family, being a good protector, and projecting an aura of quiet dominance as well as physical and mental strength. When males cannot or will not strive to accomplish legitimate modes of doing gender, they develop alternative modes to accomplish the same result, such as engaging in crime (Merton's innovation mode of adaptation, discussed in Chapter 4). In lower-class cultures, this often involves violent confrontations over status issues because taking matters into one's own hands is seen as the only way to obtain "juice" (masculine status) on the street. Violent and criminal behavior can thus be used as a resource for accomplishing masculinity—for "doing gender."

Messerschmidt (2002) also theorizes about "doing femininity" among gang girls and women engaging in "bad girl femininity." Although violence is defined culturally as masculine behavior,

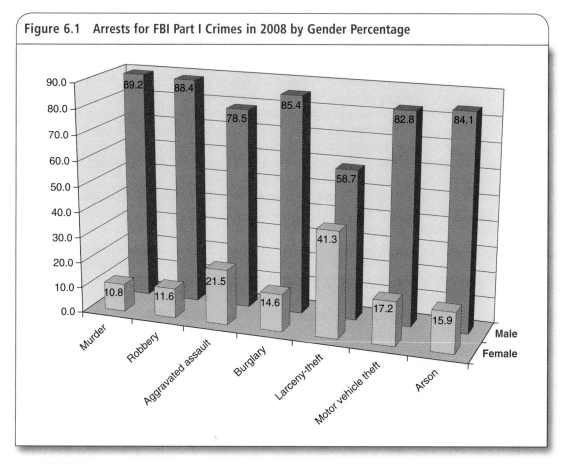

Figure 6.1 Arrests for FBI Part I Crimes in 2008 by Gender Percentage

SOURCE: FBI (2009a).

Messerschmidt asserts that if females engage in it, they are "not attempting increasingly to be masculine, but, rather, were engaging in physical violence authentically as girls and as a legitimate aspect of their femininity" (p. 465). Gang females do not consider themselves as masculine, then, but rather as "bad." Gang girls are emphatic about their feminine identity and are "very fussy over gender display (clothes, hair, makeup) and, thus, for the most part display themselves as feminine in 'culturally approved' ways" (p. 464).

Messerschmidt (2002) wants to show that gender is fluid and context specific, and that there is no incompatibility between "acting bad" and femininity. Girls and women fight to defend friends, the "hood," and "their man" from the poaching efforts of other females. He also wants to emphasize that the construction of a "badass" image is not a concern for gang females as it is among gang males, and that it is very much subordinate to constructing a feminine image. When females fight, they are contextually "doing masculinity" just as males contextually "do femininity" when "comforting and nurturing a fellow gang member" (p. 473).

Other Explanations

Both Hagan's and Messerschmidt's theories place sole reliance on socialization and gender roles to explain differences in male/female offending. Others attempt to explain it by saying that the genders differ in exposure to delinquent peers, that males are more influenced by them than females are, and that females have greater inhibitory morality (Mears, Ploeger, & Warr, 1998). Cullen and Agnew (2011) feel that statements such as these amount to nothing more than saying boys will be boys and girls will be girls, because they beg the questions of *why* males are more exposed to and more influenced by delinquent peers than females, and why females have a stronger sense of morality. A standard answer to these questions is that girls are more closely supervised than boys, but controlling for supervision level results in the same gender gap in offending (Gottfredson & Hirschi, 1990), and a meta-analysis of 172 studies found a non-significant tendency for girls to be *less* strictly supervised than boys (Lytton & Romney, 1991).

Bernard, Snipes, and Gerould (2010) note that socialist feminists (such as Joan Huber, noted earlier) have "argued that the natural reproductive differences between the sexes [ultimately] underlie male-female differences" (p. 290). These feminists agree with Dianna Fishbein (1992) that "[c]ross cultural studies do not support the prominent role of structural and cultural influences of gender-specific crime rates as the type and extent of male versus female crime remains consistent across cultures" (p. 100). They argue that because the magnitude of the gender gap varies across time and space and yet still remains constantly wide at all times and in all places, biological factors *must* play a large part (Bennett et al., 2005; Campbell, 2009). If only social factors accounted for gender differences, there should be a set of cultural conditions under which crime rates would be equal for both sexes (or even higher for females), but no such conditions have ever been found (Bernard et al., 2010). Feminists who include biological thinking in their theories assert that sex differences in dominance and aggression are seen in all human cultures from the earliest days of life and are observed in all primate and most mammalian species, and no one would evoke socialization to explain sex differences in these instances (Campbell, 2006; Hopcroft, 2009).

Neuroscientists have long shown that hormones organize the brain in male or female directions while we are still in our mother's womb (Amateau & McCarthy, 2004), and that this process organizes male brains in such a way that males become more vulnerable to the various traits associated with antisocial behavior (Ellis, 2003). Doreen Kimura (1992) tells us that that males and females come into this world with "differently wired brains," and these differences "make it almost impossible to evaluate the effects of experience [the socialization process] independent of physiological predisposition" (p. 119). Sarah Bennett and her colleagues (2005) agree, and explain the pathways from sex-differentiated brain organization to antisocial behavior:

> Males and females vary on a number of perceptual and cognitive information-processing domains that are difficult to ascribe to sex-role socialization. . . . [T]he human brain is either masculinized or feminized structurally and chemically before birth. Genetics and the biological environment in utero [in the womb] provide the foundation of gender differences in early brain morphology, physiology, chemistry, and nervous system development. It would be surprising if these differences did not contribute to gender differences in cognitive abilities, temperament, and ultimately, normal or antisocial behavior. (p. 273)

According to many theorists, the major explanation is gender differences in fear and empathy (females being higher on both), and that these traits are undergirded by testosterone and oxytocin (the "cuddle chemical"). Males have significantly more testosterone and females significantly more oxytocin (Hermans,

Putman, & van Honk, 2006). Higher testosterone equals less fear, and higher oxytocin equals greater empathy (Campbell, 2008; MacDonald & MacDonald, 2010). Shelly Taylor (2006) among others has shown how these neurohormones work against each other, and that their levels are ultimately linked to sex-differentiated evolutionary selection for nurturing behavior. No one claims that these substances are major risk (testosterone) factors for or protective (oxytocin) factors against committing criminal acts, only that they are major factors underlying gender differences in the propensity to commit such acts.

Anne Campbell's Staying Alive Hypothesis

Why do "differently wired brains" exist in the first place? Sex differences do not arise without an evolutionary reason behind them. Biologists note that sex differences in aggression and dominance seeking are related to *parental investment* (time and resources devoted to parental care), not biological sex per se. It is parental investment that provokes evolutionary pressures for the selection of the mechanisms that underlie these behaviors. In some bird and fish species, males contribute greater parental investment (e.g., incubating the eggs and feeding the young), and females take more risks, are more promiscuous and aggressive in courtship, have higher testosterone levels, and engage in violent competition for mates (Barash & Lipton, 2001). In these species, sex-related characteristics are the opposite of those found in species in which females assume all or most of the burden of parenting (the vast majority of species).

Anne Campbell (1999), who describes herself as a liberal evolutionary feminist, has attempted to account for the gender ratio problem using the logic of evolutionary theory in her **staying alive hypothesis.** Her hypothesis is based on the traits of nurturing and fear coupled with male status concerns. Campbell argues that because the *obligatory* parental investment of females is greater than that of males, and because of the infant's greater dependence on the mother, a mother's presence is more critical to offspring survival than is a father's. She notes that offspring survival is more critical to female reproductive success (the passing of one's genes to subsequent generations—the ultimate "goal" of all life forms) than to male reproductive success. Because of the limits placed on female reproductive success by long periods of gestation and lactation, females have more investment tied up in children they already have than males, whose reproductive success is only limited by access to willing females. We are reminded again that we humans are adapted to seek sexual pleasure, not reproductive success per se. Reproduction was simply a more common outcome of sexual activity in pre-contraceptive times.

Campbell (1999) argues that because offspring survival is so enormously important to their reproductive success, females have evolved a propensity to avoid engaging in behaviors that pose survival risks. The practice of keeping nursing children in close proximity in ancestral environments posed an elevated risk of injuring the child as well as herself if the mother placed herself in risky situations. Thus it became adaptive for females to experience many different situations as fearful. There are no sex differences in fearfulness *unless* a situation contains a significant risk of physical injury, and it is this fear that accounts for the greater tendency of females to avoid or remove themselves from potentially violent situations, and to employ low-risk strategies in competition and dispute resolution relative to males. Average differences in fear levels are strong and consistently found regardless of how fear is measured (Campbell, 2009). Females do engage in competition with one another for resources and mates, of course, but it is rarely violent competition. Most of it is decidedly low-key, low risk, and chronic as opposed to emotionally tense, high-risk, and acute male competition.

Campbell's theory also focuses on gender differences in status striving. Campbell (1999) shows that when females engage in crime, they almost always do so for instrumental reasons, and their crimes rarely involve risk of physical injury. There is no evidence, for instance, that female robbers crave the additional payoffs of dominance that male robbers do, or seek reputations as "hardasses." Campbell notes that while women do aggress and do steal, "they rarely do both at the same time because the equation of resources and status reflects a particularly masculine logic" (p. 210).

⊠ Evaluation of Feminist Theories

In common with critical theorists, feminist theorists have generally been content to focus on descriptive studies or on crime-specific "mini-theories" such as Chesney-Lind's *criminalizing girls' survival* that have worked well to explain female-specific offending.

Although Hagan's and Messerschmidt's theories offer some interesting insights regarding family dynamics and how gender is perceived in different social contexts, they have not moved the discipline forward. Messerschmidt's theory seems to be a rehashing of the old subcultural theories, in which "doing gender" is substituted for male status striving stressed by subculturists such as Albert Cohen in the 1950s. The assertion in Hagan's power-control theory that middle-class children will have higher rates of antisocial behavior than lower-class children is contrary to all that we know about the relationship between social class and crime, especially serious crime (Walsh, 2011b). However, Hagan (1989) admits that his theory best addresses minor misbehaviors such as smoking, drinking, and fighting, which leaves unaddressed the serious violent crimes that most strongly differentiate male and female offending. It is thus difficult not to agree with the feminists who argue that qualitative studies of specific crimes are the best way to study female crime, although, of course, this does not address the gender ratio problem.

If forced to boil down the reasons for the universal sex difference in criminal behavior to their bare minimum, differences in empathy and fear would be strong candidates. Empathy and fear are the natural enemies of crime for fairly obvious reasons:

> Empathy is other-oriented and prevents one from committing acts injurious to others because one has an emotional and cognitive investment in the well-being of others. Fear is self-oriented and prevents one from committing acts injurious to others out of fear of the consequences to one's self. (Walsh, 2011b, p. 124)

Numerous studies show highly significant differences in average levels of fear and empathy between males and females of all ages (reviewed in Walsh, 2011b).

Campbell's (1999) staying alive/high-fear hypothesis is about why females commit so little crime, not why some females commit it, but it does address the gender ratio problem. Because of its biological underpinnings, it may not be acceptable to many traditional feminists, but only 4 of the 27 commentators on her target article argued that strictly social theories better accounted for gender differences in crime. Campbell's hypothesis must be augmented with cultural factors, though, because we sometimes do see females committing more serious crimes than males. For many years, African American females have had higher homicide rates than white males (Barak, 1998).

This does not negate the basic gender ratio argument because *within* the African American community, the gender ratio is generally higher than it is in the white community; e.g., there is a bigger gap between the homicide rates of black males and females than there is between the homicide rates of white males and females (Barak 1998).

Policy and Prevention: Implications of Feminist Theories

The policy recommendations of feminist theory depend on which variety of feminism we examine. Marxist feminists seem more concerned with defeating capitalism than patriarchy; socialist feminists are more concerned with patriarchy, but have no love for capitalism; radical feminists want to abolish gender, and liberal feminists are less radical reformers who want only to abolish patriarchy (Lanier & Henry, 2010). Liberal feminist reformers have been successful in moving women into what had formerly been "male" occupations, but how the policies of other forms of feminism (abolishing capitalism and gender) could be implemented, as well as their desirability, are open to what would be highly contentious discussion. There are all sorts of other recommendations in between, the major one being the reform of our patriarchal society. Other recommendations include the more equal (less paternalistic) treatment of girls and boys by juvenile authorities, increased educational and occupational choices for women so that those in abusive relationships can leave them, more day care centers, and so forth. Feminist theory suggests that gender sensitivity education in the schools and workplaces may lead men to abandon many of their embedded sexist ideas pertaining to the relationship between the sexes.

Feminist criminologists have impacted the criminal justice system more strongly than they have criminology. The efforts of feminists to fight gender stereotypes have moved women into previously all-male occupations such as police, probation/parole, and corrections officer positions, and have placed more female judges on the bench and more lawyers before the bar. This has led to a greater understanding of female victims and their plight. Feminist criminologists and other feminist activists have been on the forefront in fighting such previously quasi-"acceptable" practices as sexual harassment, stalking, date rape, and child pornography. It was feminists who long pushed for mandatory arrests for domestic violence and achieved it, although this has led to an increase in female as well as male arrests. Feminists also fought for other reforms of such laws as the spousal exception to rape (up until the 1980s it was not legally possible for a man to rape his wife), and for rape shield laws, protecting rape victims' sexual history from examination in a rape trial. Overall, feminist criminology has been quite successful in pushing for a more just, less patriarchal, and more sensitive criminal justice system.

SUMMARY

- Critical criminology is a generic term encompassing many different theoretical positions united by the common view that society is best characterized by conflict and power relations rather than by value consensus.
- Marxist criminologists follow the theoretical trail of Karl Marx, who posited social theories based on two conflicting classes in society, the bourgeoisie and the proletariat. While some modern Marxists tend to romanticize criminals as heroic revolutionaries, Marx considered them "social scum" who preyed upon the working class. Marx is credited with introducing the terms primitive *rebellion* and *alienation* into criminology's vocabulary.
- Willem Bonger is credited with being the first Marxist criminologist. He was concerned with two opposite "social sentiments": altruism and egoism. The sentiment of altruism is killed in a capitalist social system because it generates competition for wealth, status, and jobs. Thus, capitalism produces egoism, which leads to criminal behavior on the part of both the poor and the rich.
- Marxists tend to view capitalism as the only cause of crime, and they insist that class and class values are generated by the material conditions of social life. Because only the material conditions of life really

matter, the only way to make any serious impact on crime is to eliminate the capitalist mode of production and institute a Marxist social order. Left realists realize that such a radical transformation is highly unlikely in modern times, and although they maintain a critical stance toward the system, they work within it in an effort to influence social policy.

- Conflict theorists share some sentiments with Marxists, but view conflict in pluralistic terms and as intrinsic to society, not something that can be eliminated. Crime is the result of the ability of powerful interest groups to criminalize the behavior of other less powerful interest groups when that behavior is contrary to their interests.

- Conflict criminological research tends to focus on the differential treatment by the criminal justice system of individuals who are members of less powerful groups such as minorities, women, and working-class whites.

- Peacemaking criminology is based on religious principles more than empirical science. It wants to make peace on crime, counsels us that we should appreciate the criminals' point of view, and wants us to be less punitive.

- Feminist criminology focuses on trying to understand female offending from the feminist perspective, which contends that women are faced with special disabilities living in an oppressive sexist society.

- The two big issues in feminist criminology are the *generalizability* problem (Do traditional theories of crime explain female as well as male offending?) and the *gender ratio* problem (What accounts for the huge gap in offending between males and females?).

- Early attempts to explain female crime from the feminist tradition emphasized the masculinization of female attitudes as women increasingly adopted "male" roles, or simply that as women moved into the workforce in greater numbers, they found greater opportunities to commit job-related crimes. Many feminists rejected both positions, pointing out that such theorizing provided ammunition for those who opposed the women's movement and that regardless of any increase in female offending, the male–female gap remains as wide as ever.

- John Hagan and James Messerschmidt have formulated theories of gender difference in antisocial behavior based on socialization and gender role theories. Other feminists maintain that we cannot understand gender differences in behavior without understanding the underlying biological differences between the sexes.

- The size and universality of the gender gap suggests to some that the most logical explanation for it must lie in some fundamental differences between the sexes rather than socialization, such as neurological and hormonal differences.

- Anne Campbell's staying alive hypothesis attempts to explain the gender ratio problem in terms of differential evolutionary selection pressures between the sexes. Female survival was more crucial to women's reproductive success than male survival was to theirs. Natural selection exerted pressure for females to be more fearful of dangerous situations, whereas for males the seeking of dominance and status, which aided their reproductive success, often placed them in such situations.

DISCUSSION QUESTIONS

1. Do you think that the "material conditions of life" largely determine what we will know, believe, and value, and how we will behave?

2. Do you believe that social conflict is inevitable? In what way is conflict a good thing (if at all)?

3. Can inequality ever be eliminated? If we can do this, what price would we pay, if any?

4. Do we really need a feminist criminology, or do the traditional theories suffice to explain both male and female criminality?

5. Explain how ultimate-level explanations of gender differences in behavior, such as Campbell's staying alive hypothesis, are or are not useful for criminologists.

USEFUL WEBSITES

Conflict Theory. www.newworldencyclopedia.org/entry/Conflict_theory

Feminist Criminology. http://en.wikipedia.org/wiki/Feminist_school_of_criminology

Karl Marx. www.historyguide.org/intellect/marx.html

Peacemaking Criminology. www.greggbarak.com/whats_new_2.html

Postmodern Criminology. http://uwacadweb.uwyo.edu/RED_FEATHER/lectures/051techcrm7.htm

CHAPTER TERMS

Alienation	Gender ratio problem	Primitive rebellion hypothesis
Bourgeoisie	Generalizability problem	Social sentiments
Class struggle	Left realists	Staying alive hypothesis
Economic marginalization hypothesis	Lumpenproletariat	Structured action theory
Emancipation hypothesis	Masculinization hypothesis	
	Power-control theory	

7

Psychosocial Theories

Individual Traits and Criminal Behavior

Little Jimmy Caine, a pug-nosed Irish American, is an emotionless, guiltless, walking id, all 5' 5" and 130 pounds of him. By the time he was 26, Jimmy had accumulated one of the worst criminal records the police in Toledo, Ohio, had ever seen: burglary, robbery, aggravated assault, rape—you name it, and Jimmy had done it. This little tearaway had been arrested for the brutal rape of a 45-year-old barmaid. Jimmy had entered an unlocked bar after closing time to find the lone barmaid attending to some cleaning chores. Putting a knife to the terrified woman's throat, he forced her to strip and proceeded to rape her. Because she was not sexually responsive, Jimmy became angry and placed her head over the kitchen sink and tried to decapitate her. His knife was as dull as his conscience, which only increased his anger, so he picked up a bottle of liquor and smashed it over her head. While the woman lay moaning at his feet, he poured more liquor over her, screaming, "I'm going to burn you up, bitch!" The noisy approach of the bar's owner sent Jimmy scurrying away. He was arrested 45 minutes later casually eating a hamburger at a fast food restaurant.

Jimmy didn't fit the demographic profile of individuals who engage in this type of crime. Although he had a slightly below-average IQ, he came from a fairly normal, intact middle-class home. However, Jimmy had been in trouble since his earliest days and had been examined by a variety of psychiatrists and psychologists. Psychiatrists diagnosed him with something

(Continued)

(Continued)

called a conduct disorder as an 8-year-old and as having an antisocial personality disorder at 18. Jimmy's case reminds us that we have to go beyond factors such as age, race, gender, and socioeconomic status to explain why individuals commit criminal acts. In this chapter, we look at many of the traits that psychologists and psychiatrists have examined to explain individual criminality. These explanations do not compete with sociological explanation; rather, they complete them.

The "Two Great Pillars of Psychology"

The title of this chapter says *psychosocial* rather than *psychological* theories because it is artificial to strictly separate social and psychological approaches. Self-control theory, for instance, focuses on a psychological trait, as do subcultural theories and Agnew's general strain theory. However, unlike these theories, the emphasis in this chapter is placed much more on the psychology than on the sociology of behavior. Psychosocial theories of criminal behavior are more interested in individual differences in the propensity to commit crimes than in environmental conditions that may push a person into committing them. Two of the most respected modern criminologists, Francis Cullen and Robert Agnew (2011), have written the following: "It has become increasingly clear that biological factors, individual traits, and social factors all have an important role to play in the explanation of crime" (p. 78). We have already looked at social factors, biological factors are examined in the next chapter, and in this chapter we look at individual traits other than those examined in previous chapters.

Early theories in the psychological tradition strongly emphasized two major traits contributing to criminal behavior—intelligence and temperament, the so-called "two great pillars of differential psychology" (Chamorro-Premuzic & Furnham, 2005, p. 352). Early theorists assumed that low intelligence hampered the ability to calculate the pleasures and pains involved in undertaking criminal activity, and that certain types of temperament made individuals possessing them impulsive and difficult to socialize.

David Wechsler (who devised many of the IQ tests in use today) defined **intelligence** as "the aggregate or global capacity of the individual to act purposefully, to think rationally, and to deal effectively with his [or her] environment" (quoted in Matarazzo, 1976, p. 79). Although some claim that IQ tests are biased, according to the National Academy of Sciences (Seligman, 1992) and the American Psychological Association's (APA) Task Force on Intelligence (Neisser et al., 1995), no study designed to detect such bias has ever done so.

Although most studies of IQ today involve looking at the brain areas and genes thought to be associated with it, environmental effects should not be neglected. The most important evidence of environmental effects on IQ is the so-called **Flynn effect.** Flynn (2007) has shown that the average IQ has increased in populations in all countries studied by approximately 3.1 points per decade from 1932 to 2000. These gains have been seen mostly in lower socioeconomic groups as the environment has become more equal (better schooling, medical care, and nutrition for all social classes). IQ gains have ceased in developed countries because they have wrung all the IQ-enhancing benefits that they can from the environment, although the Flynn effect is still evident in developing countries. Flynn claims that the direct genetic effect on IQ is only about 36% (as opposed to much higher estimates claimed by others), with 64% resulting from the indirect effects of genes interacting with the environment (p. 90).

The gene–environment interplay results in what Dickens and Flynn (2001) call the "multiplier effect." That is, genes are usually matched with environments ("high-IQ genes" with advantaged environments and "low-IQ genes" with disadvantaged environments) and multiply what may have been a small genetic advantage or disadvantage into a large advantage or disadvantage over time. In former times when societies were more unequal, a person born into lower-class conditions was not able to realize his or her full genetic potential, so environmental factors simply reinforced the advantage or disadvantage of genetic inheritance in a kind of "the rich get richer and the poor get poorer" fashion. Dickens and Flynn go on to say that across the time in which the Flynn effect has been working, the better environments to which successive generations have been exposed has allowed "the potency of environmental factors [to stand] out in bold relief" (p. 351).

The IQ–Crime Connection

A number of reviews of the IQ–crime relationship find it to be robust (Ellis & Walsh, 2003; Lynham, Moffitt, & Stouthamer-Loeber, 1993). It is stronger than often indicated because most IQ studies lump together boys who commit only minor delinquent acts during their teenage years with boys who will continue to seriously and frequently offend into adulthood. Casual and less serious offenders differ from non-offenders by only about 1 point, while serious persistent offenders differ from non-offenders by about 17 points (Gatzke-Kopp, Raine, Loeber, Stouthamer-Loeber, & Steinhauer, 2002; Moffitt, 1993). Simple arithmetic tells us that pooling these two groups hides the magnitude of IQ differences between non-offenders and serious offenders if the latter have lower IQs than the former.

Most IQ studies look at *full-scale* IQ (FSIQ), which is obtained by averaging the scores on *verbal* (VIQ) and *performance* (PIQ) IQ subscales. While most people have VIQ and PIQ scores that closely match, criminal offender populations are almost always found to have significantly lower-than-average VIQ scores, but not lower PIQ scores, compared to non-offenders. This PIQ>VIQ discrepancy is called **intellectual imbalance.** As L. Miller (1987) remarks, "This PIQ>VIQ relationship [is] found across studies, despite variations in age, sex, race, setting, and form of the Wechsler [IQ] scale administered, as well in differences in criteria for delinquency" (p. 120). A literature review found that overall, VIQ>PIQ boys are underrepresented in delinquent populations by a factor of about 2.6, and PIQ>VIQ boys are overrepresented by a factor of about 2.2 (Walsh, 2003).

The research on intellectual imbalance provides another example of how the role of IQ in understanding criminal behavior may be underestimated if we rely solely on full-scale IQ rather than looking deeper into the effects of verbal IQ only, or of PIQ>VIQ imbalance.

The most usual explanation for the IQ–delinquency link is that it works via poor school performance, which leads to dropping out of school, and then associating with delinquent peers (Ward & Tittle, 1994). The idea that IQ influences offending via its effect on school performance was supported in 89% of 158 studies based on official statistics and 77.7% based on self-reports (Ellis & Walsh, 2000). On the other hand, all 46 studies (100%) exploring the link between grade point average (GPA) and antisocial behavior found it to be significant. Actual performance measures of academic achievement such as GPA are probably better predictors of antisocial behavior than IQ. Academic achievement is a measure of the kind of intelligence IQ measures plus many other personal and situational characteristics such as ambition, conscientious study habits, and supportive parents.

IQ is related to a wide range of life outcomes that are themselves related to criminal and antisocial behavior such as poverty, lack of education, and unemployment. The data presented in Table 7.1 come from 12,686 white males and females in the National Longitudinal Study of Youth (NLSY). This study began in 1979 when subjects were 14 to 17 years old, and data were collected in 1989 when the subjects were 24 to 27 years old. The bottom 20% on IQ had scores of 87 and below; the top 20% had scores of 113 and above. Note the large ratios between the two groups on all outcomes. For instance, for every 1 high-IQ subject ever interviewed in jail or prison there were 31 low-IQ subjects ever interviewed in jail or prison.

Table 7.1   **The Impact of High and Low IQ on Selected Life Outcomes**

| Social Behavior | IQ Level | | Ratio |
	Bottom 20%	Top 20%	
Dropped out of high school	66%	2%	33.0:1
Living below poverty level	48%	5%	9.6:1
Unemployed entire previous year*	64%	4%	16.0:1
Ever interviewed in jail or prison	62%	2%	31.0:1
Chronic welfare recipient	57%	2%	28.5:1
Had child out of wedlock**	52%	3%	17.3:1

SOURCE: NLSY data taken from Herrnstein and Murray (1994).

*Males only **Females only

Temperament and Personality

It is obvious that low intelligence alone cannot explain criminal behavior. Most individuals with a below-average IQ do not commit crimes, and many people with an above-average IQ do. Many early psychological criminologists saw criminal behavior as a result of the interaction of low intelligence and a particular kind of temperament. As we have seen, IQ and temperament are given prominent roles as factors influencing how a person copes with strain, and thus how insulated he or she is from criminal behavior, as discussed in Robert Agnew's general strain theory in Chapter 4.

Temperament is a person's habitual mode of emotionally responding to stimuli and is largely a function of genes governing physiological arousal patterns, although arousal systems are fine-tuned by experience (Rothbart, Ahadi, & Evans, 2000). Temperamental components include *mood* (happy/sad), *activity level* (high/low), *sociability* (introverted/extraverted), *reactivity* (calm/excitable), and *affect* (warm/cold). These components make it easy or difficult for others to like us and to get along with us. Temperamental differences in children make some easy to socialize and others difficult. Children who throw temper tantrums and reject warm overtures from others may adversely affect the quality of parent–infant interactions regardless of their parents' temperaments, and thus lead to poor parent–child attachment and all the negative consequences that result. Numerous studies have shown that parents, teachers, and peers respond to children with bad temperaments negatively, and that such children find acceptance only in association with others with similar dispositions (Caspi, 2000).

Personality is an individual's set of relatively enduring and functionally integrated psychological characteristics that result from his or her temperament interacting with cultural and developmental experiences. There are many different components of personality, which psychologists call **traits,** some of which are associated with the probability of committing antisocial acts and some of which protect against doing so. People differ only on the strength of these traits; they are not characteristics that some people possess and others do not.

Any discussion of personality must acknowledge the role of the father of psychoanalysis, Sigmund Freud. Freud offered a broad sweeping theory of personality, and although he wrote little about crime, his ideas stimulated many criminologists.

Early psychological theorists never pondered what mental processes might intervene between criminal behaviors and their assumed causes. Just how does "feeblemindedness," "atavism," or anything else influence persons to commit criminal acts? If all people are hedonistic, why do only some commit crimes? If criminals are feebleminded, why don't all low-IQ people commit crimes? The psychological answer to such questions is that individuals possess different personalities, and these different personalities lead them to respond differently to identical situations.

According to Freud, the basic human personality consists of three interacting components, each having separate purposes: the id, ego, and superego. The *id* is the biological raw material of our temperament and personality; it represents our drives and instincts for acquiring life-sustaining necessities as well as life's pleasures. Like a spoiled child, the id demands instant gratification of its desires and does not care if the means used to satisfy them are appropriate or injurious to the self or to others. The id obeys what Freud called the "pleasure principle," but because it lacks the ability to engage in the hedonistic calculus, it is often dangerous to itself as well as to others. The selfish, immoral, uncaring, antisocial id is the only aspect of the personality we are born with, so in a Freudian sense we might say that we are all Lombroso's "born criminals," as discussed in Chapter 3.

The ego and the superego are formed from the raw material of the id in the process of socialization. With the correct moral training, energy from the id is used to form the *ego,* or the aspect of the personality we think of as "me" or "I." The ego obeys the reality principle because it realizes that the desires and demands of the id are necessary, but they must be satisfied in socially appropriate ways if one is to avoid negative consequences. It is the ego that performs the hedonistic calculus. It does not deny the pleasure principle; rather, it simply adjusts it to the demands of reality. Freud (1923/1976) analogized the interaction of the ego and the id in terms of a rider and a horse. The horse (the id) supplies the raw locomotive power, while the rider (the ego) supplies the goals and the direction (p. 77).

The *superego* strives for the ideal, and is thus just as irrational as the id. It represents all the moral and social rules (the "dos and do nots") internalized by the person during the process of socialization, and may be summed up as the conscience. Because many urges have

▲ **Photo 7.1**   Sigmund Freud, the father of Personality Psychology

been defined as wrong or sinful, the superego tries to suppress all the normal urges arising from the id. It is the ego's function to sort out the conflict between the antisocial demands of the id and the overly conformist demands of the censorious superego. The normal personality is one in which the ego is successful in working out compromises between its irrational partners. An abnormal personality results when either the id or the superego overwhelms the ego, resulting in psychic energy being drained from the weaker components to strengthen the stronger component. If the id is "in command" of the personality, the result is a conscienceless and impulsive individual who seeks to satisfy personal needs regardless of the expense to others.

Personality Traits Associated With Criminal Behavior

Perhaps the trait most often linked to criminal behavior is **impulsiveness,** which is the tendency to act without giving much thought to the consequences. Although impulsiveness, low constraint, and low self-control are somewhat different constructs, they are similar enough to be treated as one. All three involve disinhibited behavior in which the actor is unable or unwilling to consider the long-term consequences of his or her actions (Chapple & Johnson, 2007). A review of 80 studies examining the relationship between impulsivity and criminal behavior found that 78 were positive and the remaining 2 were non-significant (Ellis & Walsh, 2000). Although impulsiveness is a potent risk factor for criminality in its own right, it becomes more potent if negative emotionality is added to the mix.

Negative emotionality is a trait that refers to the tendency to experience many situations as aversive, and to react to them with irritation and anger more readily than with positive affective states (McGue, Bacon, & Lykken, 1993). The trait is central to Robert Agnew's general strain theory and is strongly related to self-reported and officially recorded criminality "across countries, genders, races, and methods" (Caspi et al., 1994). People who are low on constraint (i.e., they are impulsive) tend to be high on negative emotionality. Low levels of a brain chemical called *serotonin* underlie both high levels of negative emotionality and impulsivity, and Caspi and his colleagues (1994) claim that low serotonin may represent a constitutional predisposition for these traits, and thus a general vulnerability to criminality. Taking advantage of the relationship between these two traits, Agnew has developed a theory featuring them, which is discussed in Chapter 9.

Sensation seeking refers to the active desire for novel, varied, and risky sensations (Zuckerman, 1990). Sensation seekers tend to be outgoing and relatively impulsive and fearless. These other traits lead socialized sensation seekers to want to work as firefighters, police officers, or any other job that provides physical activity, variety, and excitement; unsocialized sensation seekers, on the other hand, may well find their kicks in carjacking and burglary. A review of the literature found that 98.4% of the studies reported a statistically significant relationship between sensation seeking and antisocial behavior (Ellis & Walsh, 2000).

Empathy is the emotional and cognitive ability to understand the feelings and distress of others as if they were your own. The emotional component allows you to "feel" the other person's pain, and the cognitive component allows you to understand that person's pain and why he or she is feeling it. Some people carry the pains of the world on their shoulders, while others couldn't care less about anyone. Most criminals fall into the "couldn't care less" category for obvious reasons—you are less likely to victimize someone if you feel and understand what the consequences may be for them (Covell & Scalora, 2002). Having a personality that is strong in empathy is thus a powerful protective factor against criminal offending.

Altruism can be thought of as the action component of empathy; if you feel empathy for someone, you will probably feel motivated to take some sort of action to alleviate their distress if you are able. As with empathy, altruism lies on a continuum, with criminals (again for obvious reasons) on the far end of it, with the least empathy. Lack of empathy and altruism is considered one of the most salient characteristics of psychopaths, the worst of the worst among criminals (Fishbein, 2001). A review of 24 studies of these traits found that 23 of them were statistically significant in the predicted direction; that is, the lower the level of empathy/altruism, the more antisocial behavior was displayed (Ellis & Walsh, 2000).

Conscientiousness is a primary trait composed of several secondary traits such as well-organized, disciplined, scrupulous, responsible, and reliable at one pole (the most conscientious), and disorganized, careless, unreliable, irresponsible, and unscrupulous at the other. It is easy to see how conscientiousness could be directly related to crime through the inability of people who lack it to follow a legitimate path to the American Dream:

> It is not merely a matter of talented individuals confronted with inferior schools and discrimina-
> tory hiring practices. Rather, a good deal of research indicates that many delinquents and crimi-
> nals are untalented individuals who cannot compete effectively in complex industrial societies.
> (Vold, Bernard, & Snipes, 1998, p. 177)

In other words, people with certain kinds of temperament do not develop the personal qualities needed to apply themselves to the long and arduous task of achieving financial success legitimately, and as a consequence, may attempt to obtain it through crime.

Agreeableness is the tendency to be friendly, considerate, courteous, helpful, and cooperative with others. Agreeable persons tend to trust others, to compromise with them, to empathize, and to aid them. This list of sub-traits suggests a high degree of concern for prosocial conformity and social desirability. Disagreeable persons simply display the opposite characteristics—suspicion of others, unfriendly, uncoop-erative, unhelpful, and lacking in empathy—which all suggest a lack of concern for prosocial conformity and social desirability. While agreeable people tend to also be conscientious, this is not always the case. A person can be very conscientious at work but thoroughly disagreeable as a person (think of the greedy, egotistical, and manipulative corporate criminal), and one can be very agreeable but be thoroughly lacka-daisical at work (think of the ritualist of anomie theory, from Chapter 4).

Agreeableness seems to be a better protective factor than conscientiousness. In Miller and Lynam's (2001) meta-analysis of 29 studies that compared prisoner with non-prisoner samples, they found moder-ate to strong relationships between agreeableness and antisocial behavior and moderate relationships between conscientiousness and antisocial behavior. Miller and Lynam describe the personality of the "typical" criminal in terms of agreeableness and conscientiousness:

> Individuals who commit crimes tend to be hostile, self-centered, spiteful, jealous, and indifferent
> to others (i.e., low in Agreeableness). They tend to lack ambition, motivation, and perseverance,
> have difficulty controlling their impulses, and hold nontraditional and unconventional values and
> beliefs (i.e., are low in Conscientiousness). (p. 780)

Conscience and Arousal

One of the basic ideas of psychosocial criminology is that different levels of physiological arousal correlate with different personality and behavioral patterns because arousal levels determine what we pay attention to, how strongly we pay attention, and the ease or difficulty of acquiring a conscience. When we ask most indi-viduals why they don't take advantage of others, they tend to reply that their consciences won't let them, yet few people are aware of how their consciences were formed. A **conscience** is a complex mix of emotional and cognitive mechanisms acquired by internalizing the moral rules of our social group during socialization.

▲ **Photo 7.2**   Arousal theory states that human beings have varying internal "thermostats," which explains why people differ on the levels of arousal or stimulation they need to feel comfortable. Those already set on "high" may attempt to avoid noise, activity, or crowds. On the other hand, those who crave stimulation might climb mountains, go to loud parties, or watch slasher movies.

People with strong consciences feel guilt, shame, stress, and anxiety when they violate, or even contemplate violating, these rules. Differences in the emotional component of conscience reflect variation in **autonomic nervous system (ANS)** arousal patterns (Kochanska & Aksan, 2004). The autonomic nervous system carries out the basic housekeeping functions of the body by funneling messages from the environment to the various internal organs so that they may keep the organism in a state of biological balance (e.g., adjusting pupil size, shivering or sweating in response to temperature, etc.). These processes regulating bodily functions occur automatically and never reach our conscious awareness. However, certain other messages that influence ANS functioning that do reach awareness are important for the acquisition of conscience via a process called classical conditioning.

Classical conditioning is a form of learning that is different from operant conditioning that we discussed in Chapter 5. Operant conditioning is active (it depends on the actor's behavior) and cognitive in that it forms a conscious association between a person's behavior and its consequences. Classical conditioning, on the other hand, is mostly passive (it depends more on the level of ANS arousal than on anything the actor does) and is visceral in nature (felt in the internal organs); it simply forms a subconscious association between two paired stimuli. You may have heard about Russian psychologist Ian Pavlov's classical conditioning experiment, in Psych 101, in which he conditioned dogs to salivate at the sound of a bell. Salivation is a natural ANS response to the expectation of food. A bell has no intrinsic properties that would make dogs salivate at its sound, but because Pavlov consistently paired the sound of the bell with food, the dogs learned (were conditioned) to associate the sound of the bell with food, and the sound itself became enough to make them salivate even when not paired with food.

We have all been conditioned in various ways to respond at the gut level to neutral stimuli via their association with unconditional stimuli. How did you feel as a child when the school bell rang for recess, or you heard the bells that announced the arrival of the ice cream truck? In both cases, you likely responded with some pleasure, not because you love the sound of bells themselves, but because they signaled something that you did love.

It is by way of these associations that we develop the "gut level" emotions of shame, guilt, and embarrassment that make up the emotional ("feeling") scaffolding of our consciences. Children must learn which behaviors are acceptable and which are not (the "knowledge" part of our conscience), most of which comes via parental teaching. Once children know the behavior expected of them, the degree to which emotions influence future behavior depend on the severity of the reprimand interacting with the responsiveness of their ANS (Pinel, 2000). Assuming an ANS that is adequately responsive to discipline, refraining from such behavior in the future is not simply a rational calculation of cost and benefits, but rather a function of the emotional component of conscience strongly discouraging it by generating unpleasant feelings.

Individuals with a readily aroused ANS are easily socialized; they learn their moral lessons well because ANS arousal ("butterflies in the stomach") is subjectively experienced as fear and anxiety. A hyperresponsive ANS generates high levels of fear and anxiety and is a protective factor against antisocial behavior.

Studies have shown that males with a *hyper*-arousable (quick to react) ANS living in high-crime environments are less involved with antisocial behavior than males living in low-crime environments with a *hypo*-arousable (slow to arouse) ANS (Brennan et al., 1997).

Individuals with a relatively unresponsive ANS are difficult to socialize because they experience little fear, shame, or guilt when they offend, even when discovered and punished. Measures of ANS arousal in childhood such as resting heart rate have enabled researchers to accurately predict which of their subjects would and would not have a criminal record at age 24 with 75% accuracy (Raine, 1997). Across a wide variety of subjects and settings, it is consistently found that antisocial individuals have a relatively unresponsive ANS. This relationship exists because a hypo-arousable ANS does not allow for adequate development of the social emotions. Having knowledge of what is right or wrong without that knowledge being paired with emotional arousal is rather like knowing the words to a song but not the music—not much good for the social choir.

The rational cognitive components of conscience are presumably stressed strongly in the socialization process of children from advantaged environments. If such children become delinquent and criminal, then it follows that because they lack the environmental factors that push children from less advantaged environments into crime, they will be less emotionally conditionable than their peers. Variance in physiological measures will therefore be more important in accounting for antisocial behavior among higher socioeconomic status (SES) individuals than in lower SES individuals, whose physiological risks are masked by psychosocial risk factors. This is consistently found, and has been called the **social push hypothesis.** As Scarpa and Raine (2003) define the social push hypothesis, "[I]f an individual lacks psychosocial risk factors that predispose toward antisocial behavior yet still exhibits antisocial behavior, then the causes of this behavior are more likely to be biologically than socially based" (p. 213). However, children from lower SES environments who are difficult to condition because of a hyporeactive ANS are at greatest risk for antisocial behavior because they are also more likely, on average, to lack the cognitive skills necessary for acquiring a conscience and less likely to be consistently taught moral rules.

Cognitive Arousal

Another form of arousal of interest to psychologists is neurological arousal, the regulator of which is the brain's **reticular activating system (RAS).** The RAS is a little finger-sized bundle of brain cells situated at the top of the spinal cord and can be thought of as the brain's filter system, determining what incoming stimuli the higher brain centers will pay attention to. Some individuals possess an RAS that is highly sensitive to incoming stimuli (augmenters), and others possess one that is unusually insensitive (reducers). Thus, in identical environmental situations, some people are under-aroused and other people are over-aroused. Both over- and under-arousal are psychologically uncomfortable. If you've ever taken your grandpa to a rap concert, or he has taken you to a chamber music recital, you'll know what I mean.

There is no conscious attempt to augment or reduce incoming stimuli; like the reactivity of the ANS, augmentation or reduction is solely a function of differential RAS physiology. Augmenters tend to be the people with a hyperactive ANS, and reducers tend to be people with a hypoactive ANS. Under-arousal of the ANS is associated with fearlessness and under-arousal of the RAS with sensation seeking. We can readily appreciate that sensation seeking and fearlessness are correlated since sensation seeking is aided by fearlessness (Raine, 1997).

Reducers are easily bored with levels of stimulation that are "just right" for most of us, and continually seek to boost stimuli to more comfortable levels. They also require a high level of punishing stimuli before learning to avoid the behavior that provokes punishment, and are thus unusually prone to criminal behavior.

A number of studies have shown that relative to the general population, criminals, especially those with the most serious records, are chronically under-aroused as determined by electroencephalograph (EEG) brain wave patterns, resting heart rate, and skin conductance (Ellis, 2003). EEG brain waves reflect the electrical "chatter" of billions of brain cells. Clinicians recognize four bands to classify EEG brain waves: alpha, beta, theta, and delta. *Beta waves* followed by *alpha waves* are the most rapid, and they signal when a person is alert and focused. *Theta waves* are emitted when the person is in a drowsy mental state, and *delta waves* are the slowest of them all. Most studies (about 75%) show that EEG readouts of criminals reveal that their brains are less often in the alert and focused range than are the brains of people in general (Ellis & Walsh, 2000).

Skin conductivity is measured by a meter attached to various parts of the body that records electrical responses to sweat. Sweat contains high levels of salt, and salt water is an excellent conductor of electricity. In temperature-controlled environments, increased sweating (even though the sweating may not be enough for the person to notice that he or she is sweating) occurs in response to emotional arousal. This is the basis of polygraph testing. The polygrapher asks suspects questions that evoke emotions such as guilt, shame, or embarrassment that are detected (or rather skin conductivity is detected) by the monitor. Because chronic criminals tend to have low levels of these emotions as well as lower levels of ANS arousal, they are least likely to show sweat responses to threatening questions. Thus, low skin conductivity and criminal behavior are expected to be related. In a review of this literature, all 19 studies found this relationship to be significant (Elis & Walsh, 2000).

⊠ Glen Walters' Lifestyle Theory

Perhaps the best-known modern psychosocial criminological theory is Glen Walters' (1990) **lifestyle theory.** As the term *lifestyle* implies, Walters believes that criminal behavior is part of a general pattern of life characterized by irresponsibility, impulsiveness, self-indulgence, negative interpersonal relationships, and the chronic willingness to violate society's rules. Lifestyle theory has three key concepts: choices, conditions, and cognition. A criminal lifestyle is the result of *choices* criminals make "within the limits established by our early and current biologic/environmental *conditions*" (Walters & White, 1989, p. 3, emphasis added). Thus, various biological and environmental conditions lay the foundation for future choices. Walters stresses impulsiveness and low IQ as the most important choice-biasing conditions at the individual level and attachment to significant others as the most important environmental condition. Walters' theory thus adds to rational choice theory by pointing to two important components of choice structuring.

The third concept, *cognition,* refers to cognitive styles people develop as a consequence of their biological/environmental conditions and the pattern of choices they have made in response to them. According to this theory, criminals display eight major cognitive features or **thinking errors** that make them what they are (Walters, 1990). Examples of criminal thinking errors are *cutoff* (the ability to discount the suffering of their victims), *entitlement* (the world owes them a living), *power orientation* (viewing the world in terms of weakness and strength), *cognitive indolence* (orientation to the present; concrete in thinking), and *discontinuity* (the inability to integrate thinking patterns). Little can be done to change criminal behavior until criminals change their pattern of thinking.

These thinking errors lead to four interrelated behavioral patterns or styles that almost guarantee criminality: *rule breaking, interpersonal intrusiveness* (intruding into the lives of others when not wanted), *self-indulgence,* and *irresponsibility.* These behavioral patterns are the result of faulty thinking patterns, which arise from the consequences (reward and punishment) of choices in early life, which are themselves influenced by biological and early environmental conditions. Figure 7.1 lays out lifestyle theory in diagrammatic form.

Figure 7.1	Diagram of Glen Walters' Lifestyle Theory							
Conditions		**Choices**		**Cognition**		**Behavior**		
Early and environmental biological experiences and personal traits	→	Choices resulting from conditions	→	Cognitive style formed by choices; "thinking errors"	→	Pattern of behavior: rule breaking, impulsiveness, egocentrism, etc.	→	CRIME

The Antisocial Personalities

Depending on whom you ask, antisocial personality disorder, psychopathy, and sociopathy are either terms describing the same constellation of traits or separate concepts with fuzzy boundaries. We use the generic term **psychopathy,** a syndrome we can define as being characterized psychologically by egocentricity; deceitfulness; manipulativeness; selfishness; and a lack of empathy, guilt, or remorse, or physiologically as a syndrome characterized by the inability to tie the brain areas associated with the social emotions and rational cognition together.

Some researchers believe that there is a subset of psychopaths (so-called *primary psychopaths*) whose behavior is biological in origin, and a more numerous group (*secondary psychopaths,* or as others call them, **sociopaths**) whose behavior is mostly the result of incompetent parenting (Walsh & Wu, 2008). Others view psychopathy not as something one is or isn't, but rather simply as a name we apply to the most serious and chronic criminal offenders. Psychiatrists apply the label **antisocial personality disorder (APD)** to such criminals. APD is defined by the American Psychiatric Association (1994) as "a pervasive pattern of disregard for, and violation of, the rights of others that begins in childhood or early adolescence and continues into adulthood" (p. 645). The criteria for diagnosing APD are both clinical and legal, but rest primarily on behavior.

Criminologists, however, generally want to define individuals according to criteria that are independent of their behavior and then determine in what ways those so defined differ from individuals not so defined. The most widely used measure of psychopathy is the **Psychopathy Checklist–Revised (PCL-R),** which was devised by Robert Hare, the leading expert in psychopathy in the world today (Bartol, 2002). Using case histories and semi-structured interviews that may last up to 2 hours, doctoral-level clinicians rate subjects on each of the 20 traits or behaviors listed below on a 3-point scale

▲ **Photo 7.3**   Serial Killer Ted Bundy nonchalantly leans on the Leon County jail wall as the indictment charging him with the murder of two Florida State University coeds is read to him. Only a psychopath would show such a disinterested attitude.

ranging from 0 to 2; a score of 30 or higher out of a possible 40 is the required cut-point for a psychopathic diagnosis. To put this number in perspective, offenders in general have a mean PCL-R score of 22, and non-offenders a score of 5 (Hare, 1996).

Hare (1993) is a leading proponent of the idea that psychopathy (which he considers to be a different construct from APD) is primarily biological in origin: "I can find no convincing evidence that psychopathy is the direct result of early social or environmental factors" (p. 170). This means that researchers have not found any environmental factors that *cause* psychopathy; it does not mean that such factors have no effect on how psychopathy is behaviorally *expressed*. Hare is among a subset of researchers that echo Cesare Lombroso, who probably had psychopaths in mind with his idea of "morally insane" born criminals, i.e., those "who appear normal in physique and intelligence but cannot distinguish good from evil" (quoted in Gibson, 2002, p. 25). Modern researchers no longer talk of criminals as evolutionary throwbacks whose behavior is "unnatural." Rather, they view psychopaths as behaving exactly as they were designed by natural selection to behave (Quinsey, 2002). This does not mean that their behavior is acceptable or that we cannot consider it *morally* pathological and punish it accordingly.

If psychopathy is a behavioral strategy forged by evolution, there must be some biological markers that distinguish psychopaths from the rest of us. One of the most consistent physiological findings about psychopaths is their greatly reduced ability to experience the social emotions of shame, embarrassment, guilt, and empathy (Scarpa & Raine, 2003; Weibe, 2004).

Social emotions modify brain activity in ways that lead us to choose certain responses over others. Feelings of guilt, shame, and empathy prevent us from doing things that might be to our immediate advantage (steal, lie, cheat) but would cost us in reputation and future positive relationships if discovered. Hundreds of studies using many different methods have revealed over and over that the defining characteristic of psychopaths is their inability to "tie" the brain's cognitive and emotional networks together, and thus to form a conscience (Walsh & Wu, 2008). This inability is perhaps the strongest area of agreement among psychopathy researchers. What it essentially means is that the amygdala (part of the brain that plays a role in emotional memories, particularly fear) and the prefrontal cortex (part of the brain considered the brain's "chief executive officer") do not have strong connections between them as they do in the brains of normal people (van Honk, Harmon-Jones, Morgan, & Schutter, 2010). This goes a long way in explaining the difficulties of psychopaths in processing emotions in socially approved ways.

Figure 7.2 shows the location of some brain areas mentioned here such as the amygdala and the reticular activating system. The prefrontal cortex is a large brain area that is located just above the eyes and extends about one third of the way from the forehead to the back of the head.

As we have noted, some theorists believe that sociopaths are different from psychopaths. Lykken (1995) colorfully describes sociopaths as "feral creatures, undomesticated predators, stowaways on our communal voyage who have never signed the Social Contract," and states that their behavior is "traceable to deviant learning histories interacting, perhaps, with deviant genetic predilections" (p. 22). One of the biggest factors contributing to sociopathy is poor parenting, which is itself a function of the increase in the number of children being born to unmarried women (Rowe, 2002).

According to a study of 1,524 sibling pairs from different family structures taken from the National Longitudinal Survey of Youth, Cleveland and his colleagues (Cleveland, Wiebe, van den Oord, & Rowe, 2000) found that, on average, single mothers have a tendency to follow an impulsive and risky lifestyle and to have a number of antisocial personality traits, be more promiscuous, and have a below-average IQ. Families headed by single mothers with children fathered by different men were found to be the family type that put offspring most at risk for antisocial behavior. Two-parent families with full siblings placed offspring at the

Figure 7.2   Schematic Image of the Brain Showing Various Parts of Interest

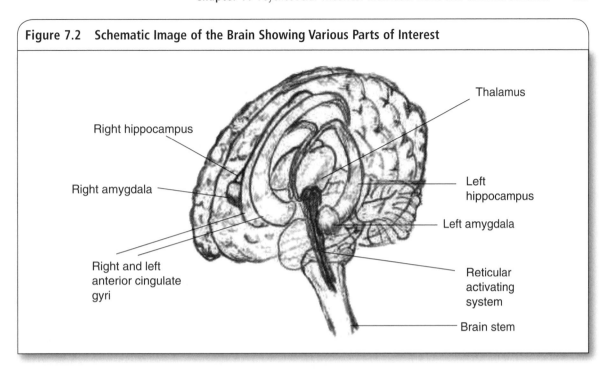

Table 7.2   Summarizing Psychosocial Theories

Theory	Key Concepts	Strengths	Weaknesses
Arousal	Because of differing ANS and RAS physiology, people differ in arousal levels they consider optimal. Under-arousal under normal conditions poses an elevated risk of criminal behavior because it signals fearlessness, boredom, and poor prospects for socialization.	Allows researchers to use "harder" assessment tools such as EEGs to measure traits. Ties behavior to physiology. Explains why some individuals in "good" environments commit crimes and why individuals in "bad" ones do not.	May be too individualistic for some criminologists. Puts all the "blame" on the individual's physiology. Ignores environmental effects.
Lifestyle	Crime is a patterned way of life (a lifestyle) rather than simply a behavior. Crime is caused by errors in thinking resulting from choices previously made, which are the result of early negative biological and environmental conditions.	Primarily a theory useful for correctional counselors dealing with their clients. Shows how criminals think and how these errors in thinking lead them into criminal behavior.	Concentrates only on thinking errors. Does talk about why they exist but pays scant attention to these reasons.
Antisocial Personality	There is a small stable group of individuals who may be biologically obligated to behave antisocially (psychopaths) and a larger group who behave similarly but whose numbers grow or subside with changing environmental conditions (sociopaths).	Concentrates on the scariest and most persistent criminals in our midst. Uses theories from evolutionary biology and "hard" brain imaging and physiological measures to identify psychopaths.	Some doubt the division of psychopath and sociopath as separate entities. While they are the scariest criminals, they are only a small proportion of all criminals.

lowest risk. Similar findings and conclusions from a large-scale British study have been reported (Moffitt & the E-Risk Study Team, 2002). Finally, Barber (2004) found that the rate of illegitimate birth was by far the strongest predictor of a composite measure of violent crime (murder, rape, assault) in his sample of 39 countries from Argentina to Zimbabwe.

The Office of Juvenile Justice and Delinquency Prevention (OJJDP) claims that delaying pregnancy until 20 to 21 years of age would lead to a 30 to 40% reduction in child abuse and neglect, and could potentially save $4 billion in law enforcement and corrections costs because offspring of teenage mothers are 2.7 times more likely than offspring of adult mothers to be incarcerated (Maynard & Garry, 1997).

Evaluation of the Psychosocial Perspective

Psychologists are always happy to point out that whatever social conditions may contribute to criminal behavior, they must influence individuals before they can affect behavior. Social factors matter and may "set the stage" for crimes, but real flesh-and-blood people commit them. Individuals are differentially vulnerable to the criminogenic forces existing in the environment because they bring different personalities to it. This person–environment interaction is captured by the old saying, "The fire that melts the butter hardens the egg." Psychologists largely take the fire (the environment) for granted and look for how the butter and eggs of our differing constitutions react to the heat of the fire. Of course, we can never take either the environment or the individual "for granted" because each affects and is affected by the other.

The relationship between IQ and criminal behavior has always been contentious. Adler et al. (2001) voice the familiar criticism that IQ tests are culturally biased, despite the findings of the National Academy of Sciences and the APA's Task Force on Intelligence cited earlier. They also cite the "debate" over whether genetics or the environment "determines" intelligence. This implies an either/or answer is possible, but since scientists involved in the study of intelligence are unanimous that all traits are *necessarily* the result of both genes and environment, it is a monumental non-debate (Carey, 2003; Flynn, 2007).

One of the most pervasive criticisms of psychological theories is that they focus on "defective" or "abnormal" personalities (Akers, 1994, p. 86). If by "abnormal" we mean a *statistical* abnormality (below or above the average on a variety of traits), however, then by definition all theories of criminality focus on abnormality. Our personalities consist of normal variation in traits we *all* possess, and these are products of the interaction of our temperaments and our developmental experiences. Lifestyle theory emphasizes that criminals are at the extremes of normal distributions of many traits, but focuses mainly on how they think as a result of their biological constitutions and their early experiences. However, there is always some risk in labeling individuals with some sort of label from psychology because we tend to think of such labels as indicating pathology (something these people "have" or "are") rather than considering them to be people with deficiencies, something they lack, such as social support, education, or a job.

The use of "hard" measuring instruments to measure ANS and RAS arousal provides us with more accurate predictions about future offending than simple "paper and pencil" methods, although we must not forget that the influences of these arousal mechanisms are strongly conditioned by the social environment. Nor should we forget that even if hard measures are better able to identify and predict, they still don't tell us how we can change criminal propensity or how we can deal with the ethical problems involved in predicting what people "might do" in the future. Predictions about human behavior are far from foolproof, and there can be many false positives (predicting something will happen and then it doesn't) as well as false negatives (predicting something won't happen and then it does).

⊠ Policy and Prevention: Implications of Psychosocial Theories

The best anti-crime policies are environmental since they are aimed at reducing the prevalence of crime in the population. But because such policies have had little impact on the crime problem in the past (Rosenbaum et al., 1998), perhaps it is wise at present to focus efforts on those who are already committing crimes rather than on conditions external to them. A variety of such programs aimed at rehabilitating offenders operate under the assumption that they are rational beings who are, however, plagued by ignorance of the long-term negative consequences of their offending behavior.

There is wide disagreement on how well rehabilitative programs work and even about what the criteria are for success. Reviews of studies with strict criteria for determining success find recidivism rates lowered by between 8 and 10% (Andrews & Bonta, 1998). These small percentages do not seem like much, but they represent many thousands of potential crimes that were not committed. Effective rehabilitation programs use multiple treatment components; are structured and focus on developing social, academic, and employment skills; use directive cognitive-behavioral counseling methods; and provide substantial and meaningful contact between treatment personnel and offenders (Sherman et al., 1997; Walsh & Stohr, 2010).

Glen Walters' theory deals with what correctional counselors call "stinkin' thinkin'" who see their task as guiding offenders to realize how destructive their thinking has been in their lives. The counselor sees offenders' problems as resulting from illogical and negative thinking about experiences that they reiterate in self-defeating monologues. The counselor's task is to strip away self-damaging ideas (such as techniques of neutralization) and beliefs by attacking them directly and challenging offenders to reinterpret their experiences in a growth-enhancing fashion. The cognitive-behavioral counselor operates from the assumption that no matter how well offenders come to understand the remote origins of their behavior, if they are unable to make the vital link between those origins and current behavioral problems, it is to no avail.

Psychopaths are poor candidates for treatment. Robert Hare (1993) states that because they are largely incapable of the empathy, warmth, and sincerity needed to develop an effective treatment relationship, treatment often makes them worse because they learn how to better push other people's buttons. Old age seems to be the only "cure" for the behaviors associated with this syndrome.

SUMMARY

- Psychosocial criminology focuses largely on intelligence and temperament as the most important correlates of criminal behavior. Low intelligence as measured by IQ tests is thought to be linked to crime because people with low IQ are said to lack the ability to correctly calculate the costs and benefits of committing crimes, and temperament is linked to crime largely in terms of impulsiveness. Intelligence is the product of both genes and environment. We concluded that IQ is probably related to crime and delinquency through its effect on poor school performance.

- Temperament constitutes a person's habitual way of emotionally responding to stimuli. The kind of temperament we inherit makes us variably responsive to socialization, although patient and caring parents can modify a difficult temperament.

- Our personalities are formed from the joint raw material of temperament and developmental processes. A number of personality traits are associated with the probability of engaging in antisocial behavior, particularly being high on impulsiveness, negative emotionality, and sensation seeking, and being low on conscientiousness, empathy, altruism, and moral reasoning.

- Classical conditioning via the ANS constitutes the emotional component of conscience and precedes the cognitive component. People differ in the responsiveness of their ANS, with those having a sluggish ANS being difficult to socialize. RAS arousal is also important to understanding criminal behavior because RAS reducers are chronically bored and seek to increase stimuli to alleviate that boredom. This may result in criminal behavior. Although people will differ greatly in their behavior depending on their innate temperaments (a function of arousal levels), their developmental and other environmental experiences also play huge parts.

- Lifestyle theory views criminal behavior as a lifestyle rather than just another form of behavior. The lifestyle begins with biological and environmental conditions that lead criminals to make certain choices, which in turn leads to criminal cognitions. The theory focuses on these cognitions, or "thinking errors." Thinking errors lead criminals into behavioral patterns that virtually guarantee criminality. The theory was devised primarily to assist correctional counselors to change criminal thinking patterns.

- Psychopaths are at the extreme end of the antisocial personality continuum. Most researchers regard the psychopathy syndrome as biological in origin, whereas some view sociopaths as formed both by genetics and the environment, with the environment playing the larger role. Many hundreds of studies have shown that psychopaths have limited ability to tie the rational and emotional components of thinking together.

- Some researchers assert that the primary cause of psychopathy is inept parenting by single mothers. Other theorists point to the fact that children born to such mothers also receive genes that predispose them to antisocial behavior from both parents in addition to being raised in an environment conducive to its expression.

DISCUSSION QUESTIONS

1. Since psychologists have long identified different temperaments as something that makes it easy or difficult to socialize children, why do you think Gottfredson and Hirschi ignored it in their self-control theory?

2. Honestly rate yourself from 1 to 10 on the traits positively associated with antisocial behavior (impulsiveness, negative emotionality, and sensation seeking), and then on the traits negatively related with antisocial behavior (conscientiousness, empathy, and altruism).

   Subtract the latter from the former. If the difference is a positive number greater than 10, or a negative number less than −10, does this little exercise correspond to your actual behavior?

3. Explain what part low arousal of the autonomic nervous system and the reticular activating system play in psychopathy.

4. How might anomie/strain theory benefit from incorporating information on IQ and conscientiousness into it?

USEFUL WEBSITES

IQ. www.psyonline.nl/en-iq.htm

IQ/Aggression Connection. www.crimetimes.org/95c/w95cp10.htm

Mental Deficiency and Crime. www.ncbi.nlm.nih.gov/pmc/articles/PMC1525117/

Mental Health. www.mentalhealth.com/

Personality Disorders. www.focusas.com/PersonalityDisorders.html

CHAPTER TERMS

Agreeableness

Altruism

Antisocial personality
    disorder

Autonomic nervous system

Classical conditioning

Conscience

Conscientiousness

Empathy

Flynn effect

Impulsiveness

Intelligence

Personality

Intellectual imbalance

Intelligence

Lifestyle theory

Negative emotionality

Psychopathy

Psychopathy Checklist–Revised
    (PCL-R)

Reticular activating system (RAS)

Sensation seeking

Social push hypothesis

Sociopaths

Temperament

Thinking errors

Traits

Biosocial Approaches

In February of 1991, Stephen Mobley walked into a Domino's Pizza store in Georgia to rob it. Once he got the money, Mobley forced store manager John Collins onto his knees and shot him execution style. In the automatic appeal to the Georgia Supreme Court to get his sentenced commuted to life in prison, his primary defense boiled down to claiming that "my genes made me do it." In support of this defense, Mobley's lawyers pointed to a Dutch study of an extended family in which for generations many of the men had histories of unprovoked violence. The researchers took DNA samples from 24 male members of the family and found that those with violent records had a marker for a mutant or variation of a gene for the manufacture of monoamine oxydase A (MAOA), an enzyme that regulates a lot of brain chemicals. Mobley's lawyers found a similar pattern of violent behavior and criminal convictions among his male relatives across the generations and requested funds from the court to conduct genetic tests on Mobley to see if he had the same genetic variant.

The court wisely denied the defense motion. Even if it were found that Mobley had the same genetic variant, it would not show that he lacked the substantial capacity to appreciate the wrongfulness of his acts or to conform to the requirements of the law. Mobley's lawyers were hoping to mitigate his sentence by appealing to a sort of genetic determinism that simply does not exist. As we shall see in this section, genes don't "make" us do anything; they simply bias us in one direction rather than another. Except in cases of extreme mental disease or defect, we are always legally and morally responsible for our behavior. Cases such as Mobley's underline the urgent need for criminologists to understand the role of genes in human behavior as that role is understood by geneticists.

⊠ The Biosocial Approach

Biosocial theories have not been popular with mainstream social scientists until fairly recently because they were interpreted as a sort of "biological determinism." This kind of thinking is rarer today as social scientists have become more sophisticated in their thinking about the interaction of biology and the environment (M. Robinson, 2004). There are still people who fear that "biological" theories can be used for racist ends, but as Bryan Vila (1994) remarks, "Findings can be used for racist or eugenic ends only if we allow perpetuation of the ignorance that underlies these arguments" (p. 329). Bigots and hate mongers will climb aboard any vehicle that gives their prejudices a free ride, and they have done so for centuries before genes were heard of.

Biosocial criminologists believe that because humans have brains, genes, hormones, and an evolutionary history, they should integrate insights from the disciplines that study these things into their theories and dismiss naïve nature *versus* nurture arguments in favor of nature *via* nurture. *Any* trait, characteristic, or behavior of *any* living thing is *always* the result of biological factors interacting with environmental factors (Cartwright, 2000), which is why we call modern biologically informed criminology *biosocial* rather than *biological.* In many ways, the early positivists were biosocial in approach because they explicitly envisioned biological and environmental interaction. Their ideas and methods were primitive by today's standards, but then, so were the ideas and methods of most sciences in the 19th century. Evolutionary ideas about the behavior of all animals (especially the human animal) were poorly understood; genes were unheard of, and the brain was still a mysterious locked black box. This has all changed with the sequencing of the human genome and with the advent of machines that enable us to see what is going on in the brain as we think and act. For these and other reasons, biosocial research into criminality is proceeding at an explosive pace. As Lilly et al. (2007) maintain, "It is clear that the time has arrived for criminologists to abandon their ideological distaste for biological theories" (p. 304).

⊠ Behavior Genetics

Behavior genetics is a branch of genetics that studies the relative contributions of heredity and environment to behavioral and personality characteristics. Genes and environments work together to develop all the traits—height, weight, IQ, impulsiveness, blood sugar levels, blood pressure, and so on—the sum of which constitutes the person.

Behavior geneticists stress that genes do not *cause* us to behave or feel; they simply *facilitate* tendencies or dispositions to respond to environments in one way rather than in another. There are no genes "for" criminal behavior, but there are genes that lead to particular traits such as low empathy, low IQ, and impulsiveness that increase the probability of criminal behavior when combined with the right environments.

Geneticists use twin and adoption studies to disentangle the relative influences of genes and environments, and they tell us that genes and environments are always jointly responsible for any human characteristic. To ask whether genes or environment is most important for a given trait is just as nonsensical as asking whether height or width is most important to the area of a rectangle. Gene expression always depends on the environment (think of identical rose seeds planted in an English garden and in the Nevada desert, and then think about where the full genetic potential of the seeds will be realized).

Behavior geneticists quantify the extent to which genes influence a trait with a measure called **heritability** (symbolized as h^2), which ranges between 0 and 1. The closer h^2 is to 1.0, the more of the variance (difference) in a trait in a *population,* not in an individual, is due to genetic factors. Since any differences among individuals can only come from two sources—genes or environment—heritability is also a measure of environmental effects ($1 - h^2$ = environmental effects). All cognitive, behavioral, and personality traits are heritable to some degree, with the traits discussed in the psychosocial section being in the .30 to .80 range (Carey, 2003).

Gene–Environment Interaction and Correlation

Gene–environment interaction and gene–environment correlation describe people's active transactions with their environment. **Gene–environment interaction (GxE)** involves the common-sense notion that people are differentially sensitive to identical environmental influences and will thus respond in different ways to them. For instance, a relatively fearless and impulsive person is more likely to seize opportunities to engage in antisocial behavior than is a fearful and constrained person.

Gene–environment correlation (rGE) means that **genotypes** and environments are related. All living things are designed to be responsive to their environments, and GxE and rGE help us to understand how by showing the indirect way that genes help to determine what aspects of the environment will and will not be important to us. In addition to furthering our understanding of the role of genes, advances in genetics have yielded enormous benefits to our understanding of the environment's role in shaping behavior; as Baker, Bezdjian, and Raine (2006) put it, "the more we know about genetics of behavior, the more important the environment appears to be" (p. 44). There are three types of G–E correlation: passive, evocative, and active.

Passive rGE is the positive association between genes and their environments that exists because biological parents provide children with genes linked to certain traits and an environment favorable for their expression. Children born to intellectually gifted parents, for instance, are likely to receive genes that lead to above-average intelligence and an environment in which intellectual behavior is modeled and reinforced, thus setting them on a trajectory that is independent (passively) of their actions.

Evocative rGE refers to the way others react to the individual on the basis of his or her behavior. Children bring traits with them to situations that increase or decrease the probability of evoking certain kinds of responses from others. A pleasant and well-mannered child will evoke different reactions than will a bad-tempered and ill-mannered child. Some children may be so resistant to socialization that parents may resort to coercive parenting or simply give up, either of which may worsen any antisocial tendencies and drive them to seek environments where their behavior is accepted.

▲ **Photo 8.1**   Former major league baseball player Jose Canseco is sworn in at a U.S. House of Representatives baseball steroids hearing. Canseco presents a fascinating case for biosocial theories. Jose had a fraternal twin brother, Ozzie, who also chose a career in baseball. However, in comparison with Jose's 462 home runs and over 1,400 RBI, Ozzie had only a "cup of coffee" in the major leagues. He came to bat only 65 times over 3 seasons and never hit a home run. Had he been an identical rather than a fraternal twin, might Ozzie have performed more like his brother? After finishing his baseball career, Jose wrote a book (*Juiced*) in which he admitted using steroids for most of his playing career and claimed that 85% of other players in his era did likewise. Because of his steroid use, many baseball experts predict Jose will never be elected to the baseball Hall of Fame, though his career numbers exceed those of many current Hall of Fame players.

Evocative rGE thus serves to magnify phenotypic differences by funneling individuals into like-minded peer groups ("birds of a feather flock together").

Active rGE refers to the active seeking of environments compatible with our genetic dispositions. Active rGE becomes more pertinent as we mature and acquire the ability to take greater control of our lives because within the range of possibilities available in our cultures, our genes help to determine what features of the environment will and will not be attractive to us. Active rGE assures us that our minds and personalities are not simply products of external forces, and that our choices are not just passive responses to social forces and situations. We are active agents who create our own environments just as they help to create us. Genes imply human self-determination because, after all, our genes are *our* genes. As Colin Badcock (2000) put it, "Genes don't deny human freedom; they positively guarantee it" (p. 71). Genes are constantly at our beck and call, extracting information from the environment and manufacturing the substances we need to navigate it. They are also what make us uniquely ourselves and thus resistant to environmental influences that grate against our natures. In short, genes do not constrain us; they enable us. This view of humanity is more respectful of human dignity than the blank slate view that we are putty in the hands of the environmental winds.

Behavior Genetics and Criminal Behavior

Although there are no genetic theories of criminal behavior per se, behavior genetic studies help us to better understand traditional criminological theories. For instance, large behavior genetic studies conducted in the United States (Cleveland et al., 2000) and the UK (Moffitt & the E-Risk Study Team, 2002) have shown that genetic factors play a large part in sorting individuals into different family structures (broken vs. intact homes), a variable often linked to antisocial behavior. In both these studies, families consisting of a divorced or never-married mother with children fathered by different men are the most at-risk family type for antisocial behavior, and families with full siblings with both biological parents present were least at risk, as noted in Chapter 7. Genes, of course, contribute to the choices people make, as well as make people easy or difficult to live with.

One of those genetic factors is almost certainly low self-control. As we saw in Chapter 5, Gottfredson and Hirschi (1990) attributed low self-control exclusively to parental treatment. However, there are now well over 100 studies that have shown rather strong links between low self-control and low levels of the neurotransmitter serotonin (Crockett, Clark, Lieberman, Tabinia, & Robbins, 2010). In other words, while we all have to be taught to control our impulses, some of us are naturally easier to teach than others. Levels of serotonin are governed both by genes and by the environment. That is, genes govern the base levels of serotonin a person has, but what is going on in the environment results in serotonin levels increasing and decreasing (J. Wright, 2011).

Unlike the relatively strong genetic influences discovered for most human traits, genetic influences on antisocial behavior are modest, especially during the teenage years. A study of 3,226 twin pairs found that genes accounted for only 7% of the variance in antisocial behavior among juvenile offenders, but 43% among adult offenders (Lyons et al., 1995). Heritability coefficients for most traits related to antisocial behavior are typically in the .30 to .80 range, and for antisocial behavior itself, two meta-analyses concluded that they are in the .40 to .58 range (Ferguson, 2010; Rhee & Waldman, 2002), with h^2 being higher in adult than in juvenile populations. What this means is that the majority of delinquents have little if any genetic vulnerability to criminal behavior, while a small minority may have considerable vulnerability. Pooling these two groups has the effect of elevating estimates of the overall influence of genes while minimizing it for those most seriously involved. Strong genetic effects on antisocial behavior are most likely to be found only among chronic offenders who begin offending prior to puberty and who continue to do so across the life course (Moffitt & Walsh, 2003).

✎ Molecular Genetics

Heritability estimates only tell us that genes are contributing to a trait, but they do not tell us which genes; only molecular genetics can tell us this. Fortunately, we can now go straight to the DNA (DNA is the acid in the nucleus of our cells that contains the instructions for particular genes, which in turn directs the manufacture of the substances we all need to survive and function) by genotyping the individuals with a simple cheek swab. We can then do an analysis of the effects of certain genes on individuals who have them and compare it with people who do not. Molecular genetic studies are being conducted with increasing frequency in criminology, with the huge National Longitudinal Study of Adolescent Health (ADD Health) study being one yielding some very important genetic findings. It is important to emphasize that any individual gene accounts only for a miniscule proportion of the variance in criminal behavior, and that it contributes to a trait linked to criminality, *not* to criminality itself, which you remember is a composite of many different traits. Genes always have *indirect* effects on behavior via the effects of the proteins they make on human traits and abilities.

All people in the world have the same genes that make them human, but we all have slight variations of them that make us all different (except for identical twins). If we didn't have these differences, the police in all those crime scene investigation shows would not be able to identify suspects by the bodily fluids left behind at crime scenes. For instance, although we all have genes that make blood, we have different blood types. Differences among individuals in behavioral traits are partially the result of what geneticists call **genetic polymorphisms.** Polymorphisms are differences in DNA sequences that code for the same gene, but which may make more or less of the substance (say, low serotonin), which leads to slightly different functional or physical traits among individuals. Let us return to Mobley's "my MAOA gene made me do it" argument in the opening vignette to illustrate how geneticists study the effects of these gene variants.

A major longitudinal study of maltreatment looking at the role of the MAOA gene showed why only about one half of abused/neglected children become violent adults (Caspi et al., 2002). The MAOA gene comes in variants that genetics call "high" and "low" activity. For a variety of reasons we cannot get into here, the low-activity version is a risk factor for a number of behavioral problems, and the high-activity version is a protective factor. Neither the genetic risk nor environmental risk factors by themselves had much effect on antisocial behavior. When combined, however, the odds of having a verified arrest for a violent crime for those with both genetic (the low variant of the MAOA gene) and environmental (maltreatment) risk factors were found to be 9.8 times greater than the odds for subjects with neither the genetic nor the environmental risk. Furthermore, although the low MAOA + maltreatment subjects were only 12% of the cohort, they were responsible for 44% of its criminal convictions.

The overall conclusion arrived at by a meta-analysis of the MAOA/maltreatment research was that their interaction is a significant predictor of antisocial behavior across all studies (Kim-Cohen et al., 2006). However, a study by Widom and Brzustowicz (2006) found that while the high-activity MAOA *allele* (an alternate form of a gene at the same location on a chromosome) buffered whites from the effects of childhood abuse and neglect as it relates to antisocial behavior later in life, it did not protect nonwhites. The authors suggest that other environmental stressors, such as the high density of antisocial others in the neighborhood, may have negated the protective power of the high-activity polymorphism among nonwhites in the study. These studies all point to the importance of studying GxE interactions—how the environment modifies the effects of genes, and how genes modify the effects of the environment. There are many other genetic polymorphisms related to antisocial behaviors being examined by biosocial criminologists, but we will encounter just one more in the next section.

⬚ **Evolutionary Psychology**

Evolutionary psychology explores human behavior using a theoretical framework and seeks to explain it with reference to human evolutionary history. Criminologists operating within the evolutionary framework explore how certain behaviors society now calls criminal may have been adaptive (such as useful in the pursuit of reproductive success, the ultimate goal of all living things) in ancestral environments. Evolutionary psychology complements genetics because it informs us how the genes of interest came to be present in the human gene pool in the first place. While genetics looks for what makes people different, evolutionary psychology focuses on what makes us all the same. Another basic difference is that evolutionary psychology looks at ultimate-level "why" questions (What evolutionary problem did this behavioral mechanism evolve to solve?), and geneticists look at proximate-level "how" questions (To what extent is this behavioral mechanism influenced by genes in this population at this time?). Ultimate causes are thus those that occurred in the past that are *ultimately* responsible for something, whereas a proximate cause is one that is most immediately responsible for causing some observed behavioral outcome.

Evolutionary psychologists agree with most criminologists that although it is morally regrettable, crime is normal behavior for which we all have the potential (Kanazawa, 2003). Evolutionary logic tells us that if criminal behavior is normal, it must have provided some evolutionary advantage for our distant ancestors. However, because modern environments are so radically different from the hunter/gatherer environments in which we evolved many of our most human traits, many traits selected for their adaptive value at the time may not be adaptive today. It is important to realize that it is the *traits* underlying criminal behavior that are the alleged adaptations, not the specific acts we call crimes (Walsh, 2009a).

Criminal behavior is a way to acquire resources illegitimately. Evolutionary psychologists refer to such behavior (whether it is defined as criminal or not) as *cheating,* and think of individual traits associated with it such as impulsiveness and aggression in terms of adaptive traits all humans share, but which also vary considerably among them. Whether exploitation occurs depends on environmental triggers interacting with individual differences and with environmental constraints. Although we all have the potential to exploit and deceive others, we are a highly social and cooperative species with minds forged by evolution to form cooperative relationships built on trust (Barkow, 2006). Cooperation is typically contingent on the reciprocal cooperation of others, and is thus a tit-for-tat strategy favored by **natural selection** because of the benefits it confers. We cooperate with our fellows because we feel good when we do, and because it identifies us as reliable and trustworthy, which confers valued social status on us.

Because cooperation occurs among groups of other cooperators, it creates niches for non-cooperators to exploit others by signaling their cooperation and then failing to follow through (Barkow, 20006). Criminal behavior may thus be viewed as an extreme form of defaulting on the rules of cooperation. But cheating comes at a cost, so before deciding to do so the individual must weigh the costs and benefits of cooperating versus defaulting. Cheating is rational (not to be confused with moral) when the benefits outweigh the costs. But if cheating is so rational, how did cooperation come to be predominant in social species? The answer is that cheating is only rational in circumstances of limited interaction and communication. Frequent interaction and communication breeds trust and bonding, and cheating becomes a less rational strategy because cooperators remember and retaliate against those who have cheated them. Ultimately, cooperation is the most rational strategy in any social species because each player reaps in the future what he or she has sown in the past.

Yet, we continue to see cheating behavior despite threats of exposure and retaliation. We do so because exposure and retaliation are threats only if **cheats** must operate within the same environment in which their reputations are known. Cheats can move from location to location, meeting and cheating a series of others who are unaware of their reputation. This is the pattern of many career criminals who move from place to place, job to job, and relationship to relationship, leaving a trail of misery behind them before their reputation catches up. This is why cheats are more likely to prosper in large cities in modern societies than in small traditional communities where the threat of exposure and retaliation is great (Ellis & Walsh, 1997).

Of course, the stability of the group and cultural dynamics must be considered. In communities where a "badass" reputation is valued by males more than anything else, criminal behavior is almost a constant thing. Even in these unstable communities containing large numbers of chronic cheats, there must be a certain level of group loyalty and cooperation.

The Evolution of Criminal Traits

There are a number of evolutionary theories of crime, all of which focus on reproductive strategies. This is not surprising because from a biological point of view, the evolutionary imperative of all living things is reproductive success. There are two ways that members of any animal species can maximize reproductive success: parenting effort and mating effort. **Parenting effort** is the proportion of reproductive effort invested in rearing offspring, and **mating effort** is that proportion of effort allotted to acquiring sexual

▲ **Photo 8.2**   A concentration on parenting effort is strongly associated with a prosocial lifestyle; a concentration on mating effort is strongly associated with an antisocial lifestyle.

partners. Because humans are born more dependent than any other animal, parenting effort is particularly important to our species. The most useful traits underlying parenting effort are altruism, empathy, nurturance, and intelligence (Rowe, 2002).

Humans invest more in parenting effort than any other species, but there is considerable variation within the species. Gender constitutes the largest division due to different levels of obligatory parental investment between the genders. Female parental investment necessarily requires an enormous expenditure of time and energy, but the only *obligatory* investment of males is the time and energy spent copulating. Reproductive success for males increases in proportion to the number of females to whom they have sexual access, and thus males have an evolved propensity to seek multiple partners. Mating effort emphasizes quantity over quality (maximizing the number of offspring rather than nurturing a few), although maximizing offspring numbers is obviously not a conscious motive of any male seeking sex. The proximate motivation is sexual pleasure, with more offspring being a natural consequence (in pre-contraceptive days) when the strategy proved successful.

Reproductive success among our ancestral females rested primarily on their ability to secure mates to assist them in raising offspring in exchange for exclusive sexual access, and thus human females evolved a much more discriminating attitude about sexual behavior (Geary, 2000). According to evolutionary biologists, the inherent conflict between the reckless and indiscriminate male mating strategy and the careful and discriminating female mating strategy drove the evolution of traits such as aggressiveness, and the lowering of trait levels (relative to female levels) such as empathy and constraint that help males to overcome both male competitors and female reluctance. The important point to remember is that *although these traits were designed by natural selection to facilitate mating effort, they are also useful in gaining nonsexual resources via illegitimate means* (Quinsey, 2002; Walsh, 2006).

The reverse is also true—traits that facilitate parenting effort underlie other forms of prosocial activity: "crime can be identified with the behaviors that tend to promote mating effort and noncrime with those that tend to promote parenting effort" (Rowe, 1996, p. 270). Because female reproductive success hinges more on parenting effort than mating effort, females have evolved higher levels of the traits that facilitate it (e.g., empathy and altruism), and lower levels of traits unfavorable to it (e.g., aggressiveness) than males. Of course, both males and females engage in both mating and parenting strategies, and both genders follow a mixed mating strategy. It is only claimed that mating behavior is more typical of males and parenting effort is more typical of females.

Empirical research supports the notion that an excessive concentration on mating effort is linked to criminal behavior. A review of 51 studies relating number of sex partners to criminal behavior found 50 of them to be positive, and in another review of 31 studies it was found that age of onset of sexual behavior was negatively related to criminal behavior (the earlier the age of onset, the greater the criminal activity) in all 31 (Ellis & Walsh, 2000). A British cohort study found that the most antisocial 10% of males in the cohort fathered 27% of the children (Jaffee, Moffitt, Caspi, & Taylor, 2003), and anthropologists tell us that there are striking differences in behavior between members of cultures that emphasize either parenting or mating strategies. Cultures emphasizing mating effort the world over exhibit behaviors (low-level parental care, hyper-masculinity, transient bonding) considered antisocial in Western societies (Ember & Ember, 1998).

Molecular genetic studies also find significant relationships between sexual and criminal behavior. A study by Beaver, Wright, and Walsh (2008) tested the evolutionary claim that the most antisocial individuals should have the largest number of sexual partners. They found that the same polymorphism of the dopamine (a neurotransmitter discussed in the next section) transporter gene (DAT1) that was significantly related to number of sexual partners was also significantly related to antisocial behavior. The reason for this is that one variant of the DAT1 gene is exceptionally efficient at clearing dopamine from the synaptic gap (see below) after it signals other neurons. This is problematic because it is dopamine that gives us pleasure when we engage in activities such as having sex, so if it is cleared too fast we are moved to seek more of the activity to get more pleasure (more dopamine). This constant seeking of activities to raise dopamine levels is the chemical basis of addiction to all sorts of things besides sex (drugs, smoking, food, gambling, alcohol, and so forth). The researchers found a positive association between number of sex partners and antisocial behavior, and that variation in the DAT1 gene explains variation in both number of sexual partners and in criminal conduct for males. Another study of 674 males found that males who had two copies (one each from their mother and father) of the same DAT1 polymorphism had significantly more sex partners (an average of 5.66) than males who had only one or no copies (an average of 2.94), as well as significantly higher delinquency scores and scores on other kinds of risky behaviors (Guo, Tong, & Cai, 2008). In other words, this particular gene variant is typically found among "people who need high levels of excitement and stimulation to activate their reward system in the same capacity as those with normally functioning reward systems" (DeLisi, Beaver, Vaughn, & Wright, 2009, p. 1189).

The Neurosciences

Whether the source of human behavior is internal or external, it is necessarily funneled through the brain, arguably the most awe-inspiring structure in the universe. Although the brain is only about 2% of body mass, it consumes 20% of the body's energy as it perceives, evaluates, and responds to its environment (Shore, 1997). This 3-pound marvel of evolutionary design is the CEO of all that we think, feel, and do. Powerful brain imaging technologies such as PET, MRI, and fMRI scans have resulted in an explosion of information on the brain over the past two decades. We are a long way from fully understanding the brain, but we cannot ignore the things known about it that are relevant to criminology. Matthew Robinson (2004) goes as far as to say that any theory of behavior "is logically incomplete if it does not discuss the role of the brain" (p. 72). As we will see, the insights criminologists can derive from neuroscience will not only buttress our theories but may also strengthen our claims for preventative *environmental* intervention.

Softwiring the Brain by Experience

All our thoughts, feelings, emotions, and behaviors are the results of networks of billions of brain cells called **neurons** communicating with one another through substances called **neurotransmitters.** There are many transmitters and other brain chemicals, but criminologists are most interested in dopamine and serotonin. Figure 8.1 shows how neurotransmitters shunt information back

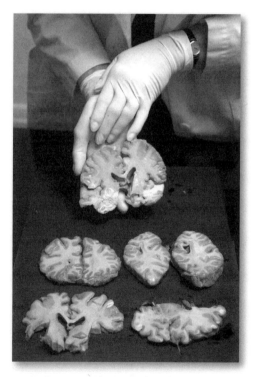

▲ **Photo 8.3**   Harkening back to the 19th century, when postmortem examinations of the brains of criminals were a frequent phenomenon, the brain of serial killer John Wayne Gacy was dissected after his execution. The attempt to locate an organic explanation of his monstrous behavior was unsuccessful.

and forth across the brain. Information from the environment is received from thousands of *dendrites* (the little finger-like projections in the figure), summated in the body of the neuron, and passed on electrically down the *axon* (the longer projection from the neuron in the figure). When the impulse reaches the end of the axon, it releases the neurotransmitters across the synaptic gap to further relay the message. The most important thing to remember here is that more "primitive" networks that control vital functions such as breathing and heart rate come "hardwired" at birth, but development of the higher brain areas depends a lot on environmental "software" downloaded after birth. The message neuroscience has for us is that *the experiences we encounter largely determine the patterns of our neuronal connections, and thus our ability to successfully navigate our lives* (Quartz & Sejnowski, 1997).

Figure 8.1   Neurons, Axons, Dendrites, and the Synaptic Process

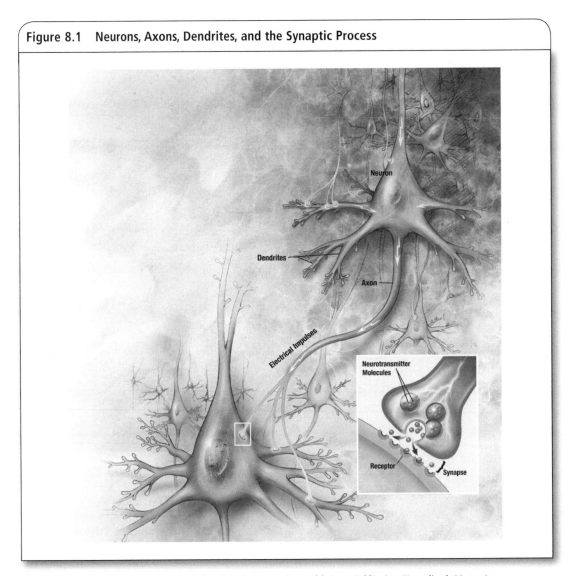

Source: National Institute on Aging website. Available at http://www.nia.nih.gov/Alzheimers/Publications/UnravelingtheMystery/

Neural networks are continually being made and selected for retention or elimination in a "use it or lose it" process governed by the strength and frequency of experience. Retention is biased in favor of networks that are most stimulated during early development (Restak, 2001). This is why bonding and attachment are so vital to human beings, and why abuse and neglect are so injurious. Hormones released by chronic stress can cause neurons to die, and children with high levels of these hormones experience cognitive and social development delays (M. Robinson, 2004). As Perry and Pollard (1998) point out, "Experience in adults *alters* the *organized* brain, but in infants and children it *organizes* the *developing* brain" (p. 36). Brains organized by stressful and traumatic events tend to relay events along the same brain pathways laid out by early events because pathways laid down early in life are more resistant to elimination than pathways laid down later in life. A brain organized by negative events is ripe for antisocial behavior.

Reward Dominance and Prefrontal Dysfunction Theories

If social animals are to function normally in their social groups, they must possess the ability to respond to signals of reward and punishment with socially appropriate approach and avoidance behavior. **Reward dominance theory** is a neurological theory based on the proposition that behavior is regulated by two opposing mechanisms, the **behavioral activating (or approach) system (BAS)** and the **behavioral inhibition system (BIS).** The BAS is associated with the neurotransmitter dopamine and with pleasure areas in the brain (Gove & Wilmoth, 2003). The BIS is associated with serotonin and with brain structures that govern memory. Neurotransmitters such as dopamine and serotonin are the chemical messengers that shunt information between neural networks. Dopamine facilitates goal-directed behavior, and serotonin generally modulates behavior (Depue & Collins, 1999).

The BAS is sensitive to reward and can be likened to an accelerator motivating a person to seek rewarding stimuli. The BIS is sensitive to threats of punishment and can be likened to a brake that stops a person from going too far too fast. The BAS motivates us to seek whatever affords us pleasure, and the BIS tells us when we have had enough for our own good. A normal BAS combined with a faulty BIS, or vice versa, may lead to a very impulsive person with a "craving brain" that can lead him or her into all sorts of physical, social, moral, and legal difficulties, by becoming addicted to pleasures such as food, gambling, sex, alcohol, and drugs (Day & Carelli, 2007).

While most of us are more or less equally sensitive to both reward and punishment (BAS/BIS balance), in some people one system might dominate the other most of the time. The theory asserts that criminals, especially chronic criminals, have a dominant BAS, which tends to make them overly sensitive to reward cues and relatively insensitive to punishment cues (Day & Carelli, 2007). Reward dominance theory provides us with hard *physical* evidence relating to the concepts of sensation seeking, impulsiveness, and low self-control we have previously discussed, since each of these traits is undergirded by either a sticky accelerator (not enough dopamine) or faulty brakes (low serotonin).

A third system of behavior control is the **fight or flight system (FFS),** chemically controlled by the adrenal hormone and neurotransmitter epinephrine (adrenaline). The FFS is that part of the autonomic nervous system that mobilizes the body for vigorous action in response to threats by pumping out epinephrine. Fear and anxiety at the chemical level are epinephrine shouting its warning: "Attention, danger ahead; take action to avoid!" Having a weak FFS that whispers rather than shouts combined with a BAS that keeps shouting, "Go get it!" and a BIS too feeble to object, is obviously very useful when pursuing all kinds of antisocial activities.

Another neurologically specific theory of criminal behavior is **prefrontal dysfunction theory.** The **prefrontal cortex (PFC)** is a part of the brain located just above the eyes that occupies about one third of the human cerebral cortex, and has been called "the most uniquely human of all brain structures" (Goldberg, 2001, p. 2).

The PFC is responsible for a number of things such as making moral judgments, planning, analyzing, synthesizing, and modulating emotions. The PFC provides us with knowledge about how other people see and think about us, thus moving us to adjust our behavior to consider their needs, concerns, and expectations of us. These PFC functions are collectively referred to as *executive functions,* and are clearly involved in prosocial behavior. If these functions are compromised in some way via damage to the PFC, the result is often antisocial behavior.

Positron emission tomography (PET) and functional magnetic resonance imaging (fMRI) studies consistently find links between PFC activity and impulsive criminal behavior. A PET study comparing impulsive murderers with murderers whose crimes were planned found that the former showed significantly lower PFC and higher limbic system activity (indicative of emotional arousal) than the latter and other control subjects (Raine et al., 1998). Cauffman, Steinberg, and Piquero (2005) combined reward dominance and PFC dysfunction theories in a large-scale study of incarcerated and non-incarcerated youths in California and found that seriously delinquent offenders have slower resting heart rates and performed poorly relative to non-delinquents on various cognitive functions mediated by the PFC.

Some Other Biosocial Risk Factors

There are numerous other biosocial risk factors, but we can only discuss a few here that we can relate back to issues in previous chapters to show how biological factors interact with the social context to produce behavior. For instance, in Chapter 6 we briefly discussed testosterone (T) in the context of male–female differences, but how about its effect among males only? Rowe (2002) discusses an interesting study of the effect of T among 4,462 males. The sample was divided into high T (upper 10%) and normal T (lower 90%) and into high and low socioeconomic status (SES). The study found that antisocial behavior more than doubled (from 14.7% to 30.1%) among low-SES/high-T males compared to low-SES/normal-T males. Among high-SES males, T levels had no effect on antisocial behavior. The interaction between T and social context is further illustrated in a longitudinal study of 1,400 boys, which found that T levels were unrelated to conduct problems for boys with "non-deviant" or "possibly deviant" friends, but conduct problems were greatly elevated among boys with high T who associated with "definitely deviant" peers (Maughan, 2005). Thus, the effects of testosterone depend quite a lot on social context, and this illustrates once again that we cannot separate biological and environmental variables and expect to understand complex behavior.

In Chapter 7, we discussed IQ without discussing biosocial factors that influence it. Exposure to noxious substances in the environment such as lead clearly reduces IQ. The IQ decrease per 1 unit increase in µg/dl (micrograms per deciliter of blood) of lead is an average of half a point (Koller, Brown, Spurfeon, & Levy, 2004). An fMRI study of predominantly black youths found that brain grey matter was inversely associated with average childhood lead concentrations in young adults (Cecil et al., 2008). The average childhood blood lead concentration of this sample was 13.3µg/dl, which is far in excess of the 2006 average of 1.5µg/dl for the general U.S. population (Bellinger, 2008). Although the grey matter lost to lead exposure was relatively small (about 1.2%), it was concentrated in vital behavior-moderating areas responsible for executive functioning and mood regulation such as the prefrontal cortex. Another study examined the relationship between childhood blood lead and verified criminal arrests. The average number of arrests for the males in the sample who were ever arrested was 5.2. The main finding of this study was that after controlling for other relevant variables, for every 5µg/dl increase in lead, there was an increase in the probability of arrest for a violent crime of about 50% (Wright, Dietrich, et al., 2008).

There are a number of neurological disorders that result from mothers drinking alcohol while pregnant, the most serious of which is **fetal alcohol syndrome (FAS).** FAS is a chronic condition resulting from an individual's prenatal alcohol exposure. Prenatal exposure to alcohol disrupts the migration and hookup of the embryo/fetus's developing neurons in brain areas such as the frontal lobes (Goodlett, Horn, & Zhou, 2005). The behavioral symptoms of FAS include low IQ; hyperactivity; impulsiveness; poor social, emotional, and moral development; and a predisposition to alcoholism (Walsh & Yun, 2011).

There are many other substances that have similar effects on neuron development and migration, because whatever the mother ingests, so does her embryo/fetus. A common risk factor is maternal smoking, which puts her fetus at risk for *hypoxia* (intermittent reduction of oxygen available to the fetus that may lead to brain cell death) (Zechel et al., 2005) as well as the toxic chemical components of tobacco (Huizink & Mulder, 2006). Cohort studies (e.g., Brennan, Grekin, & Sarnoff, 1999) consistently find that maternal smoking during pregnancy predicts criminal behavior in offspring independent of other factors. A review of a number of such studies found significantly increased risk for fetal tobacco-exposed individuals versus nonexposed individuals for various forms of antisocial behavior across diverse contexts and independent of other factors such as maternal SES and IQ (Wakschlag, Pickett, Cook, Benowitz, & Leventhal, 2002).

Table 8.1 summarizes the key concepts and strengths and weaknesses of biosocial perspectives and theories.

Table 8.1	Summarizing Biosocial Perspectives and Theories		
Theory	**Key Concepts**	**Strengths**	**Weaknesses**
Behavior and Molecular Genetics	Genes affect behavior in interaction with environmental influences. Heritability estimates the relative contribution of genetic and environmental factors and traits affecting criminality. All individual traits are at least modestly influenced by genes.	Looks at both the genetic and environmental risk factors for criminal behavior. Understanding genetic contributions and also identifies the contributions of the environment. Provides "hard" evidence.	Requires twin samples and/or adoptees, which are difficult to come by. However, technology now enables us to go straight to the DNA in molecular genetics. Expensive and requires cooperation of lab scientists.
Evolutionary Psychology Perspective	Human behavior is rooted in evolutionary history. Natural selection has favored victimizing tendencies in humans, especially males. These tendencies arose to facilitate mating effort, but are useful in pursuing criminal behavior as well. Criminals emphasize mating effort over parenting effort more than males in general.	Ties criminology to evolutionary biology. Mating effort helps to explain why males are more criminal than females and why criminals tend to be more sexually promiscuous than persons in general. Emphasizes that crime is biologically "normal" (although regrettable) rather than pathological.	Gives some the impression that because crime is considered "normal," it is justified or excused. Makes assumptions about human nature that may or may not be true. While recognizing that culture is important, it tends to ignore it.

(Continued)

Table 8.1	(Continued)		
Theory	**Key Concepts**	**Strengths**	**Weaknesses**
Neuroscience Perspective	Whatever its origin, all stimuli are channeled through the brain before given expression in behavior. The development of the brain is strongly influenced by early environmental experiences, especially those involving nurturance and attachment.	Shows how environmental experiences are physically "captured" by the brain. Emphasizes the importance of nurturing for optimal development of the brain. Uses sophisticated technology and provides "harder" evidence.	High cost of neuroimaging studies. Studies difficult and expensive to conduct but are getting cheaper all the time. The "hardness" of the data may lead us to accept findings too uncritically.
Reward Dominance Theory	Behavioral activating (BAS) and behavioral inhibiting system (BIS) are dopamine and serotonin driven, respectively. Among criminals the BAS tends to be dominant over the BIS. This BIS/BAS imbalance can lead to addiction to many things, including crime.	Explains why low serotonin is related to offending (low serotonin = low self-control). Explains why criminality persistent in some offenders, because they develop a taste for the "thrill of it all."	The neurological underpinnings of the BAS and BIS have been difficult to precisely identify. Nevertheless, it is a model used in many disciplines interested in human behavior.
Prefrontal Dysfunction Theory	Frontal lobes control long-term planning and temper emotions and their expressions. Criminals have frontal lobes that fail to function the way they do in most people, especially in terms of inhibiting actions that harm others.	Explains why moral reasoning is inversely related to involvement in persistent criminality. Explains why criminality has been linked to frontal lobe damage and to abnormal brain waves.	Dysfunction of the prefrontal lobes remains difficult to measure, even with fMRI scans. Same sampling difficulties noted for the neurosciences in general.

⊠ Evaluation of the Biosocial Perspective

Lilly et al. (2007) note that the most dramatic developments in science come most often from new observational techniques (think of the telescope and microscope) rather than new developments in theory. Criminologists now have access to new observational techniques in the form of DNA and neuroimaging data. "Never before has the sublime interplay between nature and nurture been available for scientific discovery" (DeLisi, 2009, p. 266), as it is today. The strength of biosocial approaches is that they take advantage of these new observational techniques and in their ability to incorporate biological concepts and findings that have been derived from these sophisticated physical measures into their theories. The main stumbling block is that such studies are more difficult to do and far more expensive than the typical social science study. If we want genetic information, we cannot simply go to the nearest high school and survey a few hundred students. Behavior genetic studies require comparing samples consisting of pairs of identical and fraternal twins and/or adoptees, and these are difficult to come by. However, new technologies have allowed us to go straight to the DNA, thus eliminating this need, but genotyping costs about $10 per individual.

It used to be difficult to make generalizations from the typical neuroimaging study because many tended to consist of a small number of known offenders matched with a control group. However, today there are some ambitious studies that are imaging anywhere from 400 to 2,000 subjects as costs continue to come down. Paus (2010) discusses four of these studies at length, including two longitudinal studies. All the studies are collecting mountains of environmental, behavioral, and cognitive data (e.g., socioeconomic status, maternal smoking, drinking, stressful life events, antisocial behavior, IQ, personality profiles, and many other things). Three of these four studies are also collecting DNA data. Thus, biosocial studies provide criminologists with "harder" evidence than they are typically able to get, and this evidence will help them to more solidly ground their theories. Biosocial analyses of many phenomena such as medical, psychiatric, and psychological problems are now all the rage in these disciplines, and many prominent sociologists and criminologists believe that this approach will prove just as useful in their disciplines. One of the most exciting advances is that some universities, such as the University of Texas at Dallas, are offering a double major in criminology and molecular genetics. In any event, as new discoveries are made in genetics and neuroscience, criminology can hardly ignore them.

⊠ Policy and Prevention: Implications of Biosocial Theories

The policies suggested by the biosocial perspective are midway between the macro-level sociological suggestions aimed at whole societies and the micro-level suggestions of psychological theories aimed at already convicted criminals. Mindful of how nurturing affects both gene expression and brain development in humans, many biosocial criminologists have advocated a wide array of "nurturant" strategies such as pre- and postnatal care for all women, monitoring infants and young children through the early developmental years, paid maternal leave, nutritional programs, and a whole host of other interventions (Vila, 1997). Some of the programs such as lead removal programs and educational programs to reduce maternal drinking and smoking should pay generous dividends in terms of reducing IQ loss and other negative factors caused by toxic lead and maternal substance abuse. Douglas Massey (2004), former president of the American Sociological Association, called for a biosocial understanding of such things:

> [B]y understanding and modeling the interaction between social structure and allostasis ["allostasis" refers to the dysregulation of stress response systems such as the ANS discussed in the last chapter in response to chronic levels of stress], social scientists should be able to discredit explanations of racial differences in terms of pure heredity. In an era when scientific understanding is advancing rapidly through interdisciplinary efforts, social scientists in general—and sociologists in particular—must abandon the hostility to biological science and incorporate its knowledge and understanding into their work. (p. 22)

Biosocial criminologists are typically in the forefront in advocating treatment over punishment, and toward this end they have favored indeterminate sentences over fixed sentences (Lanier & Henry, 1998). Pharmacological treatments in conjunction with psychosocial treatments have proven to be superior to psychosocial treatment alone for syndromes (alcoholism, drug addiction, etc.) associated with criminal behavior (M. Robinson, 2009). Of course, there are always dangers in seeking simple medical solutions to complex social problems. Requiring sex offenders to take anti-androgen treatment to reduce the sex drive raises both medical and legal/ethical issues regardless of how effective the treatment is. Prescribing selective serotonin reuptake inhibitors such as Prozac and Zoloft helps to curb low self-control and irritability, but there is always the temptation to treat everyone the same regardless of their serotonin levels.

One of the greatest successes of biosocial science was its pivotal role in the United States Supreme Court's outlawing of the juvenile death penalty. In writing the majority opinion in *Roper v. Simmons* (2005), Justice Anthony Kennedy noted the neurobiological evidence for the physical immaturity of the adolescent brain, which was brought to the Court's attention by the American Medical and Psychological Associations (Walsh & Hemmens, 2011). Thus, the biosocial approach can serve to advance arguments both for prevention rather than punishment and for punishments that takes into consideration valid physically identifiable differences among people.

Matt Robinson (2009), a sociologist who has spent much of his career researching crime prevention, states that

> [s]ince biosocial criminology meaningfully integrates perspective and theories from the biological and social sciences, the approach offers much hope in the area of crime prevention. At the very least, biosocial crime prevention should be far more effective than those strategies currently utilized. (p. 243)

Biosocial studies provide information about *both* environmental and biological risk factors and, as such, are "more likely to refine social policies by better specification of environmental factors than to divert funds from environmental crime prevention strategies" (Morley & Hall, 2003). In other words, they will enable us to better pinpoint environmental factors that may prove fruitful in our crime prevention efforts.

SUMMARY

- Behavior geneticists study the genetic underpinning of traits and characteristics in populations by calculating heritability coefficients. There are no genes "for" any kind of complex human behavior; genes simply bias trait values in one direction or another. This view is respectful of human dignity in that it implies self-determinism because our genes are *our* genes.

- Gene–environment interaction tells us that the impact an environmental situation (e.g., living in a crime-ridden neighborhood) has on us depends on who we are, and gene–environment correlation tells us that who we are is a product of our unique genotype and the environments we find ourselves in.

- Genes have practically no influence on juvenile delinquency, probably because of the high base rate of delinquency; i.e., almost all males commit some acts of delinquency, and thus there is little difference in behavior to be explained by either genes or environment. There are genetic effects for chronic and serious delinquents, but these few individuals tend to get "lost" in studies that combine them with those who limit their offending to adolescence. Adult criminality is much more influenced by genes. One of the reasons that we find only modest genetic effects in criminality when the traits that underlie it are strongly influenced by genes is that parents may have control over their children's behavior but little or none over their underlying traits.

- Evolutionary psychology focuses on why we have the traits we do, and is more interested in their universality than in their variability. Crime is viewed as a normal but regrettable response to environmental conditions. By this it is meant that many human adaptations forged by natural selection in response to survival and reproductive pressures are easily co-opted to serve morally wrong purposes.

- In common with all sexually producing species, humans are preeminently concerned with our own survival and reproductive success. The traits designed to assist males in their mating efforts include many that can also assist them to secure other resources illegitimately; in contrast, traits designed to assist females in their parenting efforts are conducive to prosocial behavior. Mating vs. parenting effort is not an either/or thing; males and females engage in both at various times in their lives. It is just that mating effort is more typical of males, and parenting effort is more typical of females.

- Socially cooperating species create niches that cheats can exploit to their advantage by signaling cooperation but then defaulting. Cheating is a rational strategy in the short term, but invites retaliation in the long term. This is why chronic criminals rarely have successful relationships with others and why they typically die financially destitute.

- Neuroscience tells us that genes have surrendered control of human behavior to the brain. Following genetic wiring to jump-start the process, the brain literally wires itself in response to environmental input. The "softwiring" of our brains is an electrochemical process that depends on the frequency and intensity of early experiences. Adverse experiences can literally physically organize the brain so that we experience the world negatively, which is why nurturing, love, and attachment are so important to the healthy development of humans.

- Reward dominance theory informs us that the brain regulates our behavior through the BIS and BAS systems (underlain by serotonin and dopamine neurotransmitters, respectively). In most people the BAS and BIS are balanced, but criminals tend to have either an overactive BAS or an underactive BIS. This means that their behavior is dominated by reward cues and relatively unaffected by punishment cues.

- Prefrontal dysfunction theory posits that the brain's prefrontal cortex (PFC) is vital to the so-called executive functions such as planning and modulating emotions. If the PFC is damaged in any way, the individual is deficient in these executive functions and tends to be impulsive.

DISCUSSION QUESTIONS

1. If it could be shown with high scientific confidence that some young children inherit genes that put them at 85% risk for developing antisocial proclivities, what do you think should be done? Should their parents be warned to be especially vigilant and to seek early treatment for their children, or would such a warning tend to stigmatize children? What are the costs and benefits of each option?

2. We know that males, especially young males, are more likely to perpetrate and be victimized by violent crimes. Provide a plausible evolutionary explanation for this.

3. How might reward dominance theory add strength and coherence to low self-control theory?

4. Explain why the traits underlying mating versus parenting effort are related to crime.

USEFUL WEBSITES

Anatomy of the Brain. www.neuroguide.com/index.html

Behavioral Genetics—Human Genome Project. www.ornl.gov/sci/techresources/Human_Genome/elsi/behavior.shtml

Crime Times. http://crimetimes.org/

Evolution's Voyage (Evolutionary Psychology). www.evoyage.com/

The Human Brain. www.fi.edu/brain/index.htm

CHAPTER TERMS

Behavioral activating system

Behavioral inhibition system

Behavior genetics

Cheats

Evolutionary psychology

Fetal alcohol syndrome

Fight or flight system

Gene–environment correlation

Gene–environment interaction

Heritability

Mating effort

Natural selection

Neurons

Neurotransmitters

Parenting effort

Polymorphisms

Prefrontal cortex

Prefrontal dysfunction theory

Reward dominance theory

Developmental Theories

From Delinquency to Crime to Desistance

Kathleen Holmes was a sweet child born to an "All American" family in Boise, Idaho. Her parents sent her to a Catholic girls' school where she did well in her studies. All seemed to be going well for Kathy until she was 16 years old, when she agreed to go to a local Air Force base with two older friends from the neighborhood to meet the boyfriend of one of the girls. The boyfriend brought along two of his friends, and the six of them partied with alcohol, drugs, and sex. It was Kathy's first time experiencing any of these things, and she discovered that she liked them all. Thus began a 9-year spiral into alcohol, drug, and sex addiction and into all the crimes associated with these conditions such as drug trafficking, robbery, and prostitution.

When she was 25 years old, she was involved in a serious automobile accident in which she broke her pelvis, both legs, and an arm, and suffered a concussion. She was charged with drunken driving and possession of methamphetamine for sale. Kathy spent 10 months recuperating from her injuries during which she was drug-, alcohol-, and sex-free. Because of her medical condition, she was placed on probation. Her probation officer (P.O.) she described as a real "knuckle-dragger" who, while brooking no nonsense, became something of a father figure to her. While she was recuperating, she was often taken care of by a male nurse she described as "nerdy but nice." Her parents, who had been estranged from her for some time, became

(Continued)

(Continued)

reacquainted with her, and her P.O. and nurse taught her to trust men again. She also occupied her time taking online college courses on drug addiction and counseling. She eventually married her "nerdy nurse" with her parents' blessing, and one of the honored guests was her P.O., the "knuckle-dragger."

Kathy's story illustrates some core ideas in this chapter. No matter how low a person sinks into antisocial behavior, he or she is not destined to continue the downward spiral. Certain so-called turning points can have a dramatic impact on a person's life. The auto accident and meeting the tough P.O. and the tender nurse would certainly qualify as significant turning points, as would most certainly the girl's marriage and the decision to continue her education. Before she became involved with "the wrong crowd," she had accumulated what is called "social capital" in the form of a good relationship with her family and a strong academic preparation. Although she spent most of her social capital, there was a sufficient amount left to get her back on the right track.

The Developmental Perspective

Most theories of crime implicitly assume that their favored causes are applicable across the life span. They neglect to account for changing social, biological, and psychological conditions. They also imply that criminal behavior is self-perpetuating once initiated and say little about the process of desisting. In contrast to these static views, **developmental theories** are dynamic in that they emphasize that individuals develop along different pathways, and as they develop, factors that were previously meaningful to them (e.g., acceptance by antisocial peers) no longer are, and factors that previously meant little to them (e.g., marriage and a career) become meaningful. These theories are concerned with the onset, acceleration, and deceleration of offending, and finally desisting altogether. This difference notwithstanding, all theories in this chapter tend to integrate (to varying degrees) social, psychological, and biological factors.

Offending in the early stages of life is known as **delinquency,** a legal term that distinguishes between juvenile and adult offenders. Except in rare instances in which a juvenile commits murder and is waived (transferred) to adult court, juvenile offenders are not referred to as criminals. Acts that are forbidden by law are called delinquent acts when committed by juveniles. The term *delinquent* comes from a Latin word meaning "to leave undone," with the connotation being that juveniles have *not done* something that they were supposed to (behave lawfully) rather than *done* something they were not supposed to. This subtle difference reflects the rehabilitative rather than punitive thrust of the juvenile justice system.

Juvenile delinquency occurs everywhere, although the extent and severity of it varies from culture to culture and across time. Some criminologists attempt to explain the rise in antisocial behavior in adolescence by the increase in peer involvement at this time, and its decline thereafter by the decreasing influence of peers and the increasing influence of girlfriends or boyfriends, spouses, children, and employers (Warr, 2002). However, this does not explain *why* the period between the increase in peer influence and its decline is so filled with antisocial behavior or *why* associations with peers so often lead to negative behavior.

To help us understand why, Aaron White (2004) provides us with four key messages from the 2003 conference of the New York Academy of Sciences (NYAS), which focused on adolescent brain development, a topic that has become very popular among criminologists lately for a variety of reasons, not the least of which is the influence it had on abolishing the juvenile death penalty as discussed in the last chapter:

1. Much of the behavior characterizing adolescence is rooted in biology intermingling with environmental influences to cause teens to conflict with their parents, take more risks, and experience wide swings in emotion.

2. The lack of synchrony between a physically mature body and a still-maturing nervous system may explain these behaviors.

3. Adolescents' sensitivities to rewards appear to be different from adults, prompting them to seek higher levels of novelty and stimulation to achieve the same feeling of pleasure.

4. With the right dose of guidance and understanding, adolescence can be a relatively smooth transition. (p. 4)

In other words, the NYAS statement is basically saying that the immature behavior of many adolescents is mirrored by the immaturity of their brains. The reshaping of the adolescent brain is initiated by the hormonal surges of puberty, a time when male levels of testosterone (a hormone that underlies dominance and aggression) are 10-plus times that of females (Ellis, 2003). Adolescents are also experiencing changes in the ratios of excitatory and inhibitory neurotransmitters. Dopamine (the "go get it" transmitter) is peaking, while inhibitory transmitters such as serotonin ("hold your horses") are reduced (Collins, 2004). All of this neurohormonal preparation leads many theorists (Spear, 2000; White, 2004) to conclude with Martin Daly (1996) that "[t]here are many reasons to think that we've been designed [by natural selection] to be maximally competitive and conflictual in young adulthood" (p. 193).

When adolescence is over and adulthood is attained around about the age of 20, the excitatory transmitters start to decrease and the inhibitory transmitters start to increase (Collins, 2004). McCrae and his colleagues (2000) report findings from five different countries showing age-related decreases in personality traits as positively related to antisocial behavior (e.g., neuroticism, risk taking) and increases in personality traits as positively related to prosocial behavior (e.g., conscientiousness, agreeableness). Likewise, Blonigen's (2010) review of a number of studies of age-related changes in personality traits related to antisocial behavior found that a reduction in negative emotionality and an increase in conscientiousness stood out as the strongest changes in personality. There were also moderate to large increases in agreeableness and self-control and small to moderate decreases in neuroticism (emotional instability) as people matured.

Many of us have seen media accounts of how the adolescent brain changes so much from puberty to the mid- to late twenties. One of the most important points is that the axons (see Figure 8.1) in the adolescent prefrontal cortex are not yet fully myelinated. *Myelin* is a fatty substance that coats and insulates axons and allows for the rapid transmission of brain messages, just like a plastic coating around electrical wires in the house prevents them from short-circuiting. Brains that are not fully myelinated result in a larger "time lapse" between the onset of an emotional event in the limbic system (the emotional part of the brain) and a person's rational judgment of it. In other words, there are *physical* reasons for the greater ratio of emotional to rational responses evidenced by many teens. The physical immaturity of the adolescent brain combined with a "supercharged" physiology facilitates the tendency to assign faulty attributions to situations and the intentions of others. A brain on "go slow" superimposed on a physiology on "fast-forward" explains why many teenagers find it difficult to accurately gauge the meanings and intentions of others, and why they tend to experience more stimuli as aversive during adolescence than they did as children and will do when they are adults (Walsh, 2002, p. 143).

Risk and Protective Factors for Serious Delinquency

Some individuals possess so many risk factors that they go beyond the normal adolescent hell-raising to commit serious crimes. A **risk factor** is something that increases the probability of offending. These factors are *dynamic,* meaning that their predictive value changes according to in which stage of the individual's development they occur, the presence of other risk and protective factors, and the immediate social circumstances. For instance, low socioeconomic status (SES) is a family risk factor, but a person with a high IQ and warm relationship with both parents is protected from the risks that low SES poses. It is typical for risk factors to cluster together. A single-parent family, for instance, is a risk factor that can lead to low SES and the financial necessity to reside in socially disorganized neighborhoods where children interact with antisocial peers. Likewise, protective factors tend to cluster together. A report, based on hundreds of studies, issued by the Office of the Surgeon General of the United States (OSGUS) (2001) indicated that a 10-year-old child with six or more risk factors is approximately 10 times more likely than a child of the same age with only one risk factor to be violent by the age of 18 (see Table 9.1). We have already examined many of these risk factors (poverty, social disorganization, low IQ, minimal parental supervision or presence of abuse, etc.), so we will only examine developmental factors here.

Among the risk factors listed by the OSGUS are ADHD/impulsivity, restlessness, difficulty concentrating, and aggression, which can be subsumed under the syndromes of **attention-deficit/hyperactivity disorder (ADHD)** and conduct disorder (CD). ADHD is a chronic neurological condition manifested as constant restlessness, impulsiveness, difficulty with peers, disruptive behavior, short attention span, academic underachievement, risk-taking behavior, and extreme boredom. Most healthy children will show some of these symptoms at one time or another, but they cluster together to form a syndrome in ADHD children (8 out of 14 symptoms are required for diagnosis) and are chronic and more severe than simple high spirits (Durston, 2003). ADHD affects somewhere between 2 and 9% of children and is 4 or 5 times more prevalent in males than in females (F. Levy, Hay, McStephen, Wood, & Waldman, 1997). Brain imaging studies have found differences (albeit small ones) in brain anatomy and physiology between ADHD and non-ADHD children (Raz, 2004; Sanjiv & Thaden, 2004).

Although the precise cause of ADHD is not known, genes play a large role, with heritability estimates in the .75 to .95 range (Schilling, Walsh, & Yun, 2011). As we might expect from our previous discussions of gene variants (polymorphisms), the genes most strongly involved in ADHD are associated with dopamine and serotonin functioning (Schilling et al., 2011). Environmental factors that play a role in the etiology of ADHD are fetal exposure to drugs, alcohol, and tobacco; perinatal complications; and head trauma (Durston, 2003). Subsequent environmental factors have no causal effect, although they may

Table 9.1	Delinquency Risk Factors by Domain		
Domain	**Early Onset (Ages 6–11)**	**Late Onset (Ages 12–14)**	**Protective Factors**
Individual	Being male ADHD/impulsivity Medical, physical problems Aggression Low IQ General offenses Problem (antisocial behavior) Substance abuse Exposure to TV violence Antisocial attitudes, beliefs Dishonesty[a]	Restlessness Difficulty concentrating[a] General offenses Risk taking Aggression[a] Being male Physical violence Antisocial attitudes, beliefs Crimes against persons Low IQ Substance abuse	Intolerant attitude toward deviance High IQ Being female Positive social orientation Perceived sanction for transgressions
Family	Low socioeconomic status Antisocial parents Poor parent–child relationship Harsh, lax, or inconsistent parenting Broken home Separation from parents Abusive parents Neglect	Poor parent–child relationship Low socioeconomic status Harsh, lax, or inconsistent parenting Poor monitoring, supervision Antisocial parents Broken home Abusive parents Family conflict[a]	Warm, supportive relationship with parents and other adults Parents' positive evaluation of child's peers Parental monitoring
School	Poor attitude, performance	Poor attitude, performance Academic failure	Commitment to school Recognition for involvement in conventional activities
Peer group	Weak social ties Antisocial peers	Weak social ties Antisocial, delinquent peers Gang membership	Friends who engage in conventional behavior
Community		Neighborhood crime, drugs Neighborhood disorganization	Stable, organized neighborhood

Source: Adapted from Office of the Surgeon General of the United States (2001).

[a]Males only

worsen symptoms. ADHD symptoms generally decline in their severity with age, although about 90% of ADHD sufferers continue to display some symptoms into adulthood (Willoughby, 2003).

A review of 100 studies found that 99 reported a positive relationship between ADHD and antisocial behavior (Ellis & Walsh, 2000). ADHD individuals are consistently found to be overrepresented in

▲ **Photo 9.1** A juvenile arrest can lead to many negative consequences, including placing a person on a path to a lifetime of crime and imprisonment.

juvenile detention centers, jails, and prisons worldwide (Rösler et al., 2004). Gudjonsson and colleagues' (Gudjonsson, Sigurddsson, Young, Newton, & Peersen, 2009) review of nine studies of ADHD rates among adult prison inmates using various diagnostic criteria in a number of countries found rates ranging from 24 to 67%, and a German study (Rösler et al., 2004) found that 45% of the inmates had some form of ADHD (there are a number of subtypes) compared to 9.4% of a control sample. Thus, ADHD is strongly related to criminal behavior, and it has been suggested that ADHD may underlie one of criminology's favored concepts—low self-control (Unnever, Cullen, & Pratt, 2003).

The probability that ADHD persons will persist in offending as adults rises dramatically if they are also diagnosed with **conduct disorder (CD).** CD is defined as "the persistent display of serious antisocial actions that are extreme given the child's developmental level and have a significant impact on the rights of others" (Lynam, 1996, p. 211). Markus Krueisi and his colleagues (1994) propose that ADHD is a product of a deficient BIS and CD is a product of an oversensitive BAS (see Chapter 8). ADHD individuals with CD thus suffer a double disability: Their dominant BAS inclines them to seek high levels of stimulation (to raise dopamine levels) while their faulty BIS leaves them with little sense of when to stop (low serotonin). ADHD and CD are found to co-occur in 30 to 50% of cases (Lynam, 1996). CD has an onset at around age 5 and is a neurological disorder with substantial genetic affects (Coolidge, Thede, & Young, 2000).

Patterns of Serious Delinquency

The developmental model of the progression of delinquency devised by Terrence Thornberry, David Huizinga, and Rolf Loeber (2004) focuses on the escalation of the seriousness of delinquent acts being committed as boys age. The model is based on three longitudinal studies that include more than 4,000 subjects followed since 1987. What has emerged from these studies is an image of three developmental pathways of offending, as noted in Figure 9.1. The *authority conflict pathway* starts before age 12 with simple stubborn behavior, followed by defiance and authority avoidance stages. Some boys in this pathway move into the second stage (defiance/disobedience) and a few move into the authority avoidance stage. At this point, some boys progress to one of the other two pathways, but many will go no further. The *covert pathway* starts after puberty and involves minor offenses in Stage 1 (the bottom, lightly shaded part of the triangle) that become progressively more serious for a few boys who enter Stage 3 (the peak of the triangle) on this pathway. The overt pathway progresses from minor aggressive acts in Stage 1 to very serious violent acts in Stage 3. The more seriously involved delinquents in the overt and covert pathways may switch back and forth between violent and property crimes. The overall take-home lesson of this model is that as boys get older, their crimes become more serious, but happily there are far fewer boys engaging in them.

Figure 9.1   Three Pathways to Boys' Disruptive Behavior and Delinquency

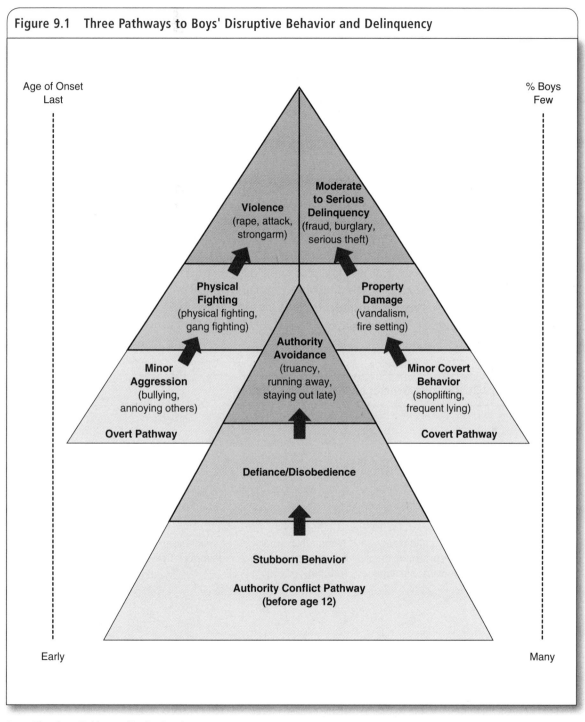

Source: Thornberry, Huizinga, and Loeber (2004).

⬚ Major Developmental Theories

As already noted, developmental theories are dynamic theories concerned with the frequency, duration, and seriousness of offending behavior from onset to desistance. All theories maintain that although a criminal career may be initiated at any time, it is almost always begun in childhood or adolescence, with only about 4% being initiated in adulthood (Elliot, Huizinga, & Menard, 1989). The duration of a criminal career may be limited to one offense or it can last well into old age, with the frequency and seriousness of offending varying widely. Onset, frequency, duration, seriousness, and desistance depend on a variety of interacting individual and situational factors that vary across the life course.

Robert Agnew's General or "Super Traits" Theory

In his general or "super traits theory," Robert Agnew (2005) identifies five life domains that contain possible crime-generating factors: *personality, family, school, peers*, and *work*. It is a developmental theory because these domains interact and loop back on one another across the life span as illustrated in Figure 9.2. Agnew suggests that personality traits set individuals on a particular developmental trajectory that influence how other people in the family, school, peer group, and work domains react to them. In other words, personality variables "condition" the effect of social variables on crime. Noting that personality traits cluster together, Agnew identifies the latent (underlying) traits of *low self-control* and *irritability* as "super traits" that encompass many of the traits we discussed in previous sections such as sensation seeking, impulsivity,

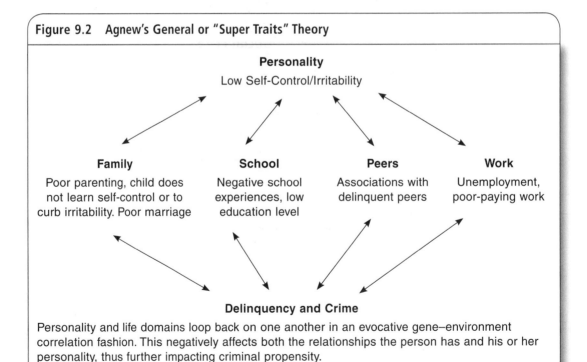

Figure 9.2   Agnew's General or "Super Traits" Theory

Personality
Low Self-Control/Irritability

Family
Poor parenting, child does not learn self-control or to curb irritability. Poor marriage

School
Negative school experiences, low education level

Peers
Associations with delinquent peers

Work
Unemployment, poor-paying work

Delinquency and Crime
Personality and life domains loop back on one another in an evocative gene–environment correlation fashion. This negatively affects both the relationships the person has and his or her personality, thus further impacting criminal propensity.

inattentiveness, and low empathy. People saddled with low self-control and irritable temperaments are likely to evoke negative responses from family members, school teachers, peers, and workmates that circle back and make those tendencies worse than they would have otherwise been. (The feedback process of evocative gene–environment correlation was discussed in Chapter 8.)

Agnew (2005) states that "biological factors [ANS and brain chemistry] have a direct effect on irritability/ low self-control and an indirect effect on the other life domains through [the impact of] irritability/low self-control [on them]" (p. 213). What Agnew calls *irritability* is analogous to the trait most psychologists call negative emotionality.

Agnew (2005) claims that the theory can explain gender, racial, and age effects in criminality, and can account for the differences between individuals who limit their offending to the adolescent years and those who offend across the life course. In terms of gender differences, Agnew says that males are more likely to inherit irritability/low self-control than females, perhaps because, in evolutionary terms, these traits have aided male reproductive success by enhancing male aggressiveness and competitiveness (p. 163). In terms of race differences, Agnew argues that African Americans are more likely to be poor and to receive discriminatory treatment, that this and other factors may significantly increase irritability, and that perceptions of poor job prospects may also lead to the adoption of an impulsive "live-for-the-day" lifestyle among some African Americans.

Agnew also notes that the immaturity of adolescent behavior is tied to the immaturity of the adolescent brain, and that adolescents tend to become more irritable because their brains are undergoing a period of intense "remodeling." At the same time adolescent brains are changing, adolescents are experiencing massive hormonal surges that tend to facilitate aggression and competitiveness. Thus, the neurological and endocrine changes during adolescence *temporarily* increase irritability/low self-control among adolescents who limit their offending to that period, while for those who continue to offend, irritability/low self-control is a *stable* characteristic.

David Farrington's Integrated Cognitive Antisocial Potential (ICAP) Theory

David Farrington's ICAP theory is based on a longitudinal cohort study of boys born in deprived areas (thus putting all boys at environmental risk) of London. The key concepts in the theory are *antisocial potential (AP),* which is a person's risk or propensity to engage in crime, and *cognition,* which is the "thinking or decision-making process that turns potential into actual behavior" (Farrington, 2003, p. 231). AP is ordered on a continuum with relatively few people with very high levels, but levels vary over time and across life events and peak in adolescence.

Farrington distinguishes between *long-term* AP and *short-term* AP. Individuals with long-term AP tend to come from poor families and tend to be poorly socialized, impulsive, sensation seeking, and have low IQ. Short-term AP individuals suffer few or any of these deficits but may temporarily increase their AP in response to certain situations or inducements. Farrington (2003) indicates that we all have "desires for material goods, status among intimates, excitement, and sexual satisfaction, and that people choose illegitimate ways of satisfying them when they lack legitimate means of doing so, or when bored, frustrated, or drunk" (p. 231).

Short-term AP may turn into long-term AP over time as a consequence of offending. This can happen if individuals find offending to be reinforcing either in material terms or in psychological terms by gaining status and approval from peers. Such outcomes lead to changes in cognition such that AP is more likely to turn into actual criminal behavior in the future. Offending can also lead to criminal labeling and incarceration, which limits future opportunities to meet one's needs legitimately. Thus, long-term AP can develop even in the absence of most or all of the risk factors (i.e., for social-situational reasons alone) said to predict chronic offending across the life course.

ICAP theory is also interested in the process of desisting from offending, which occurs for both social and individual reasons and at different rates according to a person's level of AP. As noted earlier, as people age they tend to become less impulsive and less easily frustrated. They also experience life changes such as marriage, steady employment, and moving to new areas, thus shifting their patterns of interaction from peers to girlfriends/boyfriends, spouses, and children. These events (1) decrease offending opportunities by shifting or reducing the frequency of routine activities such as men drinking with male peers, (2) increase informal controls in terms of having family and work responsibilities, and (3) change cognition in the form of reduced subjective rewards of offending because the costs of doing so are now much higher than before (the risk of losing their hard-earned stake in conformity). Potential peer approval becomes the potential disapproval of spouses (Farrington, 2003).

Terrie Moffitt's Dual Pathway Developmental Theory

Terrie Moffitt's **dual pathway developmental theory** is based on findings from an ongoing longitudinal study of a New Zealand birth cohort (in its 36th year as of 2011). It has been called "the most innovative approach to age–crime relationships and life-course patterns" (Tittle, 2000, p. 68). The data available to Moffitt and her colleagues come from collaborative efforts by scientists in sociology, criminology, psychology, medicine, genetics, and neuroscience, enabling them to test biosocial hypotheses.

It has long been known that the vast majority of youth who offend during adolescence eventually desist and that only a small number of them continue to offend in adulthood. Moffitt calls the former adolescent-limited (AL) offenders and the latter life course–persistent (LCP) offenders. LCP offenders are individuals who begin offending prior to puberty and continue well into adulthood, and who are saddled with neuropsychological and temperamental deficits that are manifested in low IQ, hyperactivity, inattentiveness, negatively emotionality, and low impulse control. These problems arise from a combination of genetic and environmental effects on brain development. Environmental risk factors include being the offspring of a single teenage mother, low SES, abuse/neglect, and inconsistent discipline. These related individual and environmental impairments initiate a cumulative process of negative person–environment interactions that result in a life course trajectory propelling individuals toward ever-hardening antisocial attitudes and behaviors.

Moffitt (1993) describes the antisocial trajectory of LCP offenders as one of

> biting and hitting at age 4, shoplifting and truancy at age 10, selling drugs and stealing cars at age 16, robbery and rape at age 22, fraud and child abuse at age 30; the underlying disposition remains the same, but its expression changes form as new social opportunities arise at different points of development. (p. 679)

This is matched by cross-situational behavioral consistency. LCP offenders "lie at home, steal from shops, cheat at school, fight in bars, and embezzle at work" (p. 679). Given this antisocial consistency across time and place, opportunities for change and for legitimate success become increasingly unlikely for these individuals. While LCP offenders constituted only 7% of the cohort, they were responsible for more than 50% of all delinquent and criminal acts committed by it (Moffitt & Walsh, 2003). Moreover, whereas AL offenders tend to commit relatively minor offenses such as petty theft, LCP offenders tend to be convicted of more serious crimes such as assault, robbery, and rape (Moffitt & Walsh, 2003).

As illustrated in Figure 9.3, AL offenders have a developmental history that places them on a prosocial trajectory that is temporarily derailed at adolescence. They are not burdened with the neuropsychological problems that weigh heavily on LCP offenders, and they are adequately socialized in childhood by competent parents. AL offenders are "normal" youths adapting to the transitional events surrounding adolescence. Their offending is a social phenomenon played out in peers groups and does not reflect any stable personal deficiencies. At least 85% of youthful offenders are adolescent limited (Moffitt, 1993).

According to Moffitt (1993), many more teens than in the past are being diverted from their prosocial life trajectories because better health and nutrition have lowered the average age of puberty, while the average time needed to prepare for participation in the economy has increased. These changes have resulted in about a 5- to 10-year **maturity gap** between puberty and entry into the job market. Thus, "adolescent-limited offending is a product of an interaction between age and historical period" (p. 692). Filled with youthful energy, strength, and confidence, and a strong desire to shed the restrictions of childhood, AL offenders are attracted to the excitement of antisocial peer groups typically led by experienced LCP delinquents. Once initiated into the group, juveniles learn the attitudes and techniques of offending through mimicking others and gain reinforcement in the form of much-desired group approval and acceptance for doing so, as social learning theorists argue.

Sampson and Laub's Age-Graded Developmental Theory

All three developmental theories we have thus far discussed posit a set of traits that set individuals on developmentally distinct pro- or antisocial pathways. However, Sampson and Laub (2005) prefer to call their **age-graded theory** a life course theory rather than a developmental theory because they deny that people are necessarily locked into developmentally distinct pathways by these traits, which of course they are not.

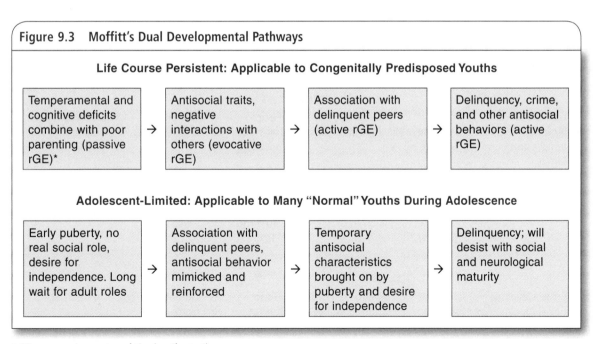

Figure 9.3   Moffitt's Dual Developmental Pathways

Life Course Persistent: Applicable to Congenitally Predisposed Youths

Temperamental and cognitive deficits combine with poor parenting (passive rGE)* → Antisocial traits, negative interactions with others (evocative rGE) → Association with delinquent peers (active rGE) → Delinquency, crime, and other antisocial behaviors (active rGE)

Adolescent-Limited: Applicable to Many "Normal" Youths During Adolescence

Early puberty, no real social role, desire for independence. Long wait for adult roles → Association with delinquent peers, antisocial behavior mimicked and reinforced → Temporary antisocial characteristics brought on by puberty and desire for independence → Delinquency; will desist with social and neurological maturity

*rGE = gene–environment correlation (see Chapter 8).

Sampson and Laub want to emphasize environmental circumstances and human agency as opposed to individual traits, although these traits are not ignored. Their theory is based on data collected from the 1930s through the 1960s by Sheldon and Eleanor Glueck (1950). Age-graded theory is essentially a social control theory extended into adulthood in that it assumes that we have to learn to be good rather than to be bad, and that ties to prosocial others are particularly important to this prosocial learning process. While the bonds to parents and school are very important during childhood, and to peers during adolescence, they become less important in adulthood when new situations offer opportunities to form new social bonds that constrain offending behavior for most people.

As with all control theories, the task of age-graded theory is not to explain why people commit crimes but rather why people do not. The theory assumes that factors such as low IQ, difficult temperament, low SES, and broken home have only indirect effects on offending via their influence on the ease with which bonding and socialization take place. The theory also places emphasis on the process of desisting from offending among those who start, and the situational factors involved in the process rather than on individual risk factors for offending.

People who bond well with conventional others build **social capital,** which is essentially a store of positive relationships built on norms of reciprocity and trust developed over time, upon which the individual can draw for support when needed. People who have opened their social capital "accounts" early in life (bonding to parents and school), even though they may spend it freely as adolescents, build quite a nest egg by the time they reach adulthood. They can then gather more interest in the form of a successful career and marriage (bonding to a career and a family of one's own). This accumulation of social capital provides people with a powerful stake in conformity, which they are not likely to risk by engaging in criminal activity. To put it in terms of Hirschi's social bonding theory discussed in Chapter 5, social capital is the result of

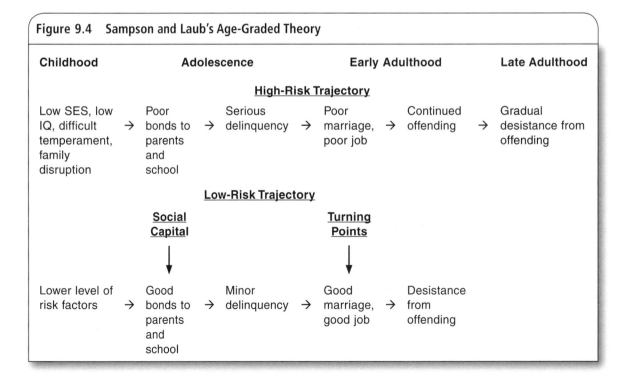

Figure 9.4   Sampson and Laub's Age-Graded Theory

Childhood	Adolescence		Early Adulthood		Late Adulthood

High-Risk Trajectory

Childhood	Adolescence	Early Adulthood		Late Adulthood
Low SES, low IQ, difficult temperament, family disruption	→ Poor bonds to parents and school →	Serious delinquency →	Poor marriage, poor job →	Continued offending →	Gradual desistance from offending

Low-Risk Trajectory

Social Capital ↓   Turning Points ↓

| Lower level of risk factors | → Good bonds to parents and school → | Minor delinquency → | Good marriage, good job → | Desistance from offending |

early attachments and commitment to parents and school, and is increased in adulthood by attachments and commitments to marriage and career.

Life is a series of transitions (or life events), which may change life trajectories in prosocial directions for persons lacking much social capital. Sampson and Laub (2005) call such events **turning points** and consider this concept the most important one in their theory (p. 14). Significant turning points include getting married, finding a decent job, moving to a new neighborhood, or entering military service. Of course, turning points are really processes rather than events, and rather than promoting change, they may accentuate antisocial tendencies, or at least leave them intact. According to Gottfredson and Hirschi (1990), the problem is that "the offender tends to convert these institutions [marriage, jobs] into sources of satisfaction consistent with his previous criminal behavior" (p. 141). In other words, offenders expand their antisocial repertoire into domestic abuse and workplace crime. Nevertheless, Laub and Sampson (2003) have shown that obtaining a good job and a good marriage does reduce offending among even previously high-rate offenders.

▲ **Photo 9.2**   From this series of family images taken over time, can you discern the behavioral and life course paths of the individuals depicted? One became a teen delinquent but then went on to be a lawabiding adult and, ultimately, a minister. One dropped out of a teen job to move away to marry a federal inmate, who had sold a number of types of drugs. Another started a career in sales but became involved in possible "Ponzi" schemes. One became a criminologist.

Although a person may be disadvantaged by the past, he or she does not have to be a prisoner to it. Age-graded theory strongly emphasizes human agency, defined as "the purposeful execution of choice and individual will," and some people freely choose a life of crime because they find it seductive and rewarding despite having full knowledge of the negative consequences (Sampson & Laub, 2005, p. 37). Nevertheless, all members of the Gluecks' (1950) original delinquent sample of 500 that Sampson and Laub were able to locate (N = 52) had desisted from offending by age 70, regardless of whether they were defined as high risk or low risk as children and regardless of the level of social capital they had managed to accumulate. This supports the long-held opinion among criminologists that all criminals "age out" of crime eventually, just as we all age out of participation in many of the activities we enjoyed as youths. Figure 9.4 above illustrates the two different life course pathways according to Sampson and Laub's theory.

Summary of Key Points, Strengths, and Differences of Developmental Theories

Theory	Key Points	Key Strengths	Key Differences
Agnew's General or Super Traits Theory	Low self-control and irritability set people on a trajectory leading to negative interactions with other in the family, in school, among peers, at work, and in marriage. These different domains interact and are in a feedback loop with one another.	Parsimoniously integrates concepts from psychology, sociology, and biology and shows how each affects and is affected by all others. Theory states that low self-control and irritability are temporarily increased during adolescence.	Does not address the process of desisting from offending. Does not explicitly address different trajectories as in Moffitt, and to a lesser extent in Farrington and Sampson & Laub.
Farrington's ICAP Theory	People have varying levels of antisocial potential (AP) due to a variety of environmental and biological factors. Few people have long-term AP, but those who do tend to offend across the life course. Short-term AP tends to occur in adolescence and can change to long-term under some circumstances.	Shows that how people think (cognition) translates AP into actual offending behavior. As with Moffitt's AL offenders, AP is said to increase temporarily during adolescence, but it can also lead to long-term AP if caught and labeled criminal because such a label limits future opportunities.	Not so much emphasis on desisting as age-graded theory, but moreso than Agnew and Moffitt. Less emphasis than Agnew and Moffitt on latent traits, but more than Sampson & Laub. Unlike Moffitt, ICAP is considered a continuum rather than a distinct two-type typology.
Moffitt's Dual Pathway Theory	There are two main pathways to offending: LCP and AL. LCP offenders have neurological and temperamental difficulties, which are exacerbated by inept parenting. LCP offenders do so across time and situations, begin prior to puberty, and continue well into adulthood. AL offenders are "normal" individuals temporarily derailed during adolescence.	Identifies two distinct pathways to offending rather than assuming all people are similarly affected by similar factors. Shows how the social bonds so important to age-graded theory are formed or not formed according to the characteristics of individuals.	Emphasizes a larger number of individual differences that affect offending behavior than all other theories. Also attempts to explain prevalence of offending with reference to the modern maturity gap. Differences between the two trajectory groups more defined than they are in other theories.

Theory	Key Points	Key Strengths	Key Differences
Sampson & Laub's Age-Graded Theory	Emphasizes the power of informal social controls across the life course. Assumes classical notions of why people commit crimes, therefore no need to dwell too much on risk factors. Turning points in life and human agency are important. These turning points are made easier if one has accumulated significant social capital.	Emphasis on the power of life events to turn trajectories around and to facilitate desistance. Also emphasis on human agency is refreshing. All offenders will eventually desist regardless of risk factors or lack of social capital.	Unlike other theories, there is little emphasis on risk factors setting people on a particular trajectory other than bonding strength. Emphasis on inhibiting (bonding, social capital) rather than facilitating factors.

⊠ Evaluation of Developmental Theories

Developmental theories offer many advantages over theories previously discussed because of their dynamic nature. It is not only consideration of the differential impact of risk factors at different junctures across the life span that distinguishes developmental theories; it is also their focus on the process of desisting from crime. Both these contributions are extremely important. Developmental theories are mostly based on longitudinal cohort data, which enables theorists to examine the links between risk factors and crime among the same individuals at every developmental stage of their lives. Longitudinal studies also enable theorists to identify causes rather than mere correlates because temporal order (which factor came first) is established among the correlates, something that cross-sectional studies cannot do.

With the exception of Agnew's theory, all developmental theories discussed in this chapter are based on longitudinal cohort data. Such data are hard to come by, and some theorists argue that cross-sectional studies (research that studies a sample at a single moment in time) are adequate and that longitudinal studies are an expensive luxury. Theorists making such claims are often those who identify a latent trait (e.g., self-control) that is considered stable throughout life; thus, cross-sectional studies capture characteristics of individuals at any one time, making multiyear studies redundant. Such a position ignores the interaction of this supposed stable latent trait with vastly different life experiences.

In short, if there is a "gold standard" for criminological theory, developmental theories would have to be it because (1) they are biosocial theories that generally integrate and consider sociological, psychological, and biological factors as a coherent whole; (2) they follow the same individuals over long periods of time, a strategy that allows for cause/effect analysis; and (3) they can identify characteristics that lead to onset, persistence, and desistance from crime in the same individuals.

Because low self-control and irritability (negative emotionality) are personality traits considered to be strongly influenced by low serotonin levels, Agnew's "super traits" theory could guide a longitudinal study in which life histories, offending records, and serotonin levels (as well as other neurohormonal substances) could be measured annually. This could tell us if at-risk individuals became even more at risk because of magnification of low self-control/high irritability due to the feedback effects of others reacting to their evocative behavior. Because neurohormonal substances are responsive to environmental contexts, significant changes in them may index important environmental changes in a person's life (Collins, 2004). We could also be informed if important turning points in subjects' lives led to decreased (or increased) offending and decreased (or increased) levels of the various measured substances.

◪ Policy and Prevention: Implications of Developmental Theories

Policies designed to prevent and reduce crime that are derived from developmental theories do not differ from those derived from other theories, but rather encompass them all and suggest a broad array of strategies. Developmental theories support the same kind of family-based, nurturant strategies supported by biosocial and social control and self-control theories. Regardless of any traits children may bring with them to the socialization process, however, these traits can be muted by patient and loving parenting, and as such, many developmental theories suggest family interventions as early as possible to help nurture bonds between children and their parents.

The *Nurse–Family Partnership* program has attempted to do this. The program began in 1978 with a sample of 400 at-risk women and girls and their infants. All mothers were unmarried, most were living in poverty, and 48% were under age 15. The women and girls were randomly assigned to four different groups, with one group (the experimental group) receiving extensive care from nurses in the form of multiple prenatal and postnatal home visitations in which the nurses gave help and advice on a variety of child care matters. The other groups received less comprehensive care for shorter periods. A 15-year follow-up study by Olds, Hill, Mihalic, and O'Brien (1998) found that the program had many beneficial outcomes for the experimental group of children and their mothers relative to the subjects in the other groups. For the mothers, there was less substance abuse, fewer subsequent illegitimate births, fewer legal difficulties, and 79% fewer verified instances of child abuse/neglect relative to the control groups. For the children, there was less substance abuse, fewer arrests, better school performance, and better all-around social adjustment.

An interesting school-based program that has been implemented in several countries and is explicitly based on the assumptions of developmental theories is the *Fast Track Project* (2005). Through multistage screening of over 10,000 kindergarten children, 891 were identified who were at high risk for antisocial behavior. These children were impulsive, had difficult temperaments, and came from unstable families living in low-income/high-crime neighborhoods. The children were divided into experimental (n = 445) and control (n = 446) groups. The experimental group was given a curriculum designed to promote social understanding, develop emotional communication skills, and improve self-control and problem-solving skills. These children were also placed in so-called *friendship groups* and *peer-pairing* designed to increase social skills and enhance friendships. Their parents received parenting effectiveness training and home visits to foster their problem-solving and general life management skills.

The program is evaluated periodically and has been found to have modest but positive outcomes. By the end of the third grade, 37% of the experimental group was judged to be free of conduct problems as opposed to 27% of the control group. By the eighth grade, 38% of the experimental group had been arrested as opposed to 42% of the control group. Although these differences are modest at best, they still reflect a good number of people saved from criminal victimization if the improvements hold in the future. If programs such as the Fast Track Project could be combined with programs like the Nurse–Family Partnership, we should see improved results since even kindergarten intervention may be too late in some cases. Yet developmental theories tell us that human life is characterized by dynamism, and that people can change at any time. This is a note of optimism for crime control/prevention strategies.

SUMMARY

- Developmental theories are dynamic and integrative and examine offending across the life course. Juvenile offending has been noted across time and cultures. A sharp rise in offending following puberty and a steady decline thereafter has been widely noted. Scientists explain the age effect with respect to the brain and hormonal processes occurring during adolescence. At puberty, a huge surge in testosterone levels is experienced, and the brain undergoes a process of intensive resculpting.

- There is a wide variety of factors that put some teens more at risk for delinquent behavior than others. The factors we focused on in this chapter were ADHD and CD, which are highly heritable. The co-occurrence of ADHD and CD is a particularly strong risk factor.

- Developmental theories follow individuals across the life course to determine the differential effect of risk factors for offending at different junctures.

- Agnew's super traits theory focuses on how low self-control and irritability (negative emotionality) interact with other life domains (school, work, marriage, etc.) across the life span to impact the probability of offending.

- Farrington's ICAP theory stresses antisocial potential and cognition and how these things are shaped in pro- or antisocial directions at different times and in different situations. Farrington distinguishes between long-term and short-term antisocial potential. Short term antisocial potential occurs primarily during adolescence.

- Moffitt's theory posits a dual-pathway model consisting of adolescent limited (AL) offenders who limit their offending to the adolescent years, and life course–persistent (LCP) offenders who offend across the life course. LCP offenders have neurological and temperamental difficulties that set them on a developmental trajectory that leads to antisocial behavior at all ages and in all social situations. AL offenders do not suffer these disabilities and have accumulated sufficient social capital that they can resettle into a prosocial lifestyle once neurological and social maturity has been reached.

- Sampson and Laub's age-graded theory is concerned with the power of informal social control (bonds) to prevent offending. Turning points in life (marriage, new job, etc.) are important in understanding the process of offenders' desistance from crime.

DISCUSSION QUESTIONS

1. Why is it important that we understand what is going on biologically during adolescence?

2. If people age out of crime as well as into it, would it be a good idea to ignore all but the most serious of juvenile crimes so that we don't risk having children gain a "criminal" reputation?

3. What is social capital and how much of it do you believe you have accumulated?

4. If only a very small number of individuals are life course–persistent offenders, shouldn't we concentrate our crime control efforts on them? If you agree with this, how do we identify them? Isn't there a danger of false-positive identification (identifying someone as likely to continue to offend well into adulthood when in fact he or she doesn't)?

5. Go to www.fasttrackproject.org/ and read more about the Fast Track Program. Identify some of the specifics of the program and report to the class.

USEFUL WEBSITES

Conduct Disorder. http://aacap.org/page.ww?section=Facts+for+Families&name=Conduct+Disorder

Developmental Psychology. http://medicine.jrank.org/pages/455/Developmental-Psychology.html

Moffitt, T. E. (1993), "Adolescent-Limited and Life-Course-Persistent Antisocial Behavior: A Developmental Taxonomy" (*Psychological Review*). www.soc.umn.edu/~uggen/Moffitt_PR_93.pdf

National Institute of Mental Heath. www.nimh.nih.gov/

Office of Juvenile Justice and Delinquency Prevention. http://ojjdp.ncjrs.org/

CHAPTER TERMS

Agnew's "super traits" theory

Antisocial potential

Attention-deficit/hyperactivity disorder (ADHD)

Conduct disorder

Delinquency

Developmental theories

Integrated cognitive antisocial potential (ICAP) theory

Maturity gap

Moffitt's dual pathway theory

Risk factor

Sampson and Laub's age-graded theory

Social capital

Turning points

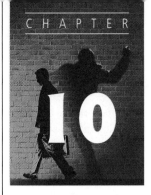

Violent Crimes

On January 8, 2011, 22-year-old Jared Lee Loughner opened fire on a political gathering ("Congress on your corner") at a Tucson, Arizona, mall, killing 6 people and wounding 13 others. The gathering had been set up outside a Safeway supermarket for the U.S. Democratic representative from Arizona, Gabrielle Giffords, who was shot in the head but not killed. Among those killed was one of Giffords' staffers and a federal judge. Loughner, who is a political radical with a hatred for religion, was arrested at the scene and, as of this writing, is awaiting trial.

In April of 1973, Edmund Kemper, a 6-foot-9, 300-pound 25-year-old hate machine, killed his mother, decapitated her, had sex with her headless body, and played darts with the head. He then invited his mother's best friend over for a "surprise" dinner to honor his mother and decapitated her also. He killed at least six other women and sexually assaulted their headless corpses. He even ate the flesh of some of his victims, cooking it into a macaroni casserole. But Kemper's biggest thrill was not sex, murder, or cannibalism; it was decapitation. In Kemper's own words, "You hear that little pop and pull their heads off and hold their heads up by the hair. Whipping their heads off, their body sitting there. That'd get me off" (quoted in Leyton, 1986, p.42).

These are extreme examples of violence in America, and although the potential for violence is in us all, the vast majority of us will never commit or be victimized by extreme violence. Nevertheless, we do seem to be just as fascinated by it as we are repelled by it.

▲ **Photo 10.1**   A victim is airlifted from the scene after Congresswoman Gabrielle Giffords and 13 others were wounded, and six people were killed, including a federal judge, when a gunman opened fire during a meet-and-greet event outside a Safeway grocery store in Casas Adobes.

✎ Violence in History

Violent crime is the use of force exercised without excuse or justifiable cause to achieve a goal at the expense of a victim. Physically harming a victim may be the motive behind a violent crime, or harm may simply be secondary to achieving the perpetrator's goal, as in a robber's desire to take a victim's money. Just as every generation in its youth seems to think it discovered sex, every generation in its maturity seems to feel that it is in the midst of an unprecedented wave of violent crime. If we had lived in earlier times, we really would have had something to complain about. Eisener's (2001) examination of the history of European murder rates showed that they dropped from a high average of 32 per 100,000 in the 13th and 14th centuries to 19, 11, 3.2, 2.6, and 1.4 per 100,000 in subsequent centuries, respectively. Eisener concludes that social control mechanisms such as the state's monopoly of power, the expansion of universal schooling, the rise of religious reform movements, and the organized discipline of the manufacturing workplace largely accounted for this decline. Eisener might also have mentioned that the high murder rates in medieval Europe had a lot to do with the combination of the habit of bearing arms, alcohol-induced quarrels, the absence of effective medical treatment for wounds, and the lack of a trusted system of justice. His data also show clearly that violence can be controlled and that violence may be the default option when controls are lacking, as social control theories maintain. This chapter examines each of the eight Uniform Crime Report (UCR) Part I violent crimes, starting with murder, including mass, spree, and serial murder.

✎ Murder

Murder is "the willful (non-negligent) killing of one human being by another" (FBI, 2010a). There were 15,241 murders in the United States in 2009, a rate of 5.0 per 100,000 of the U.S. population, which is about half of the all-time high rate of 10.2 in 1980. New Orleans and Gary, Indiana, had the highest murder rates (each with 51.7) followed by Detroit (40.1). The rates in these cities are higher than Europe's in the 13th century. Of the known offenders, 89.7% were males, 49.3% were black, 48.7% were white, and 2.0% were of other races. Note that most Hispanics—about 94%—are classified as white for FBI recording purposes. About 90% of known murders are intraracial and intrasexual. About 14% of the murders in 2009 took place within the family, as indicated by Figure 10.1. Most of the family relationships were not biological; i.e., victims were spouses, stepchildren, and half-siblings. Note that only murders in which the victim–offender relationship was known are included in this statistic.

As high as the U.S. murder rate is, it is much lower than that of many other countries. According to the World Health Organization's homicide data for young people 10 to 29 years of age, the United States is about in the middle of the pack with a rate of 11 homicides per 100,000. Colombia had the highest rate (84.4) and Japan had the lowest (0.4). In other words, for every youthful murderer in the United States, there were 7.7 in Colombia, and for every youthful murderer in Japan, there were 27.5 in the United States. Countries with low murder rates tend to be politically stable and wealthy democracies with all of the social control mechanisms Eisener describes above. France, Germany, and the United Kingdom all have rates below 1.0 per 100,000. Countries with high rates tend to be third world or developing countries, or countries experiencing rapid social, political, and economic changes, which are the conditions that Durkheim maintained led to anomie and high crime rates (see Chapter 4). Examples of such countries are South Africa (with a rate of 51 per 100,000) and Russia (18 per 100,000) (Krug, Dahlberg, Mercy, Zwi, & Lozano, 2002).

In the United States, as is the case in other countries, young people ages 18 through 24 are most likely to be killed and to be killers. Young black males are about 9 times more likely to be murdered than young white males, and young black females are about 6 times more likely to be murdered than young white females (U.S. Bureau of Justice Statistics, 2005). Males are far more likely to murder than females. When females kill males, it is typically a spouse, ex-spouse, or boyfriend in a self-defense situation (Mann, 1990). Female/female murder is very rare around the world. Martin Daly and Margo Wilson (2000) examined data on same-sex non-relative murders and found that female/female homicide constituted only 2.5% of the total number of murders for a number of different cultures around the world. Even going back to England in the 13th century, cases of females murdering other females accounted for only 4.9% of the total murders (Given, 1977).

Legal Categories of Murder

First-degree murder, or **aggravated murder,** is defined as the intentional unlawful killing of one human being by another with "premeditation and deliberation." In other words, the killer had a purposeful and evil intention (malice) to kill the victim, and had thought about and planned it (premeditation) with full awareness of the consequences (deliberation). First-degree murder is the only kind for which a convicted murderer can be executed in states that have the death penalty, although a special type of murder called felony murder may also carry that penalty, particularly when an agent of the criminal justice system is the victim. **Felony murder** does not require the intention to kill, but only the intention to commit some other felony such as robbery. If a murder occurs during the commission of the felony, it is considered a felony murder. For example, if a burglar rapes a woman in her home and then kills her, he could be charged with two counts of aggravated murder (murder in the commission of a rape and murder in the commission of a burglary) even though there is only one victim and even though he entered the victim's home only with the intention of burglarizing it.

Voluntary manslaughter (called second-degree murder in some states) is the intentional killing of another human being without premeditation and deliberation. It is murder committed in response to a mistaken belief that self-defense required the use of deadly force, or in response to adequate provocation while in the heat of passion. The term *passion* can mean fear, anger, or outrage, and can result from such provocations as being threatened in a bar or catching one's spouse in bed with someone else. In other words, voluntary manslaughter is usually charged when the suspect had temporarily been in such a high state of emotional arousal that his or her rational faculties were impeded.

Involuntary manslaughter is a criminal homicide where an unintentional killing results from a reckless act. In such cases, the defendant is charged with consciously disregarding a substantial risk that he or she should have known would put others in danger of losing their lives. The most obvious example of this is driving under the influence of alcohol or some other drug. **Negligent manslaughter** is an unintentional

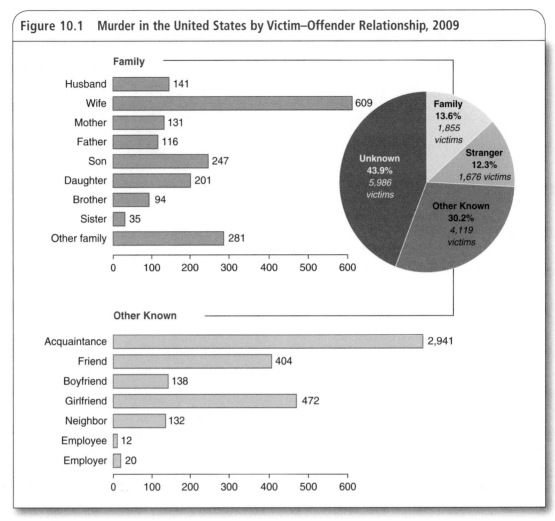

Figure 10.1 Murder in the United States by Victim–Offender Relationship, 2009

SOURCE: FBI (2010a).

homicide that is charged when a death or deaths arise from some negligent act that carries a substantial risk of death to others. Note that involuntary manslaughter is charged when defendants do something they should not have (drive drunk), and negligent homicide is charged when they neglect to do something they should have. Doing something they should not have can range widely, from driving an unsafe vehicle that should have been fixed, to failing to comply with safety regulations in a work setting.

Mass and Spree Murder

Although murder is usually considered the most serious crime a person can commit, we can comprehend murder for revenge or for personal gain because such motives have rational if not moral elements. The gruesome bloodlust of killers like Edmund Kemper, however, baffle and terrify us because they lack any objectively rational motivation with which we can identify. We can all imagine circumstances in which we might act violently, and even kill someone, but few of us can imagine ourselves becoming mass, spree, or serial killers.

Mass murder is the killing of several people at one location that begins and ends within a few minutes or hours, usually with the death of the killer by suicide. **Spree murder** is the killing of several people at different locations over a period of several days. Research suggests that the time frame involved is the only factor that differentiates mass and spree killers, and that both mass and spree killers are different from the serial killer.

Mass murderers are divided into two types: (1) those who choose specific targets the killers believe to have caused them stress (e.g., disgruntled workers who choose their workplace), and (2) those who attack targets having no connection with the killer but who belong to groups the killer dislikes such as prostitutes or people of a particular race. Such people rarely "just snap" and kill people at random. Most mass murderers are motivated by a hatred that simmers until some specific incident provides the flame that brings it to a boil. Marc Lepine, for instance, had been denied admission into the engineering school at University of Montreal and sought revenge on the woman who had taken his place in a "man's profession" by killing 14 women and wounding 13 (Fox & Levin, 2001, p. 119). Colin Ferguson, who killed 6 people and wounded 19 others on a Long Island commuter train in 1993, hated whites and claimed that "black rage" at what he saw as society's mistreatment of blacks led him to his rampage (Schmalleger, 2004). Then there was Jiverly Wong, who shot and killed 13 people and wounded 4 others at an immigration center in Binghamton, New York, in 2009 and then shot himself. Wong apparently disliked Americans because he felt they were prejudiced toward him (although most of his victims were fellow immigrants) and was embarrassed by his poor English (Lester, 2010). All three men methodically and selectively chose their victims on the basis of some alleged wrong done to them by a group their targeted victims represented.

Spree killers move from victim to victim in fairly rapid succession, and like mass murderers make little effort to hide their activities or avoid detection, as if driven by some frenzied compulsion. Spree killing is rare, but spree-killing teams are even rarer and are typically composed of a dominant leader and a submissive lover. The most recent spree-killing team is the sniper team of John Muhammad and Lee Malvo who killed 13 random individuals and wounded 6 others in the Washington, D.C., area in 2002. Although the team killed without regard to race or gender, Muhammad belonged to the Nation of Islam, which may have provided ideological impetus for his vicious spree (Hurd, 2003).

There are relatively few generalities we can make about mass and spree killers. Most research on these individuals comes from interviews with their family and friends, and sometimes notes left by the offender, because they typically commit suicide or are killed by police at or near the scene of the crime. However, reviews of the literature (Fox & Levin, 2001; Palermo, 1997) note several commonalities shared by mass and spree murderers:

- They are typically white males with an age range broader than that of serial killers.
- African Americans are overrepresented in terms of their proportion in the population.
- They have previously displayed impulsive, violent, frustrated, depressed, alienated, and antiauthoritarian behavior arising from a deep sense of having been wronged.
- They tend to have a morbid fascination with guns and to own many of them.
- Their behavior at the time of the crime, as well as the fact that it is typically committed in public places, makes it obvious that they are unconcerned about their own death, leading some researchers to view this type of murder as an elaborate suicide attempt.

- They seem to contemplate committing murder and prepare for the act, although the time and place is not generally pre-established.
- Like crime in general, spree and mass murders increased considerably in the United States during the 1960s through 1990s.
- The average age of all mass murderers and spree murderers over the past 40 years is 29.

Serial Murder

Serial murder is the killing of three or more victims over an extended period of time with a "cooling off" period in between kills (Hickey, 2006). Unlike mass and spree killers who almost invariably use guns, serial killers tend to favor "hands on" killing, often with torture. In 2007, the FBI (2008) held a symposium about serial killers with 135 experts that included law enforcement officers, mental health specialists, and criminologists who specialize in serial killing research. This group of experts put together seven commonly believed myths about serial killers, listed below.

Myth 1: Serial killers are all dysfunctional loners. The truth is that most serial killers have families, homes, and gainful employment; some have even been churchgoers and law enforcement officers. They hide in "plain sight," which is why their neighbors are shocked to hear that such a "nice man" was a serial killer and why they are so often overlooked by law enforcement.

Myth 2: Serial killers are all white males. The reality is that white males are somewhat underrepresented among serial killers in proportion to their numbers in the population, and African Americans are overrepresented. Hickey (2006) claims that approximately 44% of the serial killers operating between 1995 and 2004 have been black, which is about 3.4 times greater than expected based on the proportion of African Americans in the population (p. 143). On the other hand, the only *known* Asian American serial killer operating in the United States during the 20th century was Charles Ng, who with his white partner, Leonard Lake, tortured and killed at least 19 people in the early 1980s (Newton, 2000).

Myth 3: Serial killers are only motivated by sex. While sex is the major motivation of many serial killers, there are many other motivations such as thrill seeking, attention getting, anger, financial gain, religious or ideological reasons for wanting to eliminate certain types of people, and—perhaps the greatest motivator—a sense of power.

Myth 4: Serial killers travel and operate interstate. While a number of serial killers travel around the country trolling for victims, most conduct their operations within a well-defined geographic comfort zone where they feel confident. This comfort zone typically centers around their home or place of employment.

Myth 5: Serial killers cannot stop killing. While most serial killers do continue to kill until they are caught or die, there are some well-known killers who may stop for years before starting again. For instance, Dennis Rader, the infamous BTK ("Bind, Torture, Kill") serial killer, murdered 10 victims from 1974 to 1991 and did not kill again. He was caught in 2005 because he wanted to tell his side of the story when someone was writing a book about the unsolved BTK killings.

Myth 6: All serial killers are insane or are evil geniuses. Most serial killers probably fit the criteria for psychopathy or antisocial personality disorder, but only a very few qualify as mentally ill (e.g., schizophrenic) or legally insane. Neither are they geniuses who are able to outsmart the law at every turn. In fact, the IQs of serial killers vary considerably. The genius myth probably derives from the published IQs of particularly infamous serial killers such as Ed Kemper (IQ = 136) and Ted Bundy (IQ = 124) (Walsh & Ellis, 2007), which should warn us not to equate high intelligence with high morality.

Myth 7: Serial killers want to get caught. This is far from true and probably derives from amateur psychoanalysis. Serial killers plan their crimes (the target selection, acquisition, control, and disposal) with great care, getting better at it with each victim. They often have such feelings of power that, contrary to wanting to get caught, they feel that they *can't* be caught.

There is also an additional myth not listed by the FBI, but one that we will address anyway because there are indeed female serial killers:

Myth 8: Serial killers are never women. When Aileen Wuornos was arrested in 1991 and charged with the shooting deaths of seven males whom had she had picked up while working as a prostitute, she was dubbed "America's first woman serial killer" because she fit the definition of a "true" serial killer as a person who kills for no rational motive (Keeney & Heide, 1995). Yet there had been a number of others before her. Nursing home proprietor Amy Archer-Gilligan, who may have murdered up to 100 patients in her charge between 1907 and 1914, like almost all female serial killers, killed for financial reasons. Thus the key distinction between male and female serial killers may be this: "There are no female counterparts to a Bundy or a Gacy, to whom sex or sexual violence is a part of the murder pattern" (Segrave, 1992, p. 5).

Prevalence of Serial Killing

No one knows exactly how many serial killers there are at any given time or how many murders per year are attributable to them. Figure 10.2 presents the number and rates of serial killers in the United States between 1795 and 2004. These figures are gross estimates, as the true number of serial killers operating in any time period can never be known. The increase in the rate of killers from 1925–1944 to 1980–2004 is probably only partly a real increase in the number of serial killers. An unknown portion of the increase may be attributed to improved law enforcement methods initiated during the latter period. The increase in other forms of homicide, including spree and mass murder, during the same period, however, leads us to conclude that part of the reported increase in serial killers represents a real increase.

A Typology of Serial Killers

Although all serial killers have a common goal of killing, they have different psychological motives. Holmes and DeBurger (1998) have provided a typology that divides serial killers into four broad types: visionary, mission-oriented, hedonistic, and

▲ **Photo 10.2**   Serial Killer Dennis L. Rader, also known as the BTK (Bind, Torture, Kill) killer, is escorted into the El Dorado Correctional Facility in El Dorado, Kansas. Rader pleaded guilty to the 10 killings dating back to 1974 and received 10 life terms.

Figure 10.2 Estimated Number and Rate per 10 Million of Serial Killers Operating in the United States From 1795 to Mid-2004

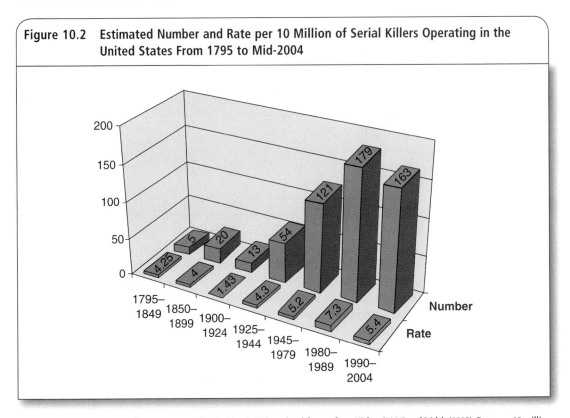

SOURCE: U.S. Justice Department figures as reported by Jenkins (1994); updated figures from Hickey (2006) and Walsh (2005). Rates per 10 million population computed by the author.

power/control. These are not definitive categories that all serial killers can be neatly assigned to; many serial killers may evidence aspects of all types at various times.

- **Visionary serial killers** are typically out of touch with reality, may be psychotic or schizophrenic, and feel impelled to commit murder by visions or "voices in my head." David Berkowitz (the "Son of Sam"), who mostly shot young lovers parked in their cars, was this type of killer. The sexual assault of victims is not usually a component of the visionary killer's pattern. The visionary killer is the stereotypical dysfunctional loner.

- **Mission-oriented serial killers** feel it is their mission in life to kill certain kinds of people such as prostitutes and homosexuals. Unlike visionary killers, mission-oriented killers do not have visions or hear voices telling them to kill. They define their own "undesirables" and set out to eliminate as many as they can. The Black Muslim group called the Angels of Death, which operated during the 1970s in California, may have killed more people than all other serial killers combined during that period. One estimate puts the total number of victims attributed to the group at 270 (Howard, 1979).

- **Hedonistic serial killers** are the majority of serial killers. They kill for thrills, engaging in perverted sexual activity. The hedonistic killer is such a self-centered psychopath that he considers someone's life less important than his sexual pleasure. Edmund Kemper is an example of this type.

He had been diagnosed as psychotic as a teenager after killing both his grandparents, but he never claimed that "voices" made him kill.

- **Power/control serial killers** gain more satisfaction from exercising complete power over their victims rather than from "bloodlust," although sexual activity is almost always involved. Like the hedonistic killer, the power/control killer frequently suffers from some form of sexual inadequacy. Ted Bundy and Jeffrey Dahmer are examples of this type of killer. Both men would keep the bodies of their victims for some time after killing them, often washing and grooming them like dolls. Dahmer went several steps further and tried to create sex slave zombies of his victims by drilling holes in their heads and pouring acid into the holes. He also cannibalized several victims.

Theories of Serial Killing

Serial killing is not the result of any single "cause" but of several risk factors interacting in various ways, and becoming a serial killer is a long drawn-out process, not a discrete event. Any one factor or combination of factors can facilitate or expedite the onset of killing for some killers but have little or no effect on others. Keep in mind that when theorists attempt to explain something as bizarre as serial killing, they are offering very broad and speculative generalities.

Stephen Leyton (1986) attempted to apply anomie/strain theory to explain serial killing across the centuries, proposing that different social classes have dominated the ranks of serial killers at different periods because classes are differentially exposed to high aspirations, and thus to crises in their social standing. During preindustrial times, only the aristocracy could aspire to greater status, and thus only they would be susceptible to the strains of anomie. In the 19th century, the middle classes had opportunities to advance their social position, and serial killers were drawn from the ranks of those who failed. In the 20th century, serial killers were drawn overwhelmingly from the ranks of ordinary working men as they aspired to the American Dream.

Significant changes in the prevalence of serial killers since the early 1960s points to some important social changes. The disinhibited counterculture that arose during the period had much to do with the increased prevalence because traditional values were questioned and rejected (Levin & Fox, 1985). Despite its rhetoric of flower power and peace, the counterculture was essentially about personal satisfaction ("Do your own thing") and not feeling bad about it ("Don't get hung up on guilt"). Following that period, many more people crossed the line to engage in a variety of aberrant behaviors because of what has been called "society's recent war against guilt" (Levin & Fox, 1985, p. 72). The late 1960s also saw an explosion of pornography, some of which depicted scenes of bondage, torture, and violent rape, which may have fed and shaped the sexual fantasies of some people, a proportion of whom subsequently acted them out (Sears, 1991).

Whatever the social factors accounting for the increased prevalence of serial killing may be, only an infinitesimally small number of people experiencing them ever kill once, let alone become serial killers. Thus, the developmental histories of serial killers should be explored. Social control theory emphasizes the role of the family, and one factor linked to serial killers is an extreme level of maternal deprivation in childhood. The increasing breakdown of the family also provides serial killers with large numbers of rootless potential victims (Saffron, 1997). The neglect, abuse, and social isolation experienced by many serial killers in their early years leave them angry and unable to relate to others in conventional ways, and eventually the resentment is detonated.

With sexuality being a central part of human life, extreme sexual dysfunction may result in deeply embedded feelings of worthlessness and powerlessness, the seeds of which may have already been implanted by childhood abuse and neglect. Kemper's fondness for sex with the dead was supposedly tied to concerns about his small penis. Serial killers may be trying to counteract feelings of inadequacy by controlling and destroying vulnerable others (e.g., women) whom they may see as the cause of their feelings. Many studies of serial murders reveal a pattern of long-standing preoccupation with fantasies devoted to sexualized violence, and that children reared in abusive homes often retreat into a private fantasy world where they can escape their fears and exert control, thus gaining in their minds that which is unavailable in reality (Carlisle, 1993).

The making of a serial killer is an extremely complicated process that may be very different for different killers. Stephen Giannangelo's (1996) *diathesis-stress* model attempts to integrate cultural, developmental, psychological, and biological concepts in his theory. The theory states that all serial killers have a congenital biological propensity (diathesis) to behave and think in ways that lead to serial killing if combined with environmental stressors such as abuse and neglect. This combination leads to self-esteem, self-control, and sexual dysfunction problems that feed back on one another and lead to maladaptive social skills, which move the person to retreat into his pornographic fantasy world. As he dwells longer and longer in this world, he enters a dissociative process in which he takes his fantasies to their moral limits. At this point, the killer seeks out victims on which to act out his fantasies, but the actual kill never lives up to his expectations or to the thrill of the hunt, so the whole process is repeated and becomes obsessive-compulsive and ritualistic, much like drug addiction. Figure 10.3 illustrates this process.

Serial Killer Profiling

Law enforcement agencies have responded to the challenge presented by serial killers with the establishment of the FBI's *Behavioral Science Unit* (BSU) in the early 1970s. The BSU, now part of the National Center for the Analysis of Violent Crime (NCAVC), has developed methods of profiling serial killers and other violent offenders. Profiling is done by extensive interviewing and formal psychological testing of incarcerated killers in order to develop a typology (the classification of offenders into different types) based on personality and other offender characteristics (FBI, 2008). Law enforcement officials have always done some sort of rudimentary profiling based on their experiences regarding "what sort of person would have done this." It is only recently that it has been elevated to an art—it cannot as yet be considered a science because profiles do not always fit the persons eventually convicted of the crimes (Canter, 2004).

Hazelwood and his FBI colleagues (Hazelwood, Ressler, Depue, & Douglas, 1987) describe the qualities of successful profilers as those who are "experienced in criminal investigation and research and possess

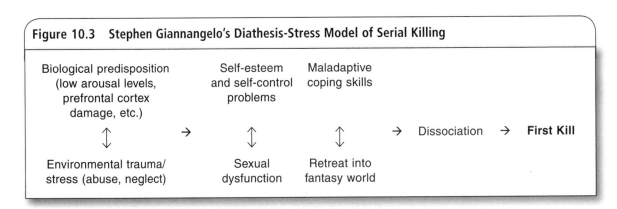

Figure 10.3   Stephen Giannangelo's Diathesis-Stress Model of Serial Killing

common sense, intuition, and the ability to isolate their feelings about the crime, the criminal, and the victim. They have the ability to evaluate analytically the behavior exhibited in the crime and to think very much like the criminal responsible."

Forcible Rape

Forcible rape is "the carnal knowledge of a female forcibly and against her will" (FBI, 2010a). This definition includes attempts to commit rape but excludes statutory rape (consensual sex with an underage female) and the rape of males. Rapes of males are classified as either assaults or other sex offenses depending on the circumstances. According to the 2010 UCR, there were 88,097 reported rapes in 2009, a rate of 56.6 per 100,000 females. Twenty-four females were also murdered during the course of the rape. By race, 65.1% of those arrested for rape were white, 32.5% were black, and the remaining 2.4% were of other races. Poor, young, unmarried, nonwhite females are disproportionately likely to be victimized, and poor, young, unmarried, nonwhite males are disproportionately likely to be perpetrators (Rand, 2009).

Some facts about rape from 20 years of victimization surveys on rape include the following:

- About half of all rapes are committed by someone known to the victim.
- The offender was armed in about 20% of the cases. Stranger rapists were more likely to be armed (29%) than were rapists known to the victim.
- Among the victims who fought their attackers or yelled and screamed, more reported that it helped the situation rather than made it worse.
- Slightly more than half of the victims report the assault to the police. Victims are more likely to report the incident if the perpetrator was armed or if they sustained physical injuries.
- As is the case with all other crimes, rape rates have fallen significantly over the last 35 years (cited in Walsh & Ellis, 2007).

Figure 10.4, from the Bureau of Justice Statistics National Crime Victimization Survey (NCVS), graphically illustrates this decline.

Rape rates vary considerably from country to country; the *reported* rape rate in the United States is typically 4 times higher than that of Germany, 13 times higher than Britain's, and 20 times higher than Japan's (Schwartz, 1995). I emphasize *reported* because determining rape rates is extremely difficult, and comparing international rape rates is more difficult still. Some countries include statutory rape in their rape reports (which would inflate the number of rapes in those countries relative to

▲ **Photo 10.3** Some males excuse their inappropriate behavior with women by adopting rape myths. For example, girls who dress provocatively or consume significant amounts of alcohol are sometimes considered to be "asking" to be victimized. Even the police and courts appear to be less sympathetic toward sex crime victims who do not take "all precautions" to avoid situations in which males might assume a female is not truly saying "No!"

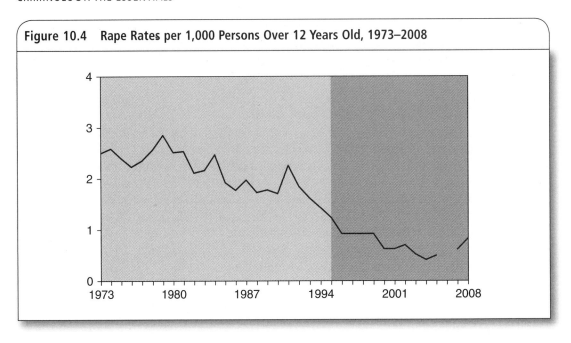

Figure 10.4   Rape Rates per 1,000 Persons Over 12 Years Old, 1973–2008

the number in the United States), others do not, and some do not differentiate between rape and other sexual offenses (which would also inflate their rates). We also have to be sensitive to the degree of stigma attached to rape victims in different cultures. For instance, Egypt, a nation of about 54 million people, reported just 3 rapes to INTERPOL in 1991, while France with approximately the same population reported 4,582 (INTERPOL, 1992, p. 54). Accusing someone of rape in some Islamic countries is not taken lightly. Proof requires the sworn eyewitness testimony of two males (or four women) of good Muslim character. As if the stigma of being raped that attaches to women in Islamic countries is not reason enough to forgo reporting the crime, if eyewitness testimony is not forthcoming, the accuser may herself be punished with 100 lashes in some Islamic countries for making a "false accusation" (Walsh & Hemmens, 2011).

There are a variety of theories about the causes of rape with different assumptions. Feminist theories assert that rape is a learned behavior and is motivated by power rather than sexual desire, that all men are capable of it, and that it is a tactic potentially used by all men to control women (Brownmiller, 1975). Social learning theorists agree with these assumptions, except that they view rapists as a psychologically unhealthy category of males, not as "normal men." Therapists who work with rapists tend to agree that all men are *potential* rapists, but assert that rape is sexually motivated and that violence (and other methods) is a tactic used to gain sexual compliance, not a goal (Mealey, 2003). Evolutionary psychologists also contend that rape is sexually motivated and that it is a maladaptive consequence of a generally adaptive behavior, i.e., males seeking as many sexual partners as possible (Thornhill & Palmer, 2000). Note that the adjectives "adaptive" and "maladaptive" are not moral statements; rather, they refer to biologists' assessment of how a behavior may have had reproductive benefits.

A study of 2,295 women from 22 universities in the United States and Canada revealed one of the strangest but most consistently reported findings about date rape: that females who self-reported a completed rape were significantly *more* likely to continue to date (27.2%) their assailants than females who self-reported an attempted, but not completed, rape (19.4%). Note that the majority of both categories, however, did not continue to date their assailants (Ellis, Widmayer, & Palmer, 2009). The authors noted that forcible tactics usually followed other unsuccessful methods of gaining compliance, such as cajoling, appeals to love, and getting the female drunk. The authors are, of course, specifically talking about date rape, and not stranger rape.

A number of 1980s surveys of high school and college students showed that a majority of males and a significant minority of females believe that it is justifiable for a man to use some degree of force to obtain sex if the woman was perceived as somehow having "led him on" (Herman, 1990). This seems to indicate that some people believe that there could be an act labeled "justifiable rape" in the same sense that there is justifiable homicide. These surveys also indicate that many believe that rape victims are often at least partially responsible for their rape because of such factors as provocative dress and lifestyle (e.g., frequenting bars) and because of the belief that "nice girls don't get raped" (Bartol, 2002, p. 295). To put this in perspective, we would not get very far claiming that we were justified in robbing a bank because the bank got us "all excited" by leading us to believe that it would give us a loan but later denied us the loan, or that its location, gaudy advertising, and apparent lack of security meant that it was "asking to be robbed."

Most studies of rapists concentrate on the violent rapist, who is usually a stranger to his victim, and who tend to have a history of other violent crimes (Freeman, 2007; Mills, Anderson, & Kroner, 2004). We know that among these subjects, violence is an important component of the sexual excitement they obtain from their crimes, just as whips and other devices are important to masochists and sadists. This pattern of preferential violence is determined by comparing penile responses of convicted rapists with those of non-rapists when exposed to sexual stimuli with strong violence content. A device called a penile plethysmograph (PPG), which is rather like a blood pressure gauge that fits around the penis, is used to measure penile response and thus the level of sexual excitement. The PPG measures the pressure of blood in the penis to ascertain how sexually excited subjects become when exposed to auditory or visual stimuli depicting various sexual situations. Violent rapists become significantly more aroused than non-rapists or date rapists when exposed to images of sexual violence (Robertiello & Terry, 2007). For instance, if a rapist achieves a 30% erection when viewing nonviolent sex and one of 80% when viewing violent sex, we can conclude both that he is more interested in violent sex than consensual sex and that he is probably a dangerous individual (Walsh & Stohr, 2010).

⬚ Robbery

Robbery is "the taking or attempted taking of anything of value from the care, custody, or control of a person or persons by force or threat of force or violence and/or putting the victim in fear" (FBI, 2010a). In 2009, there were 408,217 reported robberies in the United States, a rate of 133 per 100,000, down almost 51% from the peak rate of 272.7 in 1991. Of those robbers arrested, 60.1% were under 25 years of age, and 89.0% were male. By race, 55.5% were black, 42.8 were white, and the remaining 1.7% were of other races. In 2009, of the robbery victims, 849 were murdered (746 males and 103 females) during the course of the robbery. A total of $508 million was taken in robberies in 2009. Figure 10.2 shows the primary locations where reported robberies took place in 2009.

The danger posed by resisting victims and the severe penalties attached to committing robbery make it a high-risk crime, which suggests that those who commit it may be among the most daring and dangerous of all criminals. Interviews of active street robbers (R. Wright & Decker, 1997) reveal them to be the least educated, most fearless, most impulsive, and most hedonistic of criminals. Obtaining legitimate work is simply not an option that robbers entertain because work would seriously interfere with their "every night is Saturday night" lifestyles.

Jacobs and Wright (1999) focus on the motivating factors and decision-making processes of robbers and find that most robbers decide to commit their crimes impulsively with very little rational thought. The timing and motivation of street robbers are largely governed by their need for money, as the following statement from one of Jacobs and Wright's (1999) subjects explains:

> [The idea of committing a robbery] comes into your mind when your pockets are low; it speaks very loudly when you need things and you are not able to get what you need. It's things that you need, things that if you don't have the money, you have the artillery to go and get it. (p. 150)

The participation of robbers in street culture leads them to be blind to legitimate opportunities to the point where many seem to fatalistically believe that they have little choice but to rob. In other words, armed robbers appear to be so overwhelmed by their emotional, financial, and drug problems, and so into the "focal concerns" of their subculture, that they perceive robbery as their only way to obtain money. Sometimes robberies are committed simply because the robber sees an opportunity he can't let pass:

> If I had $5,000, I wouldn't do [a robbery] like tomorrow. But if I got $5,000 today and I seen you walkin' down the street and you look like you got some money in your pocket, I'm gonna take a chance and see. It's just natural. . . . If you see an opportunity, you take that opportunity. It doesn't matter if I have $5,000 in my pocket. (Jacobs & Wright, 1999, p. 150)

While most criminals do not specialize in any one crime, robbery does tend to be favored by the most daring of them. There are a number of attractions of robbery in comparison with other crimes. Burglary takes time and requires the burglar to find buyers for stolen property, and burglars never know who or what they might run into inside a house. Drug selling means dealing with a lot of people, the risk of being robbed oneself, and—most importantly—coping with the temptation of being one's own best customer. Robbery, on the other hand, allows the robber to pick the time, place, and victim at leisure, and then complete the job in a matter of minutes and seconds. It is the perfect crime for those with a pressing and constant need for fast cash to feed a hedonistic lifestyle and who enjoy the adrenaline rush that the crime affords them. Robbery, and flaunting the material trappings signaling its successful pursuit, is seen ultimately as a campaign for respect and status in the street culture in which most robbery specialists participate. As James Messerschmidt (1993) puts it,

> The robbery setting provides the ideal opportunity to construct an "essential" toughness and "maleness"; it provides a means with which to construct that certain type of masculinity—hardman. Within the social context that ghetto and barrio males find themselves, then, robbery is a rational practice for "doing gender" and for getting money. (p. 107)

With the exception of rape, robbery is the most "male" of all crimes, but women do engage in it. A favorite ploy for female robbers is to appear sexually available (e.g., as a prostitute) to a male victim and then, either alone or with the help of an accomplice, rob him. Female robbers will seldom rob males without an accomplice, but will practice their "art" mostly on other females. Much like their male counterparts, female robbers eschew legitimate work and prefer their hedonistic "money for nothing" lifestyles (J. Miller, 1998).

⬚ Aggravated Assault

Aggravated assault is "an unlawful attack by one person upon another for the purpose of inflicting severe or aggravated bodily injury" (FBI, 2010a). As opposed to simple assault, aggravated assault is one in which a weapon such as a knife or gun is used, although sometimes the use of hands and feet can result in a charge of aggravated assault. There were 806,834 aggravated assaults reported in 2009. This rate of 262.8 per 100,000 is a 40.5% drop from the all-time high of 441.8 in 1992. Blunt objects or "other dangerous weapons" were used in 33.5% of the cases, and 26.9% involved personal weapons such as hands and feet; firearms were used in 20.9% of the cases, knives or other cutting instruments in 18.7%. About 40% of those arrested for aggravated assault were under the age of 25. A total of 79.3% were male, 63.5% were white, 33.9% were black, and 2.6% were of other races.

In a sense, any case of aggravated assault can be viewed as a murder that never happened. Each incident of aggravated assault carries the potential threat of becoming a murder, because without the speedy access to modern medicine we enjoy today, many aggravated assaults would have turned into murders. A highly disproportionate number of aggravated assaults take place exactly where other kinds of crime take place— in socially disorganized neighborhoods—and much of it involves drug activity and disputes (Martínez, Rosenfeld, & Mares, 2008). Many other aggravated assaults are over trivial matters of "face saving" that

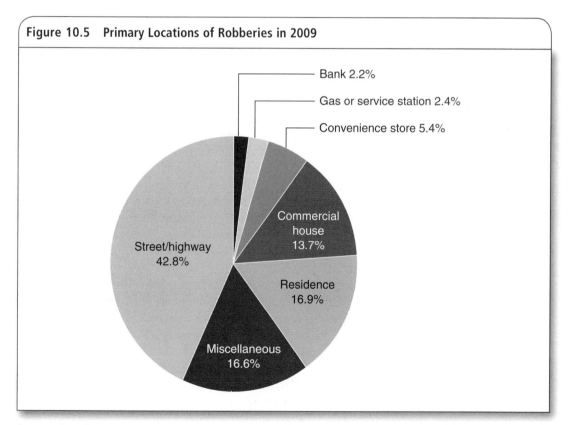

Figure 10.5   Primary Locations of Robberies in 2009

Bank 2.2%
Gas or service station 2.4%
Convenience store 5.4%
Commercial house 13.7%
Residence 16.9%
Street/highway 42.8%
Miscellaneous 16.6%

Source: FBI (2010a).

occur when someone feels that someone else has "dissed" him. As with homicides, a fair number of aggravated assaults are "victim precipitated," meaning that the victim initiated the incident that resulted in his own victimization.

⬚ Theories of Violence

Multiple murder is extreme violence requiring its own theories. Here we discuss theories of "ordinary, everyday" kinds of violence that occur much more frequently. Let us be clear that peaceful cooperation defines our species more so than violent conflict; Charles Darwin wrote about cooperation 3 times more often than about competition in *The Descent of Man* (Levine, 2006). The great majority of humans are not attracted to violence, want to avoid it, and generally strongly condemn it, but that does not mean it is not a part of our evolutionary baggage. Most humans in modern Western societies probably go from cradle to grave without ever committing a serious act of violence (Collins, 2010), but we owe this state of affairs to the fact that they inhabit societies in which law enforcement and the judiciary are largely respected and trusted; violence emerges on a large scale when these things are absent.

Researchers in different disciplines ask different questions about violence, but they all add something of value to the overall picture. Sociologists might ask what it is about the social structure of a society or the norms of a subculture within it that leads to high rates of violence. Psychologist might ask what personality features, situations, or developmental experiences increase the risk of violence. Geneticists might inquire about the mix of genetic and environmental factors associated with violence in a particular population at a particular time. Neuroscientists will focus on questions about the brain structures and neurotransmitters associated with violence. Finally, evolutionary theorists will want to know why humans have a propensity for violence in the first place and what adaptive purposes it served and serves.

According to Cao, Adams, and Jensen (1997), the subculture of violence thesis formulated by criminologists Wolfgang and Feracutti in 1967 "remains *the* definitive argument for society's role in creating violent criminal behavior" (p. 367). A **subculture of violence** is one in which the norms, values, and attitudes of its members legitimize the use of violence to resolve conflicts. This thesis reminds us that violence is not evenly distributed among all groups and in all locations in society, and it is necessary to find out why it is more prevalent in some areas and among some groups than in and among others. Wolfgang and Feracutti (1967) reasoned that "by identifying the groups with the highest rates of homicide, we should find in the most intense degree a subculture of violence" (p. 153). They found such a culture in Philadelphia's black community in the mid-1950s where the homicide rate for young black males was 27 times higher than for young white males, and the female rate was 23 times greater black than for white females, and they based their theory on studying that community (p. 152).

Subcultural norms in such areas dictate that one is expected to settle matters "like a man" and to take care of one's own beefs (i.e., don't involve the power structure). "Taking care of business" often involves violence in a subculture where it is not viewed as illicit. It is part of the street code that "emerges where the influence of the police ends and where personal responsibility for one's safety is felt to begin" (E. Anderson, 1999, p. 33). The successful application of aggression—as a manifestation of subcultural values *and* as a disavowal of mainstream cultural values—is a source of pride (E. Anderson, 1999).

Subcultures of violence have always existed and have also been referred to as honor subcultures. **Honor subcultures** are defined as "communities in which young men are hypersensitive to insult, rushing to defend their reputation in dominance contests" (Mazur & Booth, 1998, p. 362). Cultural norms that allowed duels over trivial matters of "honor" were common among the most polished and cultivated "gentlemen" of

Europe and the United States until about the middle of the 19th century. Such duels over "matters of honor" enhanced the duelists' reputations and provided them with public validation of their self-worth; i.e., they were "doing gender" (Baumeister, Smart, & Boden, 1996). Only with the establishment of modern law was dueling as a way to settle disagreements brought into disrepute. Far from acting in pathological ways, males in our modern honor subcultures are acting in historically and evolutionarily normative ways. This does not make such behavior morally acceptable, but it makes it understandable.

Neuroscience and evolutionary research supports the sociological research at different levels of analysis. Neuroscientists are interested in the effects on the brain of living in such cultures. As we saw in Chapter 8, the brain *physically* captures our experiences by molding and shaping neuronal circuitry in ways that make our behavior adaptive in the environments in which we find ourselves. The neurological literature is consistent with the evolutionary literature in suggesting that impulsiveness is the proximate behavioral expression of a brain wired by consistent exposure to violence (Niehoff, 2003).

If our brains develop in violent environments, we expect hostility from others and behave accordingly. By doing so, we invite the very hostility we are on guard against, thus confirming our belief that the world is a dangerous and hostile place and setting in motion a vicious circle of negative expectations and confirmations.

When evolutionary biologists explore the behavioral repertoire of any species, their first question is, "What is the adaptive significance of this particular behavior?" With regard to violence, they want to know how it was adaptive in evolutionary environments, what its function is, and what environmental circumstances are likely to evoke it. Evolutionary biologists assume that violence evolved to solve some set of adaptive problems; if it didn't solve some such problems, it wouldn't be part of our behavioral repertoire. However, just because it is considered "natural" and just because evolutionary accounts show how it can be useful, it certainly does not mean that it is good. There are many other natural things we would like to avoid, such as disease, death, and earthquakes.

But in what ways can violence be considered adaptive? Violence (at least credible threats of violence) is intimately related to male reproductive success in almost all animal species through its role in attaining status and dominance, and thus access to more resources and to more females; remember, this is what evolution by natural selection is all about. Reacting violently when some brute tries to steal your bananas, your cave, or your wife could be very useful in evolutionary environments when you couldn't just call 9-1-1 to have the police settle your problem. Having a reputation for violence would be even better because others would be aware of it and avoid your bananas, your cave, and your wife in the first place. In other words, in environments in which one is expected to take care of one's own beefs, violence or the threat of violence works to let any potential challenger know that it would be in his best interests to avoid you and your resources and look elsewhere. All this is why a "badass" reputation is so valued in honor subcultures, why those with such a reputation are always looking for opportunities to validate it, and why it is craved to such an extent that "[m]any inner city young men . . . will risk their lives to attain it" (Anderson, 1994, p. 89).

The evolutionary point of view shares the sociological and neuroscience point of view that a major long-term factor in violence instigation is how much violence a person has been exposed to in the past. As Gaulin and Burney (2001) explain, when many acts of violence are observed, "there is a feedback effect; each violent act observed makes observers feel more at risk and therefore more likely to resort to preemptive violence themselves" (p. 83). Inner-city children witness a lot of violence. In one study in Chicago, 33% of school children had witnessed a homicide and 66% had witnessed a serious assault (Osofsky, 1995). Another study found that 32% of Washington, D.C., children and 51% New Orleans children had been victims of violence, and 72% of Washington, D.C., and 91% of New Orleans children had witnessed violence (Osofsky, 1995). Witnessing and experiencing so much violence cannot help but stamp on the brain circuitry of these children that the world is a dangerous place in which one must be prepared to protect one's interests.

Violence and Inequality

It is frequently noted that impulsivity and discounting the future are maladaptive; that is, they are poor ways of adjusting to the conditions one finds oneself in. Margo Wilson and Martin Daly (1997) suggest, however, that discounting the future

> may be a "rational" response to information that indicates an uncertain or low probability of sur-
> viving to reap delayed benefits, for example, and "reckless" risk taking can be optimal when the
> expected profits from safer courses of action are negligible. (p. 1271)

In other words, when the young perceive little opportunity for legitimate success and when many people they know die at an early age, living for the present and engaging in risky violence to obtain resources make excellent evolutionary sense. People are naturally designed to compete for the status and resources necessary for survival and reproductive success by whatever means are available to them in the cultural environments in which they live.

Wilson and Daly (1997) tested their assumption with homicide, income inequality, and life expectancy data from the 77 neighborhoods in Chicago for the years 1988 through 1995. They hypothesized that neighborhoods with the lowest income levels and the shortest life expectancies (excluding homicides) would have the highest homicide rates. Life expectancy (effects of homicide mortality statistically removed) ranged from 54.3 years in the poorest neighborhood to 77.4 years in the wealthiest, and the attending homicide rates ranged from 1.3 in the wealthiest to 156 per 100,000 in the poorest, a huge 120-fold difference. Wilson and Daley appeal to evolutionary logic to interpret these data, viewing it as reflecting escalations of risky competitive tactics that make sense from an evolutionary point of view given the conditions in which people in disadvantaged and disorganized neighborhoods live. As Bob Dylan sang, "When you ain't got nothin', you got nothin' to lose."

Remember not to confuse an *explanation* of the facts with a *moral evaluation* of them. Wilson and Daly are saying that natural selection has equipped us to respond to high levels of inequality and expectations of a short life by creating risky, high-stakes male/male competitions that all too frequently result in violence, including homicide. From a moral point of view, this is obviously something to be condemned, but to the extent that such contingent responses are the products of natural selection, they are not pathological from the point of view of evolutionary biology.

SUMMARY

- Murder rates have been significantly higher in the past than they are today, primarily because of the lack of effective law enforcement and adequate medical attention. Homicide trends in the United States have fluctuated wildly over the years, and the country is situated somewhere in the middle in terms of its homicide rate. In the United States, the typical perpetrator and victim of homicide is a young black male living in an urban center. Female/female homicide is very rare worldwide.

- Spree, mass, and serial murder have increased dramatically since the 1960s, especially serial murders. Serial murder is the murder of three or more victims over an extended period of time.
- The FBI addresses a number of myths that are prevalent in the media about serial killers, i.e., that they are limited to white males who are either insane or geniuses and who travel the country looking for victims and really want to get caught. African Americans are overrepresented in the ranks of serial killers relative to their numbers in the population, and females are even more underrepresented than they are among other kinds of criminals. Asian Americans are also underrepresented among serial killers.
- A popular typology of serial killers contains visionary, mission-oriented, hedonistic, and power/control types. Visionary killers are usually psychotic, and mission-oriented killers feel that it is their duty to rid the world of people they consider undesirable. Hedonistic killers (the most common) kill for the pure joy of it, while power/control killers get more satisfaction from exerting complete control over their victims.
- There have been attempts to explain serial killing using traditional criminological theories. The diathesis-stress model integrates biological, psychological, and sociological variables, and it posits that serial killers have a biological disposition to kill that is exacerbated by severe environmental stress during childhood.
- Poor, young, unmarried, nonwhite females are disproportionately likely to be victimized by rape, and poor, young, unmarried, nonwhite males are disproportionately likely to be perpetrators. Feminist theories maintain that all men have a propensity to rape and that the act is about power, not sex. Social learning and feminist theory assert that rape is the result of male socialization, while evolutionary theorists maintain that it is a maladaptive consequence of male reproductive strategy.
- Robbery is a violent crime, and robbers tend to be the most impulsive, hedonistic, daring, and dangerous of all street criminals, as well as the least educated and least conscientious. Robbery is also considered an excellent way to prove a certain kind of "manliness" in certain urban areas.
- Aggravated assault is the most frequently committed of the violent Part I crimes. Each such incident carries the threat of ending up as a criminal homicide, and but for speedy access to medical treatment, many of them would have done so.
- Different disciplines study violence from different perspectives. The subculture of violence thesis is the most popular sociological theory for the influence of values and attitudes on violence. Evolutionary theories augment this model by showing why and how violence is part of the human behavioral repertoire, and neuroscience shows how constant exposure to violence is physically captured in the brain.

DISCUSSION QUESTIONS

1. Why is female/female homicide so rare, and why is this the case around the world?

2. Explain why homicide rates have dropped dramatically across the centuries.

3. What do you think of the idea that rape is only about violence and not sex?

4. Look up a famous female serial killer and discuss differences in motives and methods for committing her crimes compared to male serial killers.

5. Giannangelo's model of serial killers maintains that without a congenital disposition for it, a person would not become a serial killer. What traits do you think would be "necessary" to possess to become a serial killer when combined with environmental stressors?

USEFUL WEBSITES

Bureau of Justice Statistics. www.ojp.usdoj.gov/bjs/

Federal Bureau of Investigation. www.fbi.gov/

Rape, Abuse, and Incest National Network. www.rainn.org/

Subculture of Violence Theory. www.criminology.fsu.edu/crimtheory/wolfgang.htm

Women Organized Against Rape. www.woar.org/

CHAPTER TERMS

Aggravated assault

Aggravated murder or first-degree murder

Felony murder

Forcible rape

Hedonistic serial killer

Involuntary manslaughter

Mass murder

Mission-oriented serial killer

Murder

Negligent manslaughter

Power/control serial killer

Serial murder

Spree murder

Subculture of violence

Robbery

Violent crime

Visionary serial killer

Voluntary manslaughter

11

Terrorism

On the morning of September 11, 2001, Americans woke to horrifying images that seared into their memories forever. Nineteen Islamic terrorists led by Mohamed Atta, a shy 33-year-old son of a wealthy Egyptian lawyer, had hijacked four airliners and used them in coordinated attacks against symbols of America's financial and military might. At 8:45 a.m., American Airlines Flight 11 with 92 people on board crashed into the north tower of the World Trade Center. Eighteen minutes later, United Airlines Flight 175 with 64 people aboard smashed into the south tower. At 9:40, American Airlines Flight 77 carrying 64 people crashed into the Pentagon. Then, at 10:00 a.m., United Airlines Flight 93 carrying 45 people crashed into a Pennsylvania field, having been prevented from accomplishing its mission (apparently to destroy the Capitol Building or the White House) by the courageous steps taken by its passengers. These terrorist acts cost the lives of close to 3,000 people from 78 different countries, making it the deadliest terrorist attack in history anywhere. The financial cost of the attacks is estimated to be close to $285 billion (Jalata, 2011). What were these people trying to accomplish by such wanton acts, and what drove them to sacrifice their own lives in the process? I hope to begin to answer these questions in this chapter.

On September 3, 2004, about 30 terrorists took over 1,000 adults and children hostage in Beslan, Russia. They held them huddled together in a gymnasium for at least 2 days in sweltering heat, telling them that they (the terrorists) had come to die in the name of Allah, and that they were going to take the hostages with them. They did not feed the hostages or allow them to use the bathrooms, and they shot and killed many of them trying to make their escape. Some children died from dehydration or from unattended wounds, while others survived by drinking their own urine and eating flowers they had brought to school for their teachers (www.rediff.com). Ultimately, 340 hostages were killed and 700 others were wounded. Although much less costly in lives and property than the 9/11 attacks, the Beslan incident provides us with an even clearer, albeit grim picture of the inhumane lengths to which terrorists will go to achieve their goals because they purposely targeted children.

▲ **Photo 11.1** Hijacked United Airlines Flight 175 crashed into the south tower of the World Tade Center and exploded at 9:03 a.m., 17 minutes after hijacked American Airlines Flight 11 crashed into the north tower.

What Is Terrorism?

What kind of hatred, hostility, fanaticism, or cause could motivate individuals to board a plane or enter a school and look into the faces of innocent men, women, and children, knowing that they were to be the instruments of their deaths? And in the case of the Beslan incident, what would possess them to mock and torture frightened little children? We may well ask, with talk show host David Letterman, speaking to a shocked nation after 9/11, "If you live to be a thousand years old . . . will that [the 9/11 attacks] make any goddamn sense?" (quoted in Feeney, 2002, p. 191). These terrorist actions lead us to suppose the idea that terrorists are subhuman creatures and that terrorism is something peculiar to the modern age. Although the average terrorist will never win any humanitarian awards, and terrorism is far more prevalent and deadly today than ever before, those suppositions are incorrect.

Terrorism has a long history; it is "as old as the human discovery that people can be influenced by intimidation" (Hacker, 1977, p. ix). The term *terrorism* itself is believed to have originated with the French Revolutionary Jacobins, who instituted France's domestic Reign of Terror, killing more than 400,000 people in the process (Simonsen & Spindlove, 2004). The earliest known terrorist group was a Jewish nationalist/religious group called the *Sicarii*. They operated around 70 A.D. using deadly savage methods against the occupying Romans and their Jewish collaborators (Vetter & Perlstein, 1991). Another early group, the *Ismailis* or *Assassins,* responding to what they considered religious oppression, carried out a reign of intimidation throughout the Islamic world from about the 11th to the mid-13th centuries (Wheeler, 1991).

The FBI defines **terrorism** as "the unlawful use of force or violence against persons or property to intimidate or coerce a government, the civilian population, or any segment thereof, in furtherance of political or social goals" (quoted in B. Smith, 1994, p. 8). Terrorism is a tactic used to influence the behavior of others through intimidation, although terrorists typically appeal to a higher moral "good," such as ethnic autonomy or some religious or political dogma, to justify the killing of innocents. They strike at innocents because the essence of terrorism is public intimidation, and the randomness of terrorist action accomplishes this better than targeting specific individuals. Victims are incidental to the aims of terrorists; they are simply instruments in the objectives of (1) publicizing the terrorists' cause, (2) instilling in the general public a sense of personal vulnerability, and (3) provoking a government into unleashing repressive social control measures that may cost it public support (Simonsen & Spindlove, 2004).

Al-Qaeda's Osama bin Laden made it clear that the latter is one of the goals of his organization. He stated that in response to terrorist attacks, the United States government will have to restrict many civil liberties its citizens enjoy: "freedom and human rights in America are doomed," he is quoted as saying, adding that the United States will lead people of the Western world "into an unbearable hell and a choking life" (quoted in Kurtz, 2002, p. 5).

Former Russian President Vladimir Putin's decision to hand over sweeping new powers to the Kremlin in the wake of the Beslan attack is evidence that bin Laden may have been right. Putin's decision was criticized by another former Russian president, Boris Yeltsin, as "The strangling of freedoms, the rollback of democratic rights—this can only mean that the terrorists have won" (Walsh & Ellis, 2007, p. 345).

Thus, while terrorist violence is immoral, it is not "senseless" because it has an ultimate purpose, and evil means are justified by the ends they seek. The terrorist attacks on trains in Madrid, Spain, on March 11, 2004 (exactly 911 days after the 9/11 attacks), which took the lives of at least 200 people, led to the fall of a conservative government that supported the U.S. military action in Iraq, and the election of a socialist government 3 days later. The new government immediately pulled Spanish troops out of Iraq, which was evidently the purpose of the bombings. Every time terrorists gain an objective they have sought, the rationality of terrorism is demonstrated along with its immorality.

Is There a Difference Between Terrorists and Freedom Fighters?

Although many people accept the cliché that one person's terrorist is another's freedom fighter, this attitude has been called "sophomoric moral relativism" (Sederberg, 1989, p. 28). Of course, not everyone agrees that we can draw a sharp line between terrorists and freedom fighters, and the label one chooses to affix to a group has as much to do with one's politics as anything else. After all, Drummond (2002) points out that four individuals defined as terrorists have actually been awarded Nobel Peace Prizes (Sean McBride, Menachem Begin, Yasser Arafat, and Nelson Mandela), and Arafat, arguably the father of modern terrorism, continued to support terrorism until the day he died.

All terrorists probably claim to be freedom fighters, or at least that their cause is righteous, but there are two important distinctions between terrorists and freedom fighters (or guerrillas) that do not imply their moral equivalence. First, freedom fighters are fighters in wars of national liberation against foreign occupiers or against oppressive domestic regimes they seek to overthrow. Terrorists are typically fighting to gain some sort of ethnic autonomy, to right some perceived inequity, or to rid the world of some perceived evil, and rarely have illusions of overthrowing the government they are fighting against. While guerrillas may occasionally use terrorist tactics against noncombatants, widespread use of such tactics will deprive them of the popular support they need, and thus they tend to confine their activities to fighting enemy combatants (Garrison, 2004).

Because of the political contexts in which they operate, guerrillas may have no choice other than armed insurgence to accomplish change because they are outside the system that oppresses them. Terrorists, on the other hand, often have access to the system, but spurn the ballot box in favor of the bullet and bomb. Of course, not all claims of injustice can be righted at the polling stations, and thus the distinctions we have made here may be overdrawn to some extent. They are real enough, however, for us to conclude that the moral conflation of *terrorist* and *freedom fighter* is probably not warranted (Garrison, 2004).

⊠ The Extent of Terrorism

Although terrorism has ancient roots, it is far more prevalent today. Of the 74 terrorist groups listed by the U.S. Department of State (2004) in 2003, only 3—the *Irish Republican Army (IRA), Euskadi Ta Askatasuna (Basque Homeland and Freedom, or ETA),* and the *Egyptian Muslim Brotherhood*—operated before 1960.

Of the 74 groups, 39 are Islamic, 18 are Marxist/Maoist, and the remaining 17 are hybrids of Marxist/Islamic groups or nationalist groups. Both the IRA and ETA have supposedly relinquished terrorism, although they have done so before and then taken up arms again.

Terrorist groups and incidents rose dramatically after the 1960s, probably because the 1960s was the high point of conflict between the United States and the Soviet Union, with each having its zones of influence and with each supporting armed opposition to the other. The United States, for instance, supported the Taliban in Afghanistan both morally and financially when that organization was fighting the Soviet Union's occupation, and President Reagan even called them freedom fighters. This support has turned around to bite the United States, since we are now engaged in a bitter struggle against the Taliban.

Terrorism is also much easier to accomplish today than ever before. The Internet, e-mail, and cell phones give terrorists ready access to information and to each other, and economic globalization has led to open borders across which terrorists can easily flow (Jalata, 2011). Modern transportation systems allow terrorists to slip in and out of areas of operation with speed and efficiency, and the same systems provide terrorists with lots of potential victims by bringing large numbers of people together in places such as airports and railway stations. Modern technology also makes the terrorist's life less complicated in that it provides easily concealed and relatively cheap weapons and explosives of great destructive power. The downing of Pan American Flight 103 over Lockerbie, Scotland, in 1988 that killed 281 people was accomplished with a small amount of Semtex explosives hidden in a cassette player. Even the enemy's technology can be turned against them. Before the advent of airplanes, it would have taken a well-equipped army to topple structures like the World Trade Center; certainly 19 terrorists armed only with box cutters could not have accomplished it.

Figure 11.1 provides data on the number of terrorist attacks and resulting fatalities worldwide in 2008 from the National Counterterrorism Center (2009). Compared to 2007 figures, the total attacks were down from 14,545 and fatalities were down from 22,508. These are tragic numbers, but to put them in perspective, we have seen that there were 15,241 murders in the United States alone in 2009 (FBI, 2010a).

Terrorism and Common Crime

Like any organization, terrorist organizations must be financed. Some funding for terrorist groups comes from governments sympathetic to their cause, or hostile to the governments against which the terrorists operate. The United States government has designated Cuba, Iran, Iraq (before the 2003 U.S.-led invasion), Libya, Sudan, Syria, and North Korea as terrorist-sponsoring nations, with Iran the most active of them (U.S. Department of State, 2004). Libya renounced terrorism in 2004, and it remains to be seen what we can say of Iraq once U.S. troops leave. Some terrorist funding comes from private sympathizers, but most of it comes from common criminal activities such as drug trafficking, extortion, and bank robbery. As William Reid (2002) sees it, terrorists cloak "themselves in a 'crusade' that is more accurately viewed as criminal behavior. Even groups that preach against capitalism spend much of their energy raising money and using money from capitalist endeavors" (p. 4). The IRA has raised vast sums of money by extorting "protection" money from the very people for whom they claim to be fighting. They have made so much money from this and other criminal activities that they have had to branch out into legitimate businesses and have launched money-laundering schemes (Dishman, 2001). Evidence suggests that many IRA groups now exist with the primary purpose of developing wealth for their members. Dishman (2001) calls the IRA "a prime example of a *mutated* terrorist group who invested significant energies into committing profit-driven criminal acts" (p. 49).

Figure 11.1 Number of Terrorist Attacks and Fatalities Worldwide in 2008

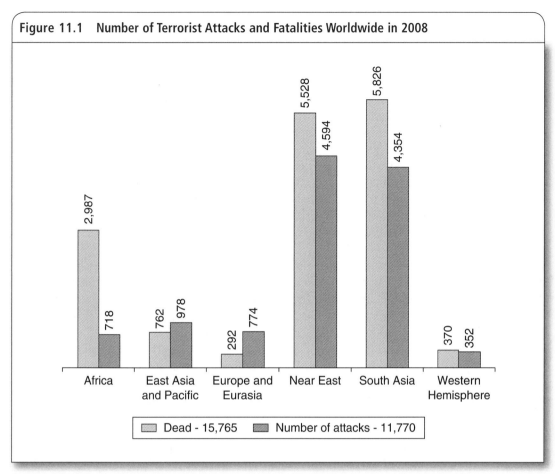

SOURCE: National Counterterrorism Center (2009).

Many Islamic groups also obtain funding from nongovernmental organizations such as charity groups; from legitimate cover businesses; and from criminal activities, particularly drug trafficking (U.S. Department of State, 2004). Terrorist groups in South America such as the Marxist/Maoist *Shining Path* of Peru make enormous profits from drug trafficking, and European groups such as Germany's Red Army Faction and Italy's Red Brigades (both Marxist oriented and now supposedly defunct) financed their activities through bank robberies and kidnapping. The widespread involvement of terrorist groups in such practices casts serious doubt on the ideological idealism they claim motivates their activities. The large amounts of money involved can corrupt the most dedicated ideologue in time, especially if fellow terrorists are lining their pockets. As Albanese and Pursley (1993) put it, "Gradually, the [criminal] activities become ends in themselves and terrorist groups begin to resemble ordinary criminal organizations hidden behind a thin political veneer" (p. 100). This is not true of groups motivated by Islamic fundamentalism, because many of the groups' members, especially its leaders, such as Osama bin Laden, often give up more than they gain in material terms.

◼ Al-Qaeda

Very few Americans knew anything about al-Qaeda or Osama bin Laden prior to the 9/11 attacks orchestrated by that group; now they are household names on par with Nazis and Adolph Hitler in a previous generation. Al-Qaeda is not a single terrorist group but rather the base (*al-Qaeda* literally means "the base") organization for a number of Sunni Muslim terrorist groups from around the world. Al-Qaeda also has cells operating in 100 countries, including the United States (Berger & Hoffman, 2010).

Al-Qaeda got its start under Osama bin Laden (see Figure 11.2) in the late 1980s and expanded dramatically in the 1990s. Bin Laden, who had fought the Russians during their invasion of Afghanistan throughout the 1980s, objected to the stationing of non-Muslim troops in Saudi Arabia, the country containing the two holiest sites in Islam—Mecca and Medina—after the first Gulf War in 1991. Because of his objections, bin Laden was exiled from Saudi Arabia and went to live in Sudan, an Islamic dictatorship. Bin Laden first built his worldwide terrorist network in Sudan, financing much of it through his vast personal fortune, as well as through the drug trade and criminal activities in a variety of countries (Simonsen & Spindlove, 2004).

After being ousted from Sudan, Osama and his henchmen moved to Afghanistan, where they found refuge and protection with the Taliban regime in power there, thanks to some extent to American support. Al-Qaeda set up terrorist training camps in Afghanistan from which terrorists were dispatched to wreak havoc around the world. Following the 9/11 attacks, President George W. Bush demanded that the Taliban turn over bin Laden to U.S. authorities for trial. When the Taliban refused the demands, American and British forces, aided by Afghan groups (mostly Shi'ite Muslims), drove the Taliban from power and scattered al-Qaeda. The group now seems to be operating primarily from Pakistan and Yemen (Berger & Hoffman, 2010). Members of al-Qaeda are virulently anti-West in general and anti-American in particular. In 1998, bin Laden issued a *fatwa* (an Islamic decree or command) and called for a *jihad* (holy war) in which he called on Muslims everywhere to kill Americans wherever they are found, whether military or civilian, man, woman, or child, and those who support Americans. In addition to the 9/11 attacks, al-Qaeda was responsible for the 1998 bomb attacks on U.S. embassies in Nairobi, Kenya, and in Dares-Salaam, Tanzania, killing or wounding more than 600 people, many of them Muslims (Primakov, 2004).

On Sunday, May 2, 2011, almost 10 years after the deadly attacks of 9/11, the 54-year-old bin Laden met justice in the form of a team of U.S. Navy SEALs who shot and killed him. This ended a frustrating manhunt spanning a decade—probably the most intense manhunt in history. Bin Laden had been hiding "in plain sight" in a luxury compound in the Pakistani city of Abbottabad, a city in which there is a large Pakistani military base. We can only speculate what influence this event will have on the immediate future of terrorism, but it was certainly good news for Americans.

Figure 11.2   The Man Who Was the World's Most Wanted Criminal

Osama bin Laden
Up to $27 Million Reward

Date of Birth: 1957
Place of Birth: Saudi Arabia
Height: 6'4"-6'6" (193-198 cm)
Weight: 160 lbs (73 kg)
Hair: Brown
Eyes: Brown
Complexion: Olive
Nationality: Saudi Arabian (citizenship revoked)
Scars/Distinguishing Characteristics: Full beard, moustache; walks with cane
Status: Fugitive

S<small>OURCE</small>: FBI (2010a).

◼ Hizballah

Hizballah ("Party of God") is the best contemporary example of a state-sponsored terrorist organization. Hizballah (sometimes spelled *Hezbollah*) ultimately owes its existence to the religious split between *Sunni*

Muslims, who believe in the legitimacy of the secular state, and *Shi'ite* Muslims, who do not. The group was organized by the Shi'ite religious leader Ayatollah Khomeini to fight the secular rule of the Shah of Iran. It emerged on the international stage after the Israeli invasion of Lebanon in 1982, which drove the Palestine Liberation Organization (PLO) out of that country. Ironically, the PLO had been the chief opponent of the Lebanese Shi'ites prior to the invasion. Hizballah fighters were sent to Lebanon by Khomeini, ostensibly to aid in the fight against Israel, but with the long-range goal of establishing an Iranian-style Islamic regime in Lebanon. Hizballah has claimed responsibility for a number of spectacular terrorist operations that helped hasten the withdrawal of American and Israeli forces from Lebanon (something no Arab army has ever accomplished). Among these actions were the bombing of the U.S. marine barracks in 1983, killing 251 American and 56 French soldiers, and the kidnapping and/or murder of several American and European citizens.

Hizballah has a sense of engaging in a sacred mission that transcends the confines of Lebanon. Much as Christian crusaders several centuries before them saw the Muslim presence in the Holy Land as an affront to Christianity, the more radical among modern Shi'ites view the existence of a Jewish state in an area they also consider holy to be an affront to Islam (Kramer, 1990). They are fiercely anti-Israeli and anti-American, viewing the United States as the decadent, drug infested, crime-ridden, sex-perverted "Great Satan" of the world. Directed and financed by Iran, Hizballah is headquartered in Lebanon and has established cells in Europe, North and South America, and Africa (U.S. Department of State, 2004).

⊠ Domestic Terrorism

Groups such as al-Qaeda are known as *transnational* terrorists because they operate in and against many countries across the globe. Groups such as the IRA and ETA are known as *domestic* terrorists because they confine operations to domestic targets (the IRA in the United Kingdom and British bases overseas, and ETA against Spanish interests). The United States also has its homegrown terrorists. Figure 11.1 clearly shows that most terrorist activity takes place far away from the United States, but the bombing of the World Trade Center in 1993 and of the Oklahoma City Federal Building in April 1995, and the horrific events of 9/11, show that the United States is not immune from it. Although the World Trade Center attacks in 1993 and 2001 were the work of foreign Islamic terrorists, the Oklahoma City bombing that took 168 lives was carried out by "All American" army veteran Timothy McVeigh and his accomplices Michael Fortier and Terry Nichols, who belonged to a Michigan militia group called the Patriots. This bombing was apparently motivated by revenge against the federal government for several operations it had undertaken against radical antigovernment groups, such as the action against the Branch Davidian religious sect in Waco, Texas, that left 78 of its members dead, as well as a number of government agents.

Perhaps the most disturbing trend in terrorism on the domestic front is "the increasing 'Americanization' of the leadership of al-Qaeda and aligned groups, and the larger number of Americans attaching themselves to these groups" (Berger & Hoffman, 2010, p. 14). Berger and Hoffman identify 63 American citizens or legal resident aliens who have been arrested and indicted, either in America or overseas, for Islamic terrorist acts from 2009 to September of 2010. When Berger and Hoffman refer to the Americanization of al-Qaeda's leadership, they are referring to people like Anwar al-Awlaki, Adnan el Shukrijumah, and Omar Hammami. The cleric al-Awlaki was raised in New Mexico and is now an important operational organizer for al-Qaeda in Yemen; Shukrijumah, who grew up in New York and Florida, is al-Qaeda's director of external operations, and Hammami, a convert from the Baptist faith from Alabama, is a key propagandist and military commander for al-Qaeda. The three are the most notorious homegrown jihadists; listed below is a sampling of

▲ **Photo 11.2** The Oklahoma City Bomber Timothy McVeigh is escorted by police and FBI agents from the Perry, OK, courthouse after his capture by an Oklahoma State Highway Patrolman. He was convicted of 11 federal offenses and sentenced to death. His execution took place on June 11, 2001.

conspiracies and attacks that took place in 2009–2010 for which American citizens or foreign nationals legally residing in the United States were responsible (Berger & Hoffman, 2010).

May 2009: James Cromitie, David Williams, Onta Williams, and Laguerre Payen were arrested for plotting to blow up Jewish centers in the Bronx, New York, and to shoot down planes at an Air National Guard Base. One was Afghan-born; the other three were African Americans who converted to Islam in prison.

June 2009: Abdulhakim Muhammed, a Muslim convert from Tennessee, killed one soldier and wounded another outside a military recruiting office in Little Rock, Arkansas.

September 2009: Afghan Najibullah Zazi was arrested after purchasing large quantities of chemicals used to make a bomb intended for detonation in the New York City subway.

September 2009: Jordanian Hosam Maher Husein Smadi was arrested in an attempt to plant a bomb in a Dallas skyscraper.

September 2009: Michael Finton was arrested after attempting to detonate a car bomb outside the Federal Building in downtown Springfield, Illinois.

November 2009: U.S. Army Major Nidal Malik Hasan opened fire on fellow soldiers in Fort Hood, Texas, killing 13 people and wounding 29 others.

December 2009: Nigerian Umar Farouk Abdulmutallab attempted to ignite an explosive device hidden in his underwear while aboard an airliner over Detroit.

May 2010: Faisal Shahzad, a Pakistani naturalized U.S. citizen, planted a car bomb in Times Square in New York. The bomb failed to detonate properly.

Many of these attacks were unsuccessful thanks to good undercover law enforcement, the amateurish efforts of poorly trained operatives whom their terrorist handlers considered expendable, or sheer good luck. The potential loss of life had some of these operations proved successful for the terrorists is staggering. As for the element of luck, we should remind ourselves that terrorists only have to get lucky once; their potential victims have to be lucky all the time.

Other Domestic Groups

We should not equate terrorism only with Islamic groups. There are, and have been in the past, a number of other groups with their own causes. Of course, they do not pose the same kinds of dire threat to the United States that groups like al-Qaeda do, but they still warrant brief discussion.

Ideological: Left-Wing. Left-wing terrorism in the United States became active during the turmoil of the 1960s. The most prominent group was the *Weather Underground* (WU). Solidly middle class, mostly white, and fiercely Marxist, the WU focused its attacks on symbols of "capitalist oppression" such as banks, corporate headquarters, and military facilities. It was thought to be defunct after the arrest of many of its leaders in the 1970s, but it renewed its robbery and bombing campaign in the 1980s. An even more radical group, the *May 19th Communist Organization* (M19CO), carried out bombing operations of U.S. military facilities and developed ties with foreign terrorist groups. The M19CO name was derived from the birthdays of North Vietnamese insurgent and later president, Ho Chi Minh, and American Black Muslim leader Malcolm X.

Another left-wing group was the *Revolutionary Armed Task Force* (RATF), forged from the alliance of the May 19th coalition and the *Black Liberation Army.* Although this group earned its terrorist credentials by bombing a number of capitalist symbols, including the FBI's headquarters in New York, much of its activity seems to be concentrated on conventional crimes such as robbery and drug trafficking. The RATF recruited from minority prisoners and parolees, especially those who see themselves as victims of a capitalist and racist America (Albanese & Pursley, 1993). The activities of these left-wing groups began to wane in the 1980s, and according to most experts, their organizations are now defunct (Council on Foreign Relations, 2004). Groups like al-Qaeda attract the kinds of people formerly attracted to the RATF.

Ideological: Right-Wing. Most right-wing American groups espouse radical libertarianism ("get the government off my back") and tend to be anti-Semites and white supremacists. An example of a right-wing group is the *Aryan Nations,* founded in the mid-1970s and headquartered in Idaho until 2001. The group espouses white supremacy, anti-Semitism, tax resistance, and radical libertarianism. The group suffered a serious blow in 2000 when it lost a $6.3 million lawsuit, which cost them the real estate they owned in Idaho as well as automobiles and other property owned by the group (Law Enforcement Agency Resource Network, 2004).

The Ku Klux Klan (KKK) is one of the oldest terrorist groups in the world, although today it is a generic name for a number of autonomous groups that range from those who never go beyond rhetoric and cross burning to those who actively practice terrorism against black churches and Jewish synagogues. At its peak, the KKK boasted a membership of 4 million, and some of its members engaged in murders, bombings, beatings, and cross burnings to intimidate blacks and white civil rights workers. The KKK shares with most other American right-wing extremist/terrorist organizations an extreme Christian fundamentalism, the advocacy of paramilitary survivalist training, and a conspiratorial view of politics. They refer to the United States government as ZOG (Zionist Occupational Government), which they say is run by Jews, liberals, and African Americans (Vetter & Perlstein, 1991). As was the case with the Aryan Nations, as a result of a lawsuit against the KKK in 2000, the organization had to "hand over all its assets and skulk off into virtual oblivion" (Walsh & Hemmens, 2011, p. 345). The poison spread by such groups is apparently still influencing some to commit acts of violence, however. On March 8, 2011, a former member of the white supremacist group the National Alliance, and wannabe member of the Aryan Nations named Kevin Harpham, was arrested and accused of leaving a bomb along a Martin Luther King Jr. Day parade route in Spokane, Washington. The bomb was apparently quite sophisticated and could have killed and wounded many people (Machetta, 2011).

▲ **Photo 11.3** The burning cross of the KKK symbolizes the violent tactics used against African Americans and others.

There are a number of groups in the United States that employ terrorist tactics that have no grand sociopolitical agenda but rather seek to resolve special issues. These groups include environmentalists seeking to protect the environment, animal rights groups seeking to protect animals, and anti-abortion groups seeking to protect the rights of the unborn. The overwhelming majority of those who align themselves with such causes are, of course, nonviolent and seek their aims through political means. However, as with any group affiliated with almost any cause, there are extremists on the fringes that turn to illegal methods to get their point across.

The *Animal Liberation Front (ALF)* and the *Earth Liberation Front (ELF)* have emerged in the past several years as major domestic terror threats, with ELF being declared by the FBI as America's number one domestic terrorist group (Consumer Freedom News, 2005), which seems like a gross exaggeration given the so-called Americanization of al-Qaeda noted above. This group has engaged in numerous acts of tree spiking, arson, sabotage of construction equipment, and other forms of vandalism, which the Law Enforcement Agency Resource Network (2004) reports have caused more than $100,000 in damage. According to the ELF website (http://earth-liberation-front .org/), it sees itself as "working to speed up the collapse of industry, to scare the rich, and to undermine the foundations of the state."

The ALF has close ties with the ELF because of the closeness of their respective agendas. Like the ELF, the ALF subscribes to the principle of "leaderless resistance," organizing itself into small autonomous cells with no centralized chain of command. This minimizes the possibility of infiltration by law enforcement (Leader & Probst, 2006). According to James Jarboe (2002), the FBI estimated that ALF/ELF had committed more than 600 criminal acts between 1996 and 2001. Despite the attention given to these two groups by the FBI, there have mercifully been no deaths attributed to the activities of ALF/ELF, although Leader and Probst see the groups as ready to turn to more violent tactics in the future.

Is There a Terrorist Personality?

Terrorists, like criminals in general, tend to be young unmarried males, although they *do not* fit the criminal profile in terms of being poorly educated relative to their peers or from single-parent families (LaFree & Ackerman, 2009). Despite self-selection for membership in terrorist groups, no study of terrorist psychology has ever produced a psychological profile, leading the majority of terrorist experts to suspect that there is no such thing as a terrorist personality (Hudson, 1999). On the other hand, the absence of a uniform terrorist personality does not mean that certain traits are not disproportionately present among those who join terrorist groups. Some scholars view terrorists as people with marginal personalities drawn to terrorist groups because their deficiencies are both accepted and welcomed by the group (Johnson & Feldman, 1992). This may apply more to domestic terrorists such as Timothy McVeigh than to Islamic terrorists.

P. Johnson and T. Feldman (1992) view terrorist groups as made up of three types of individuals: (1) the *charismatic leader,* (2) the *antisocial personality,* and (3) the *follower.* The charismatic leader is socially alienated, narcissistic, arrogant, and intelligent, with a deeply idealistic sense of right and wrong. The terrorist group provides a forum for his narcissistic rage and intellectual ramblings, and the subservience of group members feeds his egoism. Antisocial individuals have opportunities in terrorist groups to use force and violence to further their own personal goals, as well as the goals of the group. For the antisocial personality, the group functions like an organized crime family, providing greater opportunity, action, and prestige than could be found outside the group (Perlman, 2002). The majority of terrorists, however, are simple followers who see the world purely in black ("them") and white ("us") and have deep needs for acceptance, which makes them susceptible to all sorts of religious, ideological, and political propaganda (Ardila, 2002).

Terrorist expert Bruce Hoffman (2010) views homegrown terrorists as disaffected self-radicalized individuals who actively seek contact with peers with similar views, and who visit terrorist websites to make contact with radical groups. Yet aside from this, there is little to set them apart from other Americans or immigrants to America. Hoffman concludes about homegrown jihadists, to the Committee on Homeland Security, that they are

1. Good students and well-educated individuals and high school dropouts and jailbirds.

2. Persons born in the U.S. or variously in Afghanistan, Egypt, Pakistan, and Somalia.

3. Teenage boys pumped up with testosterone and middle-aged divorcees. The only common denominator appears to be a newfound hatred for their native or adopted country, a degree of dangerous malleability, and a religious fervor justifying or legitimizing violence that impels these very impressionable and perhaps easily influenced individuals toward potentially lethal acts of violence. (pp. 38–39)

⊠ Theories of Terrorism

We have a tendency to think of terrorism as being caused by religious fanaticism and political radicalism. While terrorism is certainly fed by these things, they are not its causes—fanaticism and radicalism have to have something to turn ideology into action. There are as many causes of terrorism as there are terrorist groups because it cannot be understood without understanding the historical, social, political, and economic conditions behind the emergence of each group. Perhaps the one generality we can make is that all groups originate in response to some perceived injustice. Although certain kinds of people may be drawn to terrorism, terrorists are not a bunch of "sicko-weirdos," running around the world killing for pleasure. If they were, we would have defeated terrorism long ago. Most terrorist groups take pains not to recruit anyone showing signs of mental instability because such people are not trustworthy and would arouse the suspicion of their intended targets (Hudson, 1999).

Some terrorism theorists believe that the bulk of terrorists are crusaders convinced of the moral rightness of their cause (White, 1998). If these theorists are right, we have to explain how "normal" people are persuaded to commit brutal acts against innocent people. When moral people are required to commit immoral acts, there must be some sort of personal transformation that makes it possible. In other words, the willingness to perform terrorist acts may reflect a process of moral disengagement more than an indication of pathological or criminal traits the individual brings to the terrorist group. If the essence of terrorism is "the complete transformation of sane human beings into brutal and indiscriminate killers" (Sprinzak, 1991, p. 58), terrorist acts may generate significant levels of guilt and doubt in the new recruit that must be resolved. Inconsistencies between attitudes and behavior (cognitive dissonance) are usually resolved by changing attitudes rather than behaviors. For terrorists, this typically means a deepening of the belief that their cause is just, the further dehumanizing of their targets ("infidels," "capitalist pigs," and so on), viewing the slaughter of innocents as "collateral damage," and any of a number of other ways that humans have of exorcizing behavior-inhibiting guilt and doubt.

Laurence Miller (2011) presents four cognitive stages by which terrorist groups form and by which individuals evolve into people capable of slaughtering innocents. The first stage is the belief that some set of conditions in their lives is unpleasant and unacceptable, such as poverty or rampant immorality. This is the "It's not right" stage. The second stage is "It's not fair," in which they perceive others as living a better life than the budding terrorists are living. The third stage is the "It's your fault" stage in which the cause of injustice has been identified in other groups. The fourth and final stage is the "You're evil" stage in which the group identified as the cause of the budding terrorist's suffering becomes dehumanized.

Differential association theory is relevant here. Many Islamic terrorists are recruited from religious schools known as **madrasas.** In these schools, all of Miller's four stages are present, and thus recruits are provided with an excess of definitions favorable to terrorism over definitions unfavorable to it. These schools teach secular subjects, but focus mostly on religious texts and stress the immorality ("It's not right") and materialism ("It's not fair") of Western life and the need to convert all infidels to Islam (Armanios, 2003). The madrasas are appealing to poor Muslim families because they offer free room and board, as well as free education. Many members of the Afghan *Taliban* (*Taliban* means "student") regime studied and trained in Pakistani madrasas stressing a strict form of Islam. Children are indoctrinated in these schools with anti-Israeli and anti-American propaganda from the earliest days of their lives (differential association's priority, duration, frequency, and intimacy). We might call socialization in the madrasas differential association and social learning theories on steroids, because the only definitions of reality they receive are that of the school, and they are amply rewarded (reinforced) for their acceptance and participation.

Suicide Bombers

A person nurtured on the hatred spouted in such schools is ideal material for recruitment as a martyr to the cause. Martyrdom brings with it the promise of immediate ascension into heaven, where the individual will find "rivers of milk and wine . . . lakes of honey, and the services of 72 virgins" (C. Hoffman, 2002, p. 305). The promise of a sexual paradise is powerful for young males in cultures that allow males to have multiple wives if they can afford them. This practice leaves other males, particularly the young, without access to women, a situation that must lead to great strains and frustrations. Young male suicide bombers take the Koran's promise of sexual paradise seriously. They often take elaborate steps to protect their genitals from damage so that they can enjoy the pleasures of paradise, and as one of them is supposed to have said, "most boys can't stop thinking about the virgins" (Victoroff & Kruglanski, 2009, p. 127). However, it is not only religious fervor and sexual frustration that motivate suicide bombers. Many see themselves as altruists, saints, and heroes striving to achieve noble political goals on behalf of their people. They are also afforded tremendous status in their communities, and their families are often handsomely financially rewarded for their offspring's sacrifice (L. Miller, 2011).

This supports rational choice theorists who say that we should look at what terrorist groups have to offer if we want to understand why individuals join them. In addition to the rewards offered to Islamic terrorists described above, for terrorist "wannabes," "terrorism can provide a route for advancement, an opportunity for glamour and excitement, a chance of world renown, a way of demonstrating one's courage, and even a way of accumulating wealth" (Reich, 1990, p. 271). Terrorism is much like organized crime in that it provides illegitimate ways to get what most of us would like to have—fame and fortune, and in the case of Islamic suicide martyrs, an eternity of sexual bliss. Terrorists also have a bonus in that they, and their comrades and supporters, see themselves as romanticized warriors fighting for a just and noble cause, and in the case of religious terrorists, the favor of their God and the promise of a rewarding afterlife.

The need to belong appears to be particularly important for Islamic immigrants to Europe and the United States, or the offspring of these immigrants. Such people often feel alienated from the host country, which they may see as decadent and godless. They feel their differences in religion, language, dress, and culture very acutely, and their humiliation and alienation can turn into hatred (LaFree & Ackerman, 2009). These are exactly the kinds of people that can be turned into terrorists by the rhetoric spouted by radical *imams* (preachers) in their mosques. This is aptly demonstrated by the recruitment of American-born sons (and sometimes daughters) of Islamic immigrants in certain mosques, as we have seen (B. Hoffman, 2010).

Law Enforcement Response and Government Policy

There are any number of ways a democracy can respond to terrorism, ranging from making concessions to military intervention. Concessions are likely only when there is moral substance to the terrorist cause, and when such concessions are reasonable. But the West cannot make any concessions to al-Qaeda and other such groups because they are not demanding any. What these groups have said over and over is that they want nothing less than the Islamification of the world, starting with the purification of existing Islamic regimes (as was done in Afghanistan by the Taliban) that do not match the terrorists' ideas of what an Islamic state should be.

Military intervention may be used when the terrorist threat is too big for civilian authorities to handle. But besides being distasteful to the democratic spirit, military intervention, even though successful in the short term, may be detrimental in the long term. The Islamic terrorist threat comes from many nations fed by a constant stream of religious hatred poisoning the minds of young Muslim men. It is estimated that more than 50,000 terrorists were trained in Afghanistan who are now scattered around the globe, quietly integrated into local communities and awaiting their orders to strike (Simonsen & Spindlove, 2004). Former United States Defense Secretary Donald Rumsfeld has expressed doubt that the West can win the broader global fight against Islamic terrorism, and wonders if the various groups "are turning out newly trained terrorists faster than the United States can capture or kill them" (quoted in Burns, 2004, p. 5). Clearly, we cannot defeat the threat by military force alone.

International law has been applied against terrorists, sometimes successfully, but often not. The principle of international law known as *aut dedere aut punire* (Latin for *either extradite or punish*) obligates countries to either extradite terrorists to the country where their crimes were committed or punish them themselves. Some countries neither extradite nor punish for one reason or another. (They may support the terrorist's cause, or they may fear reprisals.)

The United States has a clear-cut policy to combat terrorism:

(1) Do not make deals with terrorists or submit to blackmail. We have found over the years that this policy works.

(2) Treat terrorists as criminals and apply the rule of law.

(3) Bring maximum pressure on states that sponsor and support terrorists by imposing economic, diplomatic, and political sanctions and urging other states to do likewise (U.S. Department of State, 1995, p. iv).

Following the September 11th attacks, President George W. Bush issued an Executive Order establishing the Department of Homeland Security, the mission of which is to detect, prevent, prepare for, and recover from terrorist attacks within the United States. *Detection* involves coordinated efforts on the part of federal, state, and local agencies to collect information in an attempt to identify terrorist activities within the United States. *Prevention* relates to the investigation of identified threats; the denial of entry of suspected terrorists and terrorist materials and supplies into the United States; and the arrest, detention, and deportation of individuals suspected of membership in foreign terrorist groups. *Preparedness* refers to nationwide efforts to prepare for and lessen the impact of any terrorist attack. *Recovery* refers to efforts to quickly restore critical infrastructure facilities (distribution systems, telecommunications, utilities), the provision of adequate medical facilities, and the removal of hazardous materials in the event of a successful terrorist attack (The White House, 2001).

The mechanisms set up by the Office of Homeland Security can only work if the men and women on the front lines of security are constantly vigilant. Security was so tight in the first months after 9/11 that this period was probably the safest in which to fly in the history of aviation. But the human tendency is to grow complacent after long periods in which nothing happens, and this is terrorism's great weapon. For instance, a U.S. government document published in 1999 predicted that al-Qaeda would retaliate "in a spectacular way" for the cruise missile attacks against their training facilities in 1998, ordered by President Clinton in retaliation for the 1998 al-Qaeda attacks on U.S. embassies. This document stated the following: "Suicide bomber(s) belonging to al-Qaeda's Martyrdom Battalion could crash an aircraft packed with high explosives (C-4 and semtex) into the Pentagon, the headquarters of the Central Intelligence Agency (CIA) or the White House" (Hudson, 1999, pp. 7–8). Al-Qaeda waited patiently for 3 years before doing almost exactly as predicted, and they did it with relative ease. Thus, all the intelligence in the world is of little use unless those in the day-to-day security trenches take it as seriously every day as they did immediately after September 11, 2001.

At the time of this writing (March 2011), many Islamic states in Africa and the Middle East are in turmoil as regimes from Egypt to Yemen are being toppled or are on the verge of toppling. If all this turmoil eventually results in real democracy in these states, it bodes well for the defeat of terrorism. On the other hand, we have seen previous promises of democracy in that region of the world that turned into Islamic dictatorships—"one man (no women allowed), one vote, one time." In power struggles such as these, the spoils usually go to the best organized, most disciplined, and most ruthless groups, and these adjectives certainly define terrorist organizations more accurately than they describe democratic organizations. If Islamic extremists with their dreams of a universal *caliphate* (a global Islamic state) manage to gain power, there are decades of more terrorism to look forward to.

SUMMARY

- Terrorism is an ancient method of intimidating the public by the indiscriminate use of violence for social or political reasons. Terrorism increased rather dramatically from the 1960s to the mid-1980s, steadily dropped off into the 1990s, and has again increased in the 21st century.
- Terrorists are different from freedom fighters or guerrillas in that terrorists usually operate against democracies, and freedom fighters act against foreign colonialists or oppressive domestic regimes. Many terrorist groups tend to evolve into organized crime groups hidden behind an ideological veneer.
- Al-Qaeda is a "base" organization for a number of Islamic terrorist groups and is the group of most concern to Americans. Hizballah, a radical pro-Iranian Islamic fundamentalist group, appears to be the most active and deadly terrorist group presently operating that is not associated with al-Qaeda.
- Although American terrorists are decidedly amateur in comparison with their foreign counterparts, there are a fair number of terrorist groups operating in the United States. The most disturbing trend is the Americanization of al-Qaeda, with a number of Islamic immigrants and Islamic converts taking up arms against the United States. Other domestic terrorist groups are divided into ideological (left- and right-wing) and special issue groups.
- There is no uniform terrorist personality, although there are certain traits found more frequently among terrorists than among the general population. The most common traits found are low self-esteem and a predilection for risk taking. Since most terrorists are not mentally ill, a process of moral disengagement is posited to explain their transition from sane human beings to killers.
- There are as many causes of terrorism as there are terrorist groups. Each group has its origins in some perceived injustice, but only a minuscule number of people react to such conditions by joining terrorist organizations.
- Democracies have considerable difficulty responding to the terrorist threat because of legal restraints. The U.S. government has a clear-cut policy of treating terrorists like common criminals, not making deals with them, and imposing sanctions against nations that sponsor terrorism. The threat to democracies from terrorism is great, and the United States responded after 9/11 with the establishment of the Department of Homeland Security.

DISCUSSION QUESTIONS

1. Do you agree or disagree that there is a moral difference between terrorists and guerillas?

2. Discuss the ways in which terrorism is rational behavior.

3. Do you think that extremist Muslims hate the United States because it has troops on soil they consider holy (Saudi Arabia)? If so, why don't we take them out of there? Or is it our fear of losing Saudi oil that keeps them there?

4. Can you conceive of any circumstances under which you would commit the kinds of terrorist acts perpetrated on 9/11 or at the Beslan School, sacrificing your own life in the bargain?

5. Go to www.cfrterrorism.org and click on "groups" on the right-hand side of the page. You will find a number of terrorist groups listed there with a variety of motives. Choose an Islamic and a Marxist group and write a short paper comparing and contrasting them on their motives and methods.

USEFUL WEBSITES

Bureau of Justice Statistics. www.ojp.usdoj.gov/bjs/. Data on U.S.-based terrorism.

Council on Foreign Relations. www.cfr.org/issue/terrorism/ri135. Very useful site for up-to-date information on international terrorism.

Intelligence Resource Program. www.fas.org/irp/threat/terror.htm. Good source for checking on intelligence related to terrorism.

U.S. Department of State. www.state.gov/s/ct/rls/crt/. Numerous terrorist-related publications.

CHAPTER TERMS

Al-Qaeda	Hizballah	Terrorism
Department of Homeland Security	Madrasas	

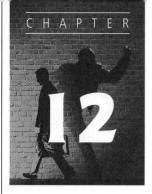

12

Property Crime

Jay Scott Ballinger was a property offender on a dark mission. He admitted in court to setting fire to between 30 and 50 churches in 11 states between 1994 and 1999. A volunteer fire-fighter was killed in one of the blazes, which also made Ballinger a murderer. He did not set these fires for profit or because he got some weird sexual kick from watching them burn; he did so on an anti-Christian mission. Ballinger, his girlfriend Angela Wood, and accomplice David Puckett traveled around the country seeking converts to satanism. To finance their travels around the Midwest and South, the trio burglarized and shoplifted, and Wood worked as a stripper.

It all came to an end when Ballinger was arrested after paramedics treating him for severe burns wondered why he had waited 2 days to seek treatment. (He had burns over 40% of his body and had to receive four skin grafts.) A police officer who remembered Ballinger's name from a previous investigation questioned him and then summoned ATF (Alcohol, Tobacco, and Firearms) agents who found fire-setting paraphernalia and satanic literature at Ballinger's home. Among the writings agents found were 50 "contracts" signed by teenagers in their own blood, pledging their souls to the devil and to do "all types of evil" for which they would be rewarded with wealth, power, and sex, the perennial male motivators. Ballinger, age 36 at the time of his arrest, was described as a misfit loner and high school dropout who was more comfortable with teens than with people his own age. He was sentenced to 42 years in prison for his multistate arson spree (Ross, 1999). As the Ballinger case shows, crimes against property can sometimes morph into something much more deadly.

⬉ What Is a Property Offense?

Although violent crime gets the lion's share of media and police attention, 87.6% (almost 9 out of 10) of the 10,339,369 offenses reported to the police in 2009 were property crimes (FBI, 2010a). Property crimes either involve the illegal acquisition of someone else's property (money or goods) or the malicious destruction of property (sometimes even including one's own if the intention is fraudulent). Just about everyone will be victimized by a property offense at some time in life, but few of us will be victimized by a serious violent crime. It is also true that while the vast majority of us will never commit a violent crime, most of us have committed, or will commit, a property offense of some kind such as pilfering items from work or shoplifting. Thus the phrase "property crime" involves everything from teens vandalizing a mailbox to a sophisticated gang stealing property worth millions from museums and mansions.

Figure 12.1 shows property crime trends in the United States over the history of the National Crime Victimization Surveys (Bureau of Justice Statistics, 2010) showing how they have been dropping dramatically over the past 35 years. Recall that the NCVS only includes offenses against individuals and households; it does not include the huge number of offenses committed against commercial establishments (stores, warehouses, factories, and so forth), thus leaving out a very large number of property offenses. It is nevertheless heartening to see such dramatic decreases in household property crimes.

⬉ Larceny-Theft

Larceny-theft is the most common property crime committed in the United States and is defined as "the unlawful taking, leading, or riding away from the possession or constructive possession of another" (FBI, 2010a). The number of larceny-thefts reported in the United States in 2009 was an estimated 6,327,230 for a rate

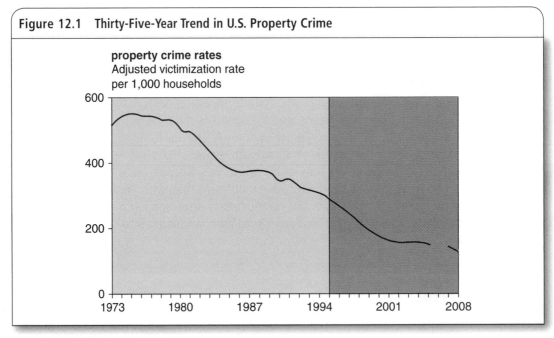

Figure 12.1 Thirty-Five-Year Trend in U.S. Property Crime

property crime rates
Adjusted victimization rate
per 1,000 households

SOURCE: Bureau of Justice Statistics (2010).

of 2,060.9 per 100,000 residents. This crime constituted 67.9% of property crimes that year. The average loss due to larceny-theft was $864, for a total national loss of approximately $5.5 billion. Of those arrested for larceny-theft in 2009, a total of 61.7% were males. Whites made up 68.1% of larceny-theft arrests, blacks 29.3%, and other races 2.7% (FBI, 2010a).

Larceny only applied to persons who achieved possession of goods belonging to others by stealth or force in early English common law; it did not cover persons who abused their victims' trust to steal from them. For instance, if person A gave person B a sheep to graze in B's field, but B killed and ate the sheep, no larceny was committed because A had voluntarily handed over the sheep to B. You can see the problems this would cause in today's society where every day numerous people hand over their money to bank tellers and their vehicles and appliances to mechanics. Lawmakers responded to this by enlarging the definition of larceny to include taking by fraud or false pretenses as well as by stealth and force. Taking by stealth has evolved into other crimes such as burglary or embezzlement, and taking by force has evolved into the crime of robbery.

Today, larceny-theft covers most types of theft that do not include the use of threats or force; excluded are theft of a motor vehicle, forgery, passing bad checks, and embezzlement. Larceny-theft includes grand theft (a felony) and petty theft (a misdemeanor), with the distinction depending on the value of the asset stolen. The cutoff value varies from state to state, but presently it is under $1,000 in every state. Whether a person is charged with grand or petty theft depends on the value of the item at the time it was stolen, not its replacement value. A stolen computer worth $100 today is a petty theft no matter how much it cost initially, but a stolen guitar bought for $100 in 1955 that is now a classic valued at $10,000 is classified as a grand theft. Because the grand theft/petty theft distinction varies across states, the UCR treats grand theft and petty theft as the same thing in its yearly larceny-theft tally. Figure 12.2 shows from where or how most incidents of larceny-theft occurred in 2009 (FBI, 2010a).

Larceny-theft is subclassified into shoplifting, pocket picking, purse snatching, thefts from motor vehicles (except for parts and accessories), theft of motor parts and accessories, theft of bicycles, and theft from buildings. Theft from motor vehicles is the most common type, and includes thefts from just about any type of motorized vehicle such as automobiles, trucks, buses, or motor homes. Purse snatching and pocket picking can easily be charged as robberies, however.

Shoplifting—theft by a person other than an employee of goods exposed for sale in a store—is the most studied of the subcategories of larceny-theft. Compulsive shoplifting is sometimes defined as a psychiatric problem (kleptomania, Greek for "stealing madness"). For instance, actress Winona Ryder was arrested in 2001 for shoplifting about $5,000 worth of goods from a Saks Fifth Avenue store in Beverly Hills (Mowbray, 2002). Ryder's "five-finger discount" is one of millions occurring each year, which cost the retail industry approximately $31 billion in 2004 ("Survey: Shoplifting Losses Mount," 2005). This loss is passed on to customers in the form of increased prices.

About 8% of shoplifters say they engage in the practice as a primary source of income (R. Moore, 1984). These are the individuals with the greatest level of expertise who know how to minimize the risk of being caught, target the most expensive items, and steal almost exclusively for resale. Most shoplifters, however, are impulsive amateurs who shoplift relatively inexpensive items on the spur of the moment and for their own gratification (Lamontagne, Boyer, Hetu, & Lacerte-Lamontagne, 2000). One self-report study is consistent with rational choice theory in that shoplifters said that they engaged in shoplifting simply because it is an easy, low-risk crime for which there are abundant opportunities (Tonglet, 2001). Some criminologists maintain that the rewards received from such crimes as shoplifting often go far beyond any material gain. They argue that a big part of the reward for engaging in crime comes from the sheer thrill of getting away with something in the face of the possibility that one could get caught, and point out that there is little

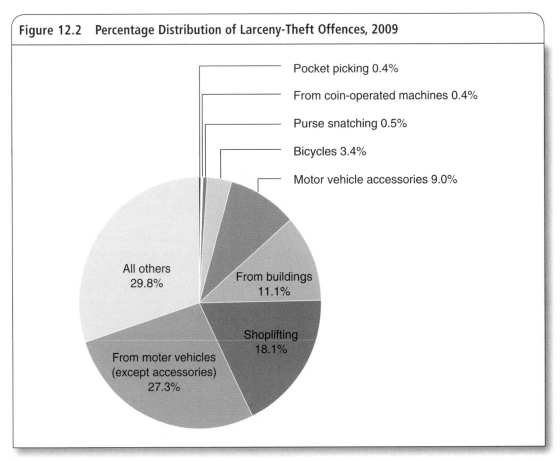

Figure 12.2 Percentage Distribution of Larceny-Theft Offences, 2009

Pocket picking 0.4%

From coin-operated machines 0.4%

Purse snatching 0.5%

Bicycles 3.4%

Motor vehicle accessories 9.0%

All others 29.8%

From buildings 11.1%

Shoplifting 18.1%

From moter vehicles (except accessories) 27.3%

Source: http://www2.fbi.gov/ucr/cius2009/offenses/property_crime/larceny-theft.html

economic gain from most crimes. The "seduction" of crime is crime itself, the conquering of fear, and the euphoric thrill of completion that is the real payoff (Young, 2003).

Burglary

Burglary is "the unlawful entry of a structure to commit a felony or theft" (FBI, 2010a). Burglary has always been considered a very serious offense under common law dating back many centuries because of the importance attached to the sanctity of the home. Victims of residential burglary experience feelings of anger, fear, and a profound sense of invasion of privacy and vulnerability in addition to financial loss. The original common-law definition involved breaking and entering at night with the intention of committing a felony, but now includes any unlawful entry whether forceful or not and regardless of the time of day. This has enabled a number of states to charge shoplifting (even of a $2 lipstick) as a burglary if the shoplifter admits that he or she entered the store intending to steal some item.

The number of burglaries reported to the police in 2009 was 2,199,125 for a rate of 729.5 per 100,000. Of these, 72.6% were residential with an average loss of $2,096. Of those arrested for burglary in 2009, a total of 94.4% were male, 66.5% were white, 31.7% were black, and the remaining 1.8% were of other races (FBI, 2010a).

Burglars and Their Motives

The "typical" burglar is a young male firmly embedded in street culture. Burglars are perhaps a little less daring than robbers since there is less chance of victim contact, injury, and identification for burglary than for robbery, and the penalties and probability of arrest are lower (the national clearance rate for burglary in 2009 was 12.5 versus 28.2 for robbery [FBI, 2010a]). Almost all of the 105 active burglars interviewed by Wright and Decker (1994) admitted to numerous other offenses they had committed. They evidenced pride in their ability to exploit the range of criminal opportunities that come their way, but most of them considered burglary their preferred crime because it offers the greatest chance of success and reward with the least amount of risk.

Demographically, Wright and Decker's (1994) burglars came from poor, run-down, and socially disorganized neighborhoods where unemployment was rife. They were poorly educated, unreliable, resistant to taking orders, and most came from single-parent homes. Consistent with Walter Miller's focal concerns concept addressed in Chapter 4, there was a strong sense of toughness and masculine independence, fate ("I had little choice but to burgle"), excitement (sexual activity, drugs, alcohol), autonomy ("As a burglar I'm my own man"), and smartness (outwitting the law; getting something for nothing).

▲ **Photo 12.1**   O. J. Simpson is taken into custody after being found guilty during his trial at the Clark County Regional Justice Center on October 3, 2008. Simpson and co-defendant Clarence "C. J." Stewart were found guilty on all charges after standing trial for crimes including felony kidnapping, armed robbery, and conspiracy related to a 2007 confrontation with sports memorabilia dealers in a Las Vegas hotel. The verdict comes 13 years to the day after Simpson was acquitted of murdering his ex-wife Nicole Brown Simpson and Ronald Goldman.

Wright and Decker (1994) state that burglars constantly need money quickly to finance their lifestyle, but many of them also reported that they found the psychic rewards of committing burglaries to be a secondary benefit, describing the act as "an adventure," "a challenge," "fun," "exciting," and "thrilling" (p. 58), which once again points to the psychological payoff for committing crimes. Given the lack of legitimate skills and general trustworthiness they see among burglars, Wright and Decker are dubious about the possibility that job creation programs could change burglars into law-abiding citizens because most burglars see burglary as being far more profitable than working. Rengert and Wasilchick (2000) also reject the notion that burglary is a default option of the jobless, claiming that many burglars give up jobs to concentrate on burglary: "Unemployment is not what caused crime. Crime caused the unemployment" (p. 47). Legitimate employment simply would not fit into these people's lives because for them, party time is all the time. Almost all of the proceeds of Wright and Decker's burglars were spent on drugs, alcohol, and sex, and legitimate jobs provide neither adequate time nor money to engage in these pleasures to the extent desired.

In 2010, an estimated 15.7% of persons arrested for burglary were females. Females overwhelmingly commit burglaries in mixed-gender teams, and thus obviously share most of the demographic characteristics of their male partners (Mullins & Wright, 2003). Mullins and Wright found that most of the women were initiated into burglary by their boyfriends and that some were coerced ("If you love me, you'll do it"). Mullins and Wright also found that female burglars capitalized on their sexuality to locate potential targets and gain access to homes that they and their partners would burglarize later. Once inside a target's home, the woman could assess potential valuables and entry points, and perhaps even discover where the victim kept spare keys. They could also elicit other important information such as the target's schedule so that the woman and her partner could be sure to enter the house when the target is not at home. As is the case with their male counterparts, most of female burglars' loot is spent on drugs, although they also tend to spend a lot on clothing.

Choosing Burglary Targets

Selecting a suitable home for a burglary is an obvious concern for burglars. Working from a routine activities framework, Mawby (2001) lists the four most important considerations in target selection as target exposure, guardianship, target attractiveness, and proximity.

- *Target exposure* refers to the visibility and accessibility of the home, i.e., isolation from other homes and easy access via side and back doors shielded by abundant trees and shrubs. Can the premises be seen by neighbors and passersby? Thus, some homes are at a much greater risk for burglary than others simply because of location and the physical properties of the premises.
- *Guardianship* refers to how well the home is protected. Does the home show signs of occupancy such as cars in the driveway, lights on, music playing; is there a burglar alarm or dog present; or is there mail in the mailbox, newspapers in the foyer, and a general silence about the place? Households in urban socially disorganized neighborhoods are not only vulnerable because they have numerous motivated offenders looking for opportunities to score, but also because the neighborhood has low levels of collective efficacy (see Chapter 4).
- *Target attractiveness* refers to signs that there should be rich pickings in the house. Previous surveillance may have revealed high-priced cars in the driveway or delivery trucks delivering expensive items.
- *Proximity* refers to the distance between the target home and the burglar's home and/or to potential sources of loot disposal, such as pawn shops.

All these concerns are relative to the "professionalism" of the burglar. High-level burglars may travel miles to a particularly attractive target after very careful surveillance and planning, but the majority of burglars are low- to mid-level opportunists who engage only in minimal, even spur-of-the-moment, planning. For these individuals, the "planning" of a burglary is little more than opportunism. Proximity is important both because burglars are most familiar with their own areas and because many of them lack transportation. As one of Wright and Decker's (1994) subjects put it, "I ain't gonna go no further than 10 blocks; that's a ways to be carryin' stuff. . . . Since I'm on foot, I got to keep walkin' back and forth until I get it all" (p. 86). As is the case with murderers, robbers, and rapists, the great majority of low- and mid-level burglars prey on residents in the same neighborhoods in which they also reside. Target exposure and attractiveness are simply making the best of a bad deal for such burglars since the pickings are pretty slim in their neighborhoods.

Guardianship is the most important consideration for low- and mid-level burglars, with many choosing homes occupied by individuals known to them such as neighbors, acquaintances, and even friends

(Mawby, 2001; Wright & Decker, 1994). Wright and Decker report the statement of one of their respondents: "I be knowin' what house I'm going to hit. It could be a friend of mine, I could be over at his house all last week, I know he got a new VCR, we be lookin' at movies. I know what time they work. I know where his wife at or he stay by himself" (p. 70). Typically the only planning such individuals do is to call and see if their intended victims are home. Some of those who victimize friends and acquaintances do report occasional pangs of guilt, but justify their actions as the result of a desperate need to get money for another drug fix. Given their willingness to criminally exploit almost anyone, including so-called friends, we can easily see why Wright and Decker characterized their sample of burglars (who obviously lack the social emotions of shame, embarrassment, empathy, and guilt) as "self-centered individuals without notably strong bonds to other human beings; their allegiance seemed forever to be shifting to suit their own needs" (p. 72).

Disposing of the Loot

The most immediate pressure facing burglars after a successful burglary is to convert the stolen goods into cash. Burglars turn to a variety of sources to dispose of the loot, including the use of fences. A **fence** is a person who regularly buys stolen property and who often has a legitimate business to cover his or her activities. Only a minority of burglars (the high-level burglar) use a fence because fences prefer to deal only with people they trust. Fences are valued by burglars who use them because it is the fastest way of getting rid of "hot" property and they can be trusted to be discreet. Fencing is a UCR Part II crime, which is formally known as receiving stolen property. Anyone knowingly buying or possessing stolen property can be charged with this crime.

Burglars without connections to a professional fence must turn to other outlets. One method is a pawn-shop, but this is not a very popular outlet for most burglars because pawnbrokers must ask for identification, take pictures of people selling to them, and possess "hot sheets" of stolen goods. Some burglars who have developed a trusting relationship with certain pawnbrokers are able to sell "off camera," but because pawn-brokers always have the upper hand in negotiations and offer very little for the goods, only 13 of Wright and Decker's (1994) sample said that they regularly used them. A more popular outlet was the drug dealer because it can entail a strict "drugs-for-merchandise" deal without involving any middleman. Others regularly sold to relatives, friends, and acquaintances because few people can resist buying merchandise at below even "fire sale" prices. Because of the high value of the property they go after, high-level burglars would never use any of these alternatives to the professional fence.

⊠ Motor Vehicle Theft

Motor vehicle (MV) theft is simply "the theft or attempted theft of a motor vehicle" (FBI, 2010a), i.e., any motorized land vehicle such as a motorcycle, bus, automobile, truck, or snowmobile, although the vast majority of thefts involve automobiles. There were an estimated 794,616 MV thefts in 2009 for a rate of 258.8 per 100,000, down 31.5% from 2000. The nationwide loss attributable to MV theft in 2009 was approximately $5.2 billion. Of those arrested for MV theft, 82.1% were males, 61.1% were white, 36.3% were black, and other races made up the remainder (FBI, 2010a). Table 12.1 presents the top 10 most stolen vehicles in 2009 and the cities with the highest theft rates. The older vehicles are stolen primarily for their parts.

Most MV thefts are committed by juveniles strictly for fun. Juveniles will spot a "cool" car with the keys in the ignition, steal it, drive it around until it runs out of gas, and then abandon it. Some of the more malicious joyriders will get an additional kick by smashing it up a little before abandoning it (Rice & Smith, 2002).

Table 12.1	**The 10 Most Stolen Vehicles and 10 Cities With Highest MV Theft Rates, 2009**
1. 1994 Honda Accord	1. Laredo, TX
2. 1995 Honda Civic	2. Modesto, CA
3. 1991 Toyota Camry	3. Bakersfield, CA
4. 1997 Ford F-150 Pickup	4. Stockton, CA
5. 2004 Dodge Ram Pickup	5. Fresno, CA
6. 2000 Dodge Caravan	6. Yakima, WA
7. 1994 Chevrolet Pickup (full size)	7. San Francisco/Oakland/Fremont, CA
8. 1994 Acura Integra	8. Visalia/Porterville, CA
9. 2002 Ford Explorer	9. Las Vegas/Paradise, NV
10. 2009 Toyota Corolla	10. Albuquerque, NM

SOURCE: National Insurance Crime Bureau (2010).

The high recovery rate of stolen vehicles (about 62%) indicates that most MV thefts are for expressive (to show off, to get some kicks) rather than for instrumental reasons (financial gain) (Linden & Chaturvedi, 2005). Motor vehicles are also obviously stolen for profit. Most vehicles stolen for profit are taken to so-called chop shops where they are stripped of their parts and accessories. These items are easily sold to auto supply stores, repair shops, and individuals who get faster delivery at a cheaper price than they would from legitimate suppliers. Other stolen vehicles may be shipped abroad where they are worth more than they are in the United States. Some professional auto thieves (called *jockeys*) even steal particularly high-value vehicles "to order" for specific customers.

The common theme that emerges from in-depth interviews with all types of criminals, including those who specialize in stealing automobiles, is their general disdain for legitimate employment and their unending pursuit of a self-indulging lifestyles (alcohol, drugs, sex, being a "sack hound" [sleeping a lot] and one's "own man," and so on). Again, this is a "rational choice," but one that is severely bounded by offenders' knowledge, cognitive and emotional skills and traits, and the context of their developmental experiences in disorganized neighborhoods. The content of a series of interviews with car thieves on probation or parole by Copes (2003) is almost identical in content to our earlier discussion of robbers and burglars. This is one example illustrating the kind of lifestyles to which the typical criminal is drawn:

> The life is mostly party. I don't think people understand that it's quite like that, but it is. In other words, you don't work. . . . When you get your money, you usually get it real fast and you have a lot of time to spend it. You can sleep all day if you want to and you can go out and get drunk, get high—you don't have to get up the next morning to go to work. (pp. 315–316)

The most serious form of MV theft is **carjacking,** defined as the theft or attempted theft of a motor vehicle from its occupant by force or threat of force. Carjacking thus involves multiple crimes for which the offender may be charged, including robbery, assault, and MV theft. Carjacking has increased significantly in recent years (by an average of 34,000 incidents per year from 1992 to 2002 [Klaus, 2004]). Media accounts of carjacking have sparked many copycat offenders: "Hey, I can steal any vehicle I want without damaging it, I get the car keys, and I can rob the owner too; what a concept!" (McGoey, 2005, p. 1).

Victims report that 93% of the carjackers were male, 3% involved a male/female team, and 3% were lone females. By race/ethnicity, 56% were African American, 21% were white, and the remainder were not identified by race. Approximately 32% of the victims of completed carjackings and 17% of attempted carjackings were injured. Two well-known victims of carjacking are singer-songwriter Marc Cohn and rapper Cam'ron, both of whom were shot and wounded in botched carjackings in 2005.

Interviews with active carjackers reveal that they are more like street robbers in terms of their demographics, motivations, and lifestyles than they are like professional car thieves (Jacobs, Topalli, & Wright, 2003). Most carjackings are motivated by the need to bankroll a hedonistic lifestyle and displays of status. Many times, carjacking is a spur-of-the-moment thing, as described by "Tall":

> I was broke. I didn't have enough bus fare. I'm walking down the street, there's a guy sitting in his car. I go ask him for change. He was going for his pocket. I just grabbed him outta his car. Why just take his change when I can take his car and get a little bit more? (Jacobs et al., 2003, p. 678)

Carjacking may also be precipitated by the victim driving around "flossing"—that is, engaging in ostentatious displays of wealth and status that others see as an affront ("dissing") to neighborhood carjackers. One carjacker described his attitudes toward flossing as follows: "This motherfucker [was] . . . flossing . . . showboating and shit. He had all that shit in that motherfucking [car] . . . He flossed his ass off . . . So we was gonna get the motherfucker" (Jacobs et al., 2003, p. 682). The sheer thrill, the rush, the dance with danger for its own sake, are also powerful motivators for carjacking: "It's a rush thing when you're pulling someone out of a car . . . I mean, I feel good" (p. 683). Some carjackers also say that they enjoy the opportunity to brutalize and humiliate their victims: "The way people look at you when they're scared and panicky and stuff . . . [I]t is funny just to see them shaking and pissing all over theyself." As another of Jacobs et al.'s respondents put it, "You get a kick out of seeing them screaming and hollering . . . especially when they all [acting as though they were] tough" (p. 684).

⬛ Arson

Arson is "any willful or malicious burning or attempting to burn, with or without intent to defraud, a dwelling house, public building, motor vehicle or aircraft, personal property of another, etc." (FBI, 2010a). Arson was added to the UCR in 1979, and there still exists a great deal of difficulty in gathering statistics from reporting agencies because of the problem of deciding whether a "suspicious" fire was arson. Only fires that have been determined by investigators to have been willfully and maliciously set are classified as arsons. For these reasons, the UCR does not provide an estimated national rate for arson, but the U.S. Fire Administration (2010) reported a total of 26,500 cases of "intentionally set structure fires" in 2009 that cost the lives of 170 people and incurred a direct dollar loss of $684 million. Almost 47% of all persons arrested for arson in 2009 were juveniles. Whites accounted for 74.8% of arson arrests and blacks for 22.7%. Males accounted for 85.5% of arson arrests.

Arson can have a variety of instrumental motivations such as financial gain, revenge, and intimidation, or expressive motivations ("thrills") that may signal psychopathology of some sort. For instance, an owner of a failing business may hire a professional arsonist (a *torch*) to burn down his or her place of business; a person may set fire to the property of another because of some perceived wrong suffered; or labor unionists may set fires in a labor dispute to intimidate management, as was the case with the massive Dupont Hotel fire in Puerto Rico in 1986, which led to the death of 97 people and injured 140 others.

Because juveniles who have reached the age of responsibility comprise only about 7% of the American population but are consistently arrested in about half of arsons cases, expressive motivations are of great interest. Juvenile fire-setting may be the result of curiosity and may never be repeated if the juvenile is caught and dealt with, but persistent fire-setters are another matter. It is worth noting that fire-setting is one of three childhood behaviors that psychiatrists have long used to predict adult violence (Vaughn et al., 2010). This is the so-called MacDonald triad of fire-setting, cruelty to animals, and enuresis (continued bed-wetting at ages significantly beyond the normal age of cessation, which is an indication of sluggish autonomic nervous system activation).

A variety of studies have shown that compared with youths in general, persistent fire-setters have higher levels of other antisocial behaviors such as hostility, they are more impulsive, and they have lower levels of sociability and assertiveness. They have also been shown to suffer more psychiatric symptoms, have higher levels of depression, and to come from families with low levels of affectionate expression and child monitoring (Brett, 2004; Hakkanen et al., 2004; Santtila, Hakkanen, Alison, & Whyte, 2003).

The most ambitious of these studies comes from a nationally representative sample of over 43,000 U.S. residents 18 years of age and over (Vaughn et al., 2010). Vaughn and his colleagues found that the 407 respondents who admitted ever setting fires on purpose to destroy someone else's property or just to see it burn differed greatly from the rest of the sample on many variables. For instance, the odds of a fire-setter forcing someone to have sex with him were over 18 times greater than the odds of a non–fire-setter ever doing so, almost 13 times the odds of a non–fire-setter for animal cruelty, and almost 20 times the odds of a non–fire-setter for ever committing a robbery.

⬛ Crimes of Guile and Deceit: Embezzlement, Fraud, and Forgery/Counterfeiting

The property crimes we have discussed thus far are "physical" crimes committed largely by "street" people. The UCR lists three Part II property crimes—embezzlement, fraud, and forgery/counterfeiting—that are committed by a demographically broader range of people than we see committing such crimes as burglary and MV theft. Some criminologists consider these crimes committed by guile and deceit to be white-collar crimes. White-collar workers certainly commit these crimes, but blue- and pink-collar workers as well as the unemployed and welfare recipients commit them also. It is for this reason that most criminologists examine them as property crimes rather than white-collar crimes.

Embezzlement is the misappropriation or misapplication of money or property entrusted to the embezzler's care, custody, or control. Embezzlement is the rarest of property crimes with only 11,756 cases being reported in the UCR for 2009. Females (51.1%) were actually arrested more often than males in that year. Whites constituted 66% and blacks 31.7% of arrests. Most embezzlers do what they do because they have some pressing financial problem, or simply because they have access to money and the ability to hide any discrepancies for some time. After being exposed, many insist that they were only "borrowing the money," and that they fully intended to pay it back.

Banks have long been embezzlement targets, and the advent of computers has made it both easier to embezzle from them and more lucrative. In the first decade of the "computer revolution" in banking, arrests for embezzlement rose 56%, with the average loss to banks per computer embezzlement crime being as high as $500,000 compared with the average loss per armed bank robbery of just over $3,000 (Rosoff, Pontell, & Tillman, 1998). A favorite method of stealing via the computer is known as the salami ("slicing off") technique whereby the embezzler will open up "phantom accounts" in his or her name and slice off a few cents from a large number of accounts whose owners are hardly likely to notice. This technique can garner the embezzler large sums of money over a period of time (Rosoff et al., 1998).

The most successful embezzler in U.S. history was Robert L. Vesco who looted close to $250 million from a variety of mutual funds while he was head of a Swiss-based investment organization. When Vesco learned that he was under investigation for criminal fraud in 1972, he fled to Costa Rica, avoiding extradition by "contributing" $300,000 to the Costa Rican president (Coleman, 1986, p. 83). In 1982, Vesco settled down in Cuba where he set up a criminal empire, becoming a middleman and dealmaker to a variety of dictators and criminal elites in the Central American/Caribbean region. Justice prevailed in the end when the Cubans arrested him in 1995 on suspicion of being a foreign agent and sentenced him to 13 years in prison for "economic crimes against the state" (Associated Press, 1996).

Fraud is theft by trick, i.e., obtaining the money or property of another through deceptive practices such as false advertising and impersonation. The FBI (2010b) reported 161,233 arrests for fraud in 2009, of which 56.1% were males. In terms of race, 67% were white, 31.2% were black, and the remainder were Asian/Pacific Islanders.

Obtaining resources by fraudulent means probably began when the first human being realized that he or she could obtain them with less risk and effort by using brains rather than brawn. Examples of fraud include dishonest telemarketing, quack medical cures, phony faith healers, "cowboy" home repair companies, price gouging, and diploma mills promising "accredited" college degrees for a lot of money and little study.

Perhaps the most successful U.S. fraudster of the 20th century is the penny stock king, Robert E. Brennan. The penny stock business is a legitimate one, but one that is almost designed for fraud. Penny stocks are shares/securities in small start-up companies in need of financing that are not listed on a recognized exchange and are sold "over the counter." Penny stock fraudsters such as Brennan purchase large blocks of virtually worthless stocks at as little as one tenth of a cent per share and aggressively sell them at a higher price. When the stock reaches a predetermined price, they dump their own shares (a practice known as "pump and dump"), leaving the hapless investors with worthless paper and the brokers with millions in ill-gotten gains. At its high point, Brennan had over 500,000 customers and a sales force of 1,200 brokers. The primary targets of Brennan's firm were the elderly, who would be called by salespersons who were instructed, "Never hang up until the customer buys or dies" (Griffin, 2002, p. 254).

Forgery is the creation or alteration of documents to give them the appearance of legality and validity with the intention of gaining some fraudulent benefit from doing so. Strictly speaking, forgery is the "false writing" of a document and *uttering* is the passing of that document to another with knowledge of its falsity with intent to defraud. One can thus commit a forgery without uttering (passing the document on) and can utter (passing a forged document on that he or she did not forge) without committing a forgery.

Counterfeiting, the creation or altering of currency, is a special case of forgery. In most states, forgery and counterfeiting are allied offenses, which is the reason that they appear that way in the UCR. Would-be counterfeiters no longer need the fine craftsmanship of an engraver who produces quality plates for professional counterfeiters. Printing currency with copiers available today has become an amateur's do-it-yourself enterprise (just feed in a $20 bill and press the button). However, such bills are far more difficult to pass today due to new technology embedded in genuine currency and new detection technology in stores. There were 49,992 arrests for forgery/counterfeiting in 2009, of which 63.3% were males, suggesting only a small gender difference in the commission of these crimes. By race, 66.7% were white and 31.7% black (FBI, 2010a).

Cybercrime: Oh What a Tangled World Wide Web We Weave

Cybercrime is the use of computer technology to criminally victimize unwary individuals or groups. Any invention that can be used by criminals to exploit others has been, but few of these have been as useful as the computer. Now even the weak and timid who would never dream of using a gun to rob or otherwise victimize someone can steal, assault, and harass from the comfort of his or her home with little or no risk involved. Everyone who enters cyberspace, uses a credit card, or has a social security number—which means just about everybody—is a potential victim of cybercrime. We have seen that conventional criminals such as robbers and burglars typically operate in their own or nearby neighborhoods, but the global reach of the Internet now allows someone in Birmingham, England, to victimize someone in Birmingham, Alabama, or vice versa. The number of offenses it is possible to classify under cybercrime are legion, ranging from terrorism (the targeting of a country's computer-run infrastructures such as air traffic control and power grid systems) to simple e-mail harassment. We can concentrate only on the most common ones here, beginning with identity theft.

▲ **Photo 12.2** There are many kinds of hackers, some of whom are purely interested in the intellectual challenge of breaking into difficult systems and do so without damaging them. Others (sometimes known as cyberpunks or virtual vandals) break into systems and implant viruses to destroy data.

Identity Theft

Identity theft is the use of someone else's personal information without their permission to fraudulently obtain goods and services. According to a Federal Trade Commission (2010) report, 279,389 complaints of identity theft were filed with the commission in 2009, although it estimates that about 9 million people actually are victims of it in some form, and that the cost to the economy is approximately $50 billion a year. Identity theft can range from a criminal's short-term use of a stolen or lost credit card to the long-term use of a person's complete biographical information (name, social security number, and other identifiers) to "clone" the victim's identity and to commit multiple crimes that may be attributed to the victim.

Criminals gain access to the personal information of others in a variety of ways.

They can steal it, buy it, or simply be given it by their unwary victims. People are continually providing confidential information to all sorts of businesses and agencies that goes into huge data banks that may be legitimately accessed by employees who may steal it, or they can be "hacked" into and information stolen. Credit card numbers can be copied during a financial transaction such as when a restaurant server takes your card for processing, or they can be surreptitiously recorded on a skimming device, which is typically a cigarette pack–size device that is run across a credit card to record the electronic information in the magnetic strip. This information is then used to make duplicate cards. Thieves can also steal new boxes of checks left in mailboxes for pickup, copy the information on them, and buy duplicate checks from mail order firms.

Another method is *phishing,* which as the name implies, involves thieves casting thousands of fraudulent e-mails into the cyberpond asking for personal information and waiting for someone to bite. Phishers may send out official-looking e-mails with a bank logo asking recipients to "update" their information or telling them that their account may have been fraudulently used and that the bank needs to "verify" their personal information. One study indicated that 40% of recipients of one fraudulent bank e-mail believed it to be real (Kshetri, 2006). A victim may also be literally scared into providing his or her information. Imagine receiving an e-mail from "Lolita Productions" telling you that your credit card has been billed $99.95 for the first two child pornography CDs and that it will be automatically billed $49.95 each month for further CDs. The message also says that if you want to cancel membership, you should e-mail back with full credit card details "for verification." Knowing the penalties for possessing child porn, you may be anxious to do anything to free yourself from the electronic embrace of Lolita Publications.

Perhaps the most notorious phishers are the so-called Nigerian frauds run by Nigerian organized crime groups. E-mails have been received by millions of people the world over. A small number of people fall for it. These people are asked to provide help in moving a large amount of cash out of some country in Africa, a good portion of which (often in the millions) they are supposed to receive for their trouble. They are first asked to send a small amount of money (perhaps $200 or less) to "cover expenses," but are suckered into sending ever-larger amounts as "complications" arise. Some of the more gullible have even been lured to Nigeria with their cash and have been killed (Baines, 1996).

Most stolen identity information is not for the personal use of the thief but for sale to others. An organization of about 4,000 individuals called the Shadowcrew stole large volumes of personal information for many years and arrogantly advertised and sold it on websites worldwide. If you wanted to buy card numbers with security codes, you could get 50 of them for $200; if you wanted the same thing complete with the original owner's social security number and date of birth, you would only have to pay $40 each (S. Levy & Stone, 2005). Leading members of the Shadowcrew were arrested by the U.S. Secret Service in 2004, effectively closing the business that authorities estimated had trafficked at least 1.5 million credit and bank cards, account numbers, and other counterfeit documents such as passports and driver's licenses (U.S. Department of Justice, 2004).

Denial of Service Attack: Virtual Kidnapping and Extortion

Denial of service (DoS) attacks occur when criminals "kidnap" a business website or threaten to kidnap it so that business cannot be conducted. DoS attacks are accomplished by overloading the computational resources of the victim's system by flooding it with millions of bogus messages and useless data. Sometimes an attack is simply malicious mischief carried out by computer-savvy disgruntled employees, customers, or just someone who has a bone to pick with the services the company provides. Other times DoS attacks are committed by criminals who demand ransom. Online gambling sites are prime targets for cyber-extortionists because a "kidnapped" website cannot accept bets and stands to lose millions. Paying the ransom is cheaper than losing the business, especially if the threat comes during peak operation times. Millions of dollars have been paid to cyber-extortionists with only a miniscule few ever reported to the police (Kshetri, 2006). Not that the police could do much anyway, as many of these attacks originate overseas.

Who Are the Hackers?

A **hacker** may be simply defined as someone who illicitly accesses someone else's computer system. Hackers may be seen as the upscale version of Albert Cohen's lower-class delinquents we met in Chapter 4 who engaged in malicious, destructive, and non-utilitarian vandalism "just for the heck of it." We do not include in this categorization people who hack into computers for instrumental or political reasons such as professional criminals or cyberterrorists.

There are many kinds of hackers, some of whom are purely interested in the intellectual challenge of breaking into difficult systems and do so without damaging them, and there are others (sometimes known as cyberpunks or virtual vandals) who break into systems and implant viruses to destroy data. Most, however, appear to be intellectual thrill seekers who enjoy the challenge of doing something illegal and getting away with it (Voiskounsky & Smyslova, 2003). There is something of a counterculture among these people analogous to graffiti artists and gang members. They take on cybernames such as Nightcrawler and Kompking and romanticize and tell stories about their accomplishments, as well as the accomplishments of "legendary" hackers. Gaining the respect of fellow hackers serves as a source of psychological reinforcement for them in ways similar to ordinary street delinquents (Kshetri, 2006). Hackers tend to be young white males, loners, "nerdy," with a high IQ, and idealistic, but also unpopular with others, prone to lying and cheating, and perhaps prone to drug or alcohol abuse (Power, 2000; Voiskounsky & Smyslova, 2003).

Software Piracy

Software piracy is illegally copying and distributing of software for free or for sale. The Business Software Alliance (BSA) has estimated that the worldwide cost of software piracy in 2004 was $31 billion, and that although the United States has the lowest piracy rate (ratio of legitimate to pirate market) in the world, it leads the world in losses to piracy—about $7 billion annually (BSA, 2005). In the United States, the illegal market is 21% of the total market, whereas in countries such as China, Vietnam, and Russia, the illegal market is around 90% of the total. The BSA estimates that worldwide for every two dollars' worth of software purchased, one dollar's worth was illegally obtained.

Software piracy is a crime, but few people see it as such unless multiple copies are made and sold for profit. Many view making copies for friends the same way they view loaning books to them—"I bought it. Shouldn't I be able to give it to whomever I please?" Having purchased something legally, they see no reason why the law should mandate that they alone should be allowed rights to it. A large survey of university employees found that this was indeed the attitude of many. Forty-four percent of the respondents said that they had obtained unauthorized copies of software, and 31% said that they had made such copies (Seale, Polakowski, & Schneider, 1998).

SUMMARY

- Property crimes constitute the vast majority of crimes committed in the United States. Larceny-theft is the most common of these crimes and is divided into misdemeanor and felony categories depending on the value of the stolen property. Shoplifting has received the most attention of all the subcategories of larceny-theft, largely because a number of celebrities supposedly suffer from a psychiatric condition known as kleptomania.
- Burglary is a more serious property offense because it violates victims' homes and could lead to personal confrontation. Studies of burglars find them to be motivated by the need to get quick and easy cash to finance a hedonistic lifestyle. They are typically members of the lower class with the focal concerns of that class, such as seeking excitement and autonomy. Female burglars typically work as auxiliaries to male partners, and they too spend much of their money on alcohol and drugs.
- Motor vehicle theft is a serious larceny often committed by juveniles for the purpose of joyriding, although many vehicles are stolen for profit. Carjacking is a relatively new method of stealing cars made "necessary" by the improvement of antitheft devices. There is a high degree of injury to the victim inherent in this crime, which is essentially a robbery.
- Arson is a particularly dangerous property crime because it can lead to deaths. Juveniles commit the majority of arsons, suggesting that it serves some expressive function for them. Compulsive fire-setting in childhood signals some very serious underlying psychological problems and is a strong predictor of later criminality.
- Crimes of guile and deceit included in the UCR property crime classification are embezzlement, fraud, and counterfeiting/forgery. There are only small gender differences in the commission of these crimes.
- Cybercrime is the use of computer technology to criminally victimize unwary individuals or groups. There are many different forms of cybercrime, ranging from the use of the Internet to steal identity information to the destruction of whole computer networks. Other offenses include denial of services "kidnapping" of firms and the very common crime of software piracy.

DISCUSSION QUESTIONS

1. Survey several friends or classmates and ask them if they have ever shoplifted and if so, how they felt afterwards. Did some feel guilty? Did others feel proud about getting away with it? What were their motivations? Did some get a thrill out of it?

2. According to the burglary researchers discussed in this chapter, would it be wise policy to implement job training and job creation programs for convicted burglars? Why or why not?

3. Go to www.theideabank.com/programs/index.php and view the various youth fire-setting intervention programs listed there. Click on two or three of the research reports listed there and report about what you learn about juvenile fire-setting and its treatment.

USEFUL WEBSITES

FTCs Identity Theft Site. www.ftc.gov/bcp/edu/microsites/idtheft/

InterFire. www.interfire.org/

Motor Vehicle Theft Statistics. www.fbi.gov/ucr/cius_04/offenses_reported/property_crime/motor_vehicle_theft.html

Property Crime Statistics. www.fbi.gov/ucr/cius_04/offenses_reported/property_crime/index.html

Property Crime Trends. www.ojp.gov/bjs/glance/house2.htm

CHAPTER TERMS

Arson	Cybercrime	Identity theft
Burglary	Embezzlement	Larceny-theft
Carjacking	Forgery	Motor vehicle theft
Counterfeiting	Fraud	

Public Order Crime

Joe Alladyce and Jared Livingston were both literally born drunk. Their mothers were heavy drinkers who continued to drink during their pregnancies, and if mothers drink, so do their fetuses. If the fetus survives this assault, it is highly likely to be born with a condition called fetal alcohol syndrome (FAS; see Chapter 8), symptoms of which include neurological abnormalities, intellectual impairment, behavioral problems, and various bodily and facial imperfections. Joe and Jared were made wards of the court and sent to a special institution where staff did their best to educate and care for them. The boys formed a bond with each other and soothed each other's feelings of anger and depression. When they were both 17, they walked away from the home and made their way to the nearest town where they robbed a liquor store and went on a drinking binge. Walking down the street in a stupor, they came across Mr. and Mrs. Whelan and little 7-year-old Angela walking toward them. Angela made a remark about their behavior and appearance and started to giggle. Enraged, Jared smashed Angela over the head with the beer bottle he was carrying, and Sam did the same thing to her father when he tackled Jared. Both boys mercilessly beat and kicked all three family members to death.

This tragic story illustrates the insidious nature of alcohol abuse. Joe and Jared didn't ask to be born with incurable disabilities, and according to many FAS experts could no more be held responsible for their actions than a blind person is for not recognizing faces. They have brains incapable of appreciating right from wrong and of linking cause and effect. Their mothers not only ruined their own lives but also the lives of their sons and the lives of surviving members of the Whelan family. There is a huge cost to society caused by what has been aptly named "the beast in the bottle" and by other substances that tear the rationality from our brains and replace it with all manner of monsters.

What Are Public Order Crimes?

Public order crimes are a smorgasbord of offenses, some of which have been variously called vice offenses, consensual offenses, victimless crimes, or even nuisance offenses. Some public order crimes are considered very seriously (the sale of drugs), and some are dismissed with a shrug of the shoulders or a look of disgust (drunken and disorderly behavior). Public order crimes are of the "moving target" type—legal in some places and at some times (prostitution in Nevada, drugs in Amsterdam, gambling in London) and illegal at other times and in other places. There is one school of thought that maintains that allowing or ignoring public order offenses can only lead to more serious crimes because it signals that nobody cares for the community (Wilson & Kelling, 1982). This so-called broken windows approach to crime control has had a major impact on policing and may be considered an approach akin to Sampson et al.'s collective efficacy concept discussed in Chapter 4.

All public order offenses cause some social harm, but whether or not the harm is great enough to warrant siphoning off criminal justice resources that could be applied to more serious crimes is a matter of debate. For instance, the debate about whether the use of mind-altering drugs should be legalized is not about the effects of these drugs—everyone realizes that they are harmful. Rather, it is about whether legalization or decriminalization would be the lesser of two evils. The notion that offenses categorized as public order offenses are "victimless" has been rejected by most criminologists today because there are always secondary victims (family members, friends, etc.) who may be profoundly harmed by the actions of the offender. The man who brings a sexually transmitted disease back to his wife after visiting a prostitute, the man who gambles away the family's money, and the woman who takes illicit drugs during her pregnancy are all causing great harm to many other people. Rather than victimless, public order crimes are better conceived of as consensual mala prohibita acts that always have the potential for causing harm to others besides the person engaging in them. The "victimless" act of drinking alcohol to excess by the mothers of Joe and Jared in the above vignette started a horrible chain of events that took three lives and ruined many others. These two mothers caused more social, financial, and emotional harm than any two burglars or thieves probably ever did.

Alcohol and Crime

Humans have a love of ingesting substances that alter their moods. We swallow, sniff, inhale, and inject with a relish that suggests that sobriety is a difficult state for us to tolerate. Alcohol has always been humans' favorite way of temporarily escaping reality. We drink this powerful drug to loosen our tongues, to be sociable, to liven up our parties, to feel good, to sedate ourselves, and to anesthetize the pains of life. Benjamin Franklin once supposedly opined that "Beer is proof positive that God loves us and wants us to be happy," and many centuries before, the ancient Sumerians and Egyptians were singing the praises of beer, wine, and the various spirits, but also warning about the consequences of excessive use (E. Burns, 2004, p. 2).

Of all the substances used to alter mood and consciousness, alcohol is the most directly linked to crime, especially violent crime (Martin, 2001). It has been estimated that at least 70% of American prison inmates (Wanberg & Milkman, 1998) and 60% of British inmates (McMurren, 2003) are alcohol and/or drug addicted. Alcohol is linked to about 110,000 deaths a year versus the "mere" 19,000 fatalities attributable to other drugs (M. Robinson, 2005), although this should be interpreted in light of the fact that many more people drink alcohol than take illicit drugs.

Police officers spend more than half of their time on alcohol-related offenses, and it is estimated that one third of all arrests (excluding drunk driving) in the United States are for alcohol-related offenses (Mustaine & Tewkesbury, 2004). About 75% of robberies and 80% of homicides involve a drunken offender and/or victim, and about 40% of other violent offenders in the United States had been drinking at the time of the offense (Martin, 2001). The U.S. Department of Health and Human Services (2009) estimates the cost of alcohol abuse to society to be a staggering $185 billion. Of the total costs to society of both drug and alcohol abuse, the National Institute on Alcohol Abuse and Alcoholism (1998) estimates that alcohol abuse accounts for 60% and drug abuse the remaining 40%.

▲ **Photo 13.1**   Many alcohol researchers compare societies in which drinking alcohol is considered normal behavior from an early age with American society, in which youth alcohol drinking is illegal. In the United States, many teens drink anyway, and binge drinking rather than moderate use is a potential problem.

The Effects of Alcohol and Context on Behavior

The effects of alcohol (or any other drug) on behavior is a function of the interactions of the pharmacological properties of the substance, the individual's physiology and personality, and the social and cultural context in which the substance is ingested. Pharmacologically, alcohol is a depressant drug that inhibits the functioning of the higher brain centers. As more and more alcohol is drunk, behavior becomes less and less inhibited as the rational cortex surrenders its control of the drinker's demeanor to the more primitive limbic system (the "emotional" brain). What's going on in the drinker's brain to cause this? Although alcohol is a brain-numbing depressant, at low dosages it is actually a stimulant because it raises dopamine levels (Ruden, 1997). Alcohol also reduces inhibition by affecting a neurotransmitter called GABA, which is a major inhibitor of internal stimuli such as fear, anxiety, and stress (Buck & Finn, 2000). In addition, alcohol decreases serotonin, reduces impulse control, and increases the likelihood of aggression (Martin, 2001). Alcohol's direct effects on the brain can thus help us to reinvent ourselves as "superior" beings: the fearful to become more courageous, the self-effacing to become more confident, and the timid to become more assertive.

As powerful a behavioral disinhibitor as alcohol is, it is not sufficient by itself to change anyone's behavior in the direction of serious law violations. Most people don't become violent or commit criminal offenses when drinking, or even when they are "over the limit." Alcohol is a releaser of behaviors that we normally keep under control but which we may be prone to exhibit when control is weakened. Hence, we may become silly, amorous, melancholic, maudlin, and even aggressive and violent when our underlying propensity to be these things is facilitated by alcohol and the social context in which it is drunk. In some social contexts, drinking may lead to violence, but not in others. Many violent incidents between strangers take place in or around drinking establishments in which both victims and perpetrators had been drinking (Richardson & Budd, 2003).

Groups of young men assembled in bars are recipes for trouble. Experimental research has shown that drinking increases fantasies of power and domination, and that men who are the heaviest drinkers are the most likely to have them (Martin, 2001). With loosened inhibitions, such fantasies might lead to males flirting with the girlfriends of males from another group and then not backing off when challenged, or interpreting some comment or gesture as threatening. If a male values his reputation as a macho tough guy, aggressive responses are more likely when his friends are present and he is looking to validate his reputation. There's an old saying among heavy drinkers: "It's not how many beers you drink; it's who you drink them with."

There are also cultural factors to be considered when evaluating the alcohol–crime relationship. Two of the major cultural factors influencing the relationship between alcohol consumption and criminal behavior are "defining a drinking occasion as a 'time-out' period in which controls are loosened from usual behavior and a willingness to hold a person less responsible for their actions when drinking than when sober by attributing the blame to alcohol" (Martin, 2001, p. 146). If one's culture defines alcohol as a good-time elixir, the unfortunate (but often subjectively experienced as enjoyable) by-product of which is a loss of control over behavioral inhibitions, then one is granted cultural "permission" to do just that.

Binge drinkers frequently consume anywhere from 5 to 10 drinks in few hours' time and are particularly likely to define drinking as a time-out period. Binge drinkers are typically college-age single young adults who drink solely to get drunk. An American study found that 40% of college students reported at least one episode of binge drinking in the previous 2 weeks (L. Johnson, O'Malley, & Bachman, 2000), and a Russian nationwide study found that almost one third of the men admitted binge drinking at least once a month (Pridemore, 2004). The cultures of both American college students and Russians in general have a high level of tolerance for engaging in heavy drinking. Richardson and Budd's (2003) British study found that 39% of binge drinkers admitted to a criminal offense in the previous 12 months, whereas 14% of other regular drinkers and 8% of occasional or non-drinkers did. The corresponding percentages for a violent crime were 17, 4, and 2. A survey of 180,455 male and 3,664 female arrestees in major U.S. cities found that 47.9% of the males and 34.9% of the females reported that they had engaged in binge drinking on at least one occasion in the 30 days preceding their arrest (Zhang, 2004).

But do heavy drinking plus social context *per se* cause increased antisocial behavior? It could well be that antisocial individuals are more prone to drink heavily and to be attracted to social contexts in which violence is most likely to occur. In this view, antisocial propensities are simply exacerbated under the influence of alcohol and social setting (Bartol, 2002). Heavy alcohol intake certainly has a greater disinhibiting effect on behavior than heavy tea intake, so alcohol-induced disinhibition may be considered a cause of antisocial acts. Likewise, violence and other antisocial behaviors are assuredly more likely to occur in a biker bar than in a tearoom, and thus social context may be considered a cause as well. But a stricter standard of causation should consider that perhaps the substance and the setting are secondary in causal importance to the traits of individuals drinking the beverage of their choice in the setting of their choice.

Drunk Driving

Traffic fatalities caused by drivers under the influence of alcohol are another evil caused by the "beast in the bottle." In state statutes, this crime is typically referred to as driving under the influence (DUI) or driving while intoxicated (DWI). According to the Insurance Information Institute (2009), 11,773 people died in alcohol-impaired crashes in 2008, down 9.7% from 13,041 in 2007. In 2008, there were 1,483,396 DUI arrests, 80.5% of whom were males and 84% of whom were white (FBI, 2009a).

Many people used to consider deaths due to drunk drivers as "accidents" rather than "crimes," and penalties were relatively light. Attitudes began to change in the United States with the founding of MADD (Mothers Against Drunk Driving) in 1980, an organization that has effectively lobbied for legislation nationwide to increase the legal drinking age and for stricter penalties for drunk drivers. MADD also lobbied to lower the blood alcohol count (BAC) level that defines intoxication from 0.10 to 0.08 grams per deciliter of blood. Every state in the union has now enacted all these measures. As we see in Figure 13.1, from the National Highway Traffic Safety Commission (2009), these combined measures reduced alcohol-related traffic fatalities by 57% from 1982 to 2008.

A 1999 nationwide study of DUI offenders found that the average BAC at arrest was 0.24, or 3 times the legal limit. The average time for DUI offenders in jail was 11 months, and for offenders sent to prison it was 49 months (Maruschak, 1999). The significant drop in alcohol-related traffic fatalities following a lowered tolerance and increased penalties shows that we can indeed legislate morality, if only in the case of drunken driving.

Alcoholism: Type I and Type II

Alcoholism is a chronic disease condition marked by progressive incapacity to control alcohol consumption despite psychological, social, or physiological disruptions. It is a state of altered cellular physiology caused by chronic consumption of alcohol that manifests itself in physical disturbances (**withdrawal** symptoms) when alcohol use is suspended. While most alcoholics do not get into serious trouble with the law, numerous theorists have hypothesized that alcoholism and criminality are linked because they share a common cause, which is probably the dysregulation of the behavioral activating system (BAS; see Chapter 8) that leads

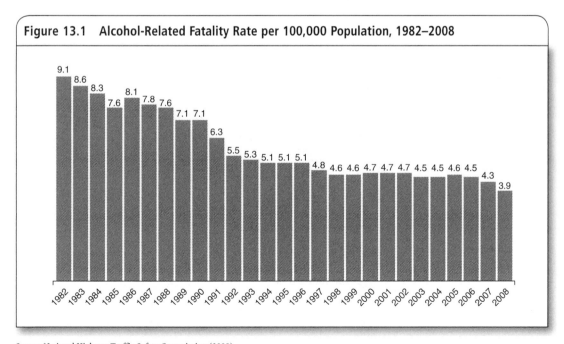

Figure 13.1   Alcohol-Related Fatality Rate per 100,000 Population, 1982–2008

Source: National Highway Traffic Safety Commission (2009).

to a "craving brain"(Gove & Wilmoth, 2003). Alcoholics have a saying that one drink is one too many and a hundred drinks are not enough. This seemingly contradictory statement tells us that a single drink activates the brain's pleasure centers and leads to such a craving for more that a hundred drinks will not satiate. Thus, both alcoholics and serious criminals are "reward dominant" in terms of their neurophysiology. Figure 13.2 shows the alcohol reward system. Alcohol stimulates the release of the "pleasure neurotransmitter" dopamine, which is made in the ventral tegmental area and then sent to the nucleus accumbens, the brain's major "pleasure center" (Oscar-Berman et al., 2009).

There are two different types of alcoholics, which can be likened to Moffitt's adolescent limited and life course–persistent offenders discussed in Chapter 9: Type I and Type II. Crabbe (2002) describes the two types in this way: "**Type I alcoholism** is characterized by mild abuse, minimal criminality, and passive-dependent personality variables, whereas **Type II alcoholism** is characterized by early onset, violence, and criminality, and is largely limited to males" (p. 449). Type II alcoholics start drinking (and using other drugs) at a very early age, rapidly become addicted, and have many character disorders and behavioral problems that *precede* their alcoholism. Type I alcoholics start drinking later in life than type II's and progress to alcoholism slowly. Type I's typically have families and careers, and if they have character defects, these are induced by their alcohol problem and are not permanent (DuPont, 1997).

Heritability estimates for Type II alcoholism are about 0.90, and about 0.40 for Type I's (McGue, 1999), indicating that environmental factors are more important to understanding Type I alcoholism than Type II alcoholism (Crabbe, 2002). The genetic influence on alcoholism reflects genetic regulation of

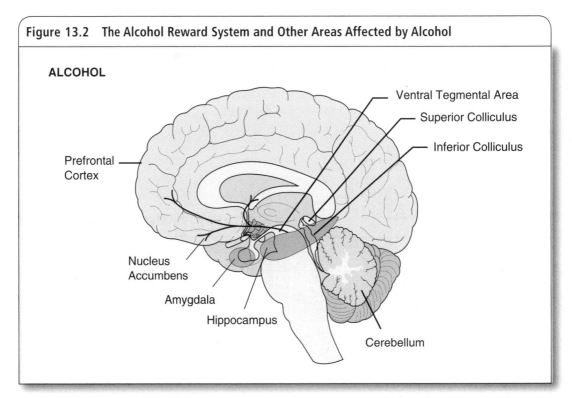

Figure 13.2   The Alcohol Reward System and Other Areas Affected by Alcohol

ALCOHOL

Ventral Tegmental Area
Superior Colliculus
Inferior Colliculus
Prefrontal Cortex
Nucleus Accumbens
Amygdala
Hippocampus
Cerebellum

SOURCE: National Institute of Drug Abuse (1996).

neurotransmitters such as GABA, dopamine, and serotonin (Buck & Finn, 2000), or their regulation by enzymes such as the MAOA enzyme discussed in Chapter 8 (Demir et al., 2002).

▧ Illegal Drugs and Crime

The Extent of the Illicit Drug Problem

Alcohol use is a legal and socially acceptable way of drugging oneself, but substances discussed in this section are not. This was not always the case, for many of these drugs have been legitimately used in religious rituals, for medical treatment, and for recreational use around the world and across the ages. Up until 1914, drugs now considered illicit were legally and widely used in the United States for medicinal purposes. Physicians and politicians were not fully aware of the dangers of addiction at the time, and many substances were openly advertised and sold as cures for all sorts of ailments and as refreshing "pick-me-ups." The most famous of these was Coca-Cola, which was made with the coca leaf (used to process cocaine) and kola nuts (hence the name) until 1903. Many patented medicines such as Cocaine Toothache Drops and Mother Barley's Quieting Syrup, used to "soothe" infants and young children, contained cocaine, morphine, or heroin.

Attitudes toward drug usage in America gradually began to change as awareness of the addictive powers of many of these substances grew. The **Harrison Narcotic Act** of 1914 was the benchmark act for changing America's concept of drugs and their use. According to Richard Davenport-Hines (2002), "By the early 1920s, the conception of the addict changed from that of a middle-class victim accidentally addicted through medicinal use, to that of a criminal deviant using narcotics (or stimulants) for pleasure" (p. 14). The Harrison Act did reduce the number of addicts (estimated at around 200,000 in the early 1900s), but it also spawned criminal black market operations (as did the Volstead Act prohibiting the production and sale of alcohol in 1919) and ultimately many more addicts (Casey, 1978, p. 11).

Figure 13.3 shows percentages of individuals participating in the 2008 National Household Survey on Drug Abuse (NHSDA) who admitted to the use of any illicit drug during the month prior to being interviewed. As with delinquency and crime, drug use rises to a peak in the age 18–20 category and then drops precipitously. The use of illicit drugs by most adolescents probably reflects experimentation (adolescent limited use), while their continued use in adulthood (life course–persistent use) reflects a far more serious antisocial situation.

Drug Addiction

All addictive drugs mimic the actions of normal brain chemistry by inhibiting or slowing down the release of neurotransmitters, stimulating or speeding up their release, preventing their reuptake after they have stimulated neighboring neurons, or breaking

▲ **Photo 13.2**   This man is damaging his brain and all those who care about him.

Figure 13.3   Illicit Drug Use in Past Month by Age, 2008

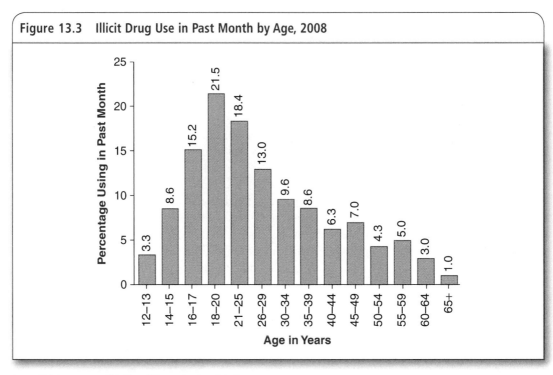

Source: U.S. Department of Health and Human Services (2009).

transmitters down more quickly. As we saw earlier (Figure 13.2), the brain has evolved pleasure centers by which Mother Nature rewards us when we do things that lead to survival and reproductive success. That is, we are neurologically rewarded when we eat, drink, have sex, reach safe havens, and enjoy the good company of others. Drugs hijack the brain's pleasure centers and produce more powerful, rapid, and predictable effects on our pleasure centers than are naturally obtained by the action of neurotransmitters in response to non–drug-induced pleasurable experiences.

People turn to illegal drugs for many of the same reasons that people turn to alcohol—to be "with it," to be sociable, to conform, to induce pleasure, to escape stress, or to escape chronic boredom. Among those who experiment with drugs, there are some who are genetically predisposed to develop addiction to their substance(s) of choice just as others are "sitting ducks" for alcoholism (T. Robinson & Berrridge, 2003).

The Drug Enforcement Administration (DEA, 2003) defines **drug addiction** as "compulsive drug-seeking behavior where acquiring and using a drug becomes the most important activity in the user's life," and estimates that 5 million Americans suffer from drug addiction (p. 13). **Physical dependence** on a drug refers to changes to the body that have occurred after repeated use of it and that necessitate its continued administration to avoid withdrawal symptoms. Physical dependence is not synonymous with addiction as commonly thought, but **psychological dependence** (the deep craving for the drug and the feeling that one cannot function without it) is synonymous with addiction.

Regardless of the type of drug, addiction is not an invariable outcome of drug usage any more than alcoholism is an invariable outcome of drinking. The DEA (2003) estimates that about 55% of today's youth have used some form of illegal substance, but few descend into the hell of addiction (Kleber, 2003). Genetic differences are undoubtedly related to a person's chances of becoming addicted given identical levels of usage and an identical period of time using.

Drug Classification

There are several drug classification schemes, which are determined by the purpose for which the classification is being made. The DEA (2003) schedule classification scheme divides chemical substances into five categories, or schedules. Schedule I substances are those that have high abuse liability and no medical use in the United States, such as heroin, peyote, and LSD. Schedule II substances have equally high (or higher) abuse liability, but have some approved medical usage, such as opium or cocaine. Schedule III and IV substances have moderate to moderately high abuse liability and are legally available with prescription, and Schedule V substances can be purchased without a prescription. The three major types of drugs defined in terms of their effects on the brain are the narcotics, stimulants, and hallucinogenics.

- **Narcotics:** Narcotic drugs are those that reduce the sense of pain, tension, and anxiety, and produce a drowsy sense of euphoria. Heroin is an example.
- **Stimulants:** The stimulants have effects opposite to those of narcotics. Stimulants such as cocaine and methamphetamine keep the body in an extended state of arousal.
- **Hallucinogenics:** Hallucinogenic drugs are mind-altering substances such as lysergic acid diethylamide (LSD) and peyote.

The Drugs–Violence Link

Illegal drugs are associated with violence in three ways: (1) systemic, (2) economic-compulsive, and (3) pharmacological (Goldstein, 1985). Systemic violence is violence associated with "doing business" (the growing, processing, transporting, and selling of drugs) in the criminal drug culture. There is so much systemic violence because the drug business is tremendously lucrative for those involved in it, and there is much competition for a slice of that business. The United Nations estimates the annual worth of the international illicit drug trade to be $400 billion (cited in Davenport-Hines, 2002, p. 11).

As with any industry, the illicit drug industry consists of several levels of business between extracting the product from the ground and selling it to the eventual consumer. Cocaine and heroin both begin as natural products grown in fields, cocaine as the coca leaf and heroin as the poppy flower. According to the U.S. Department of State (2005), the number of acres used for coca cultivation in 2004 in South America was 60,787, down from 552,763 in 2001. This huge reduction was accomplished mainly by the aerial spraying of the coca crop with herbicides. On the down side, it was also reported that Afghanistan had 510,756 acres (798 square miles) devoted to cultivating poppies, up from a mere 4,164 acres under cultivation during the last full year (2001) of the Taliban regime.

After the crop has been picked, the raw material must be processed, packaged, and smuggled via various "pipelines" into the countries in which the customers for the product reside. Figure 13.4 shows the trafficking routes for cocaine and heroin from points of origin to eventual destination. Once it arrives at its destination, it is "cut" (mixed with various other substances) to increase its volume and then distributed to street-level outlets for sale to drug users. Profit is made along each step of the way. In 2000, for instance, a kilogram (about 2.2 pounds) of heroin cost an average of $2,720 in Pakistan but sold for an average of

Figure 13.4   Illegal Drug Marketing, From Grower to Market

Grower	→	Processor	→	Transporter	→	Wholesaler	→	Retailer
Farmers plant and harvest poppy, coca, and marijuana crops.		Chemicals such as motor oil, sulfuric acid, kerosene, and insecticides are used to refine product.		Smugglers use planes, boats, trucks, and many other methods to get product to wholesaler.		Organized criminal groups cut product and distribute it to dealers.		Seller deals directly with consumer in crack houses or on street.

$129,380 on the streets in the United States, and a kilogram of coca base cost an average of $950 in Colombia in 1997 and sold for $25,000 in the United States (Davenport-Hines, 2002, p. 16).

Systemic violence and other criminal activity begins with the bribery and corruption of law enforcement officials and political figures, or their intimidation and assassination, in the countries where raw materials are grown and through which the processed product is transported. On the streets of the United States, systemic violence is most closely linked with gang battles over control of territory (control of drug markets). Goldstein and his colleagues (Goldstein, Brownstein, Ryan, & Belluci, 1989) found that just over one half of a sample of 414 murders committed in New York in 1988 were drug related, with 90% of those drug-related crimes involving cocaine.

Economic-compulsive violence is violence associated with efforts to obtain money to finance the high cost of illicit drugs. The drugs most associated with this type of activity are heroin and cocaine because they are the drugs most likely to lead to addiction among their users and are the most expensive (Parker & Auerhahn, 1998). Crimes committed to obtain drug money run the gamut from shoplifting, robbery, and prostitution, to trafficking in the very substance the addict craves. A study of newly incarcerated drug users found that 72% claimed that they committed their latest crime to obtain drug money (Lo & Stephens, 2002).

Pharmacological violence is violence induced by the pharmacological properties of the drug itself. Violence induced by illicit drugs is rare compared with violence induced by alcohol, the legal drug. A criminal victimization survey found that less than 5% of victims of violent crimes perceived their assailants to be under the influence of illicit drugs versus 20% who perceived them to be under the influence of alcohol (Parker & Auerhahn, 1998).

What Causes Drug Abuse?

Sociological explanations of drug abuse mirror almost exactly the explanations for crime. Erich Goode illustrates the almost indistinguishable explanations offered for the causes of crime and drug abuse in his book *Drugs in American Society* (2005). In anomie terms, drug abuse is a retreatist adaptation of those who have failed in both the legitimate and illegitimate worlds, and drug dealing is an innovative adaptation. In social control terms, drug abusers lack social bonds; in self-control terms, drug abuse is the hedonistic search for immediate pleasures; and in social learning terms, drug abuse reflects differential exposure to individuals and groups in which it is modeled and reinforced. Goode favors conflict theory most as an explanation. As the rich get richer, the poor get poorer, and economic opportunities are shrinking for the uneducated and the unskilled, drug dealers have taken firm root among the increasingly demoralized, disorganized, and politically powerless "underclass." He notes that most members of this class do not succumb to addiction, but enough do "to make the lives of the majority unpredictable, insecure, and dangerous" (p. 77). Goode maintains that conflict theory applies "more or less exclusively to heavy chronic, compulsive use of heroin or crack" (p. 74).

Does Drug Abuse Cause Crime?

Figure 13.5 shows the percentage of adult arrestees in some of the largest cities in the United States who tested positive for illicit drugs in 2007 through 2009. Clearly, these data show that illicit drug abuse is strongly *associated* with criminal behavior, but is the association a *causal* one? There are three possible explanations for the connection: (1) Drug use causes high rates of offending; (2) high rates of offending cause drug use; and (3) there is no causal connection, i.e., certain individuals are predisposed to high levels

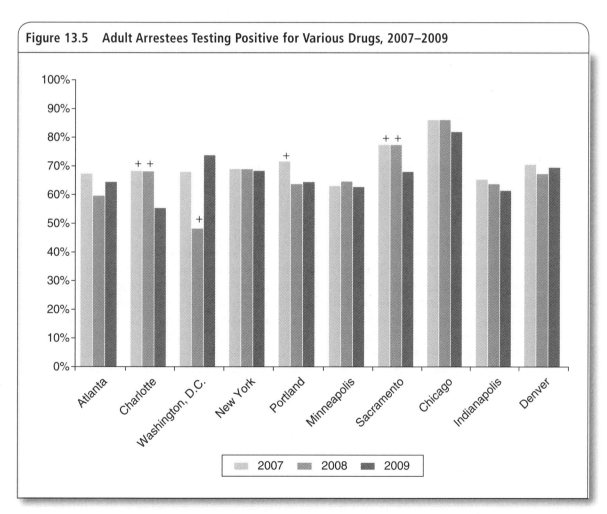

Figure 13.5   Adult Arrestees Testing Positive for Various Drugs, 2007–2009

SOURCE: Office of Drug Control Policy (2010).

of involvement in both drugs and crime. A large body of research indicates that drug abuse does not appear to *initiate* a criminal career, although it does increase the extent and seriousness of one (McBride & McCoy, 1993; Menard, Mihalic, & Huizinga, 2001). Drug abusers are not "innocents" driven into a criminal career by drugs, although this might occasionally be true. Rather, chronic drug abuse and criminality are part of a broader propensity of some individuals to engage in a variety of deviant and antisocial behaviors (Fishbein, 2003; McDermott et al., 2000). The reciprocal (feedback) nature of the drugs–crime connection is explained by Menard et al. (2001) as follows:

> Initiation of substance abuse is preceded by initiation of crime for most individuals (and therefore cannot be a cause of crime). At a later stage of involvement, however, serious illicit drug use appears to contribute to continuity in serious crime, and serious crime contributes to continuity in serious illicit drug use. (p. 295)

It is clear that the use of illicit drugs is very harmful to individuals, to their families, and to society. What is even clearer, however, is that the "War on Drugs," just like the war on alcohol during prohibition, is the cause of more harm than it prevents. Most countries have abandoned their own wars on drugs today and reverted to harm-reduction policies (i.e., policies aimed solely at minimizing harm). The United States is among only a handful of countries that reject harm-reduction programs favored by agencies such as the World Health Organization (WHO) and the United Nations Children's Fund (UNICEF) (Wodak, 2007). Such programs involve syringe exchange programs, drug substitution programs (such as methadone for heroin), and—most importantly—decriminalizing of drug usage. Yes, such practices do upset many people's sense of morality, but it does seem to be the practical thing to do if our goal is to reduce the overall social harm of drugs in our society. According to drug expert Alex Wodak, "No country which has started harm reduction programs has ever regretted that decision and then reversed their commitment" (p. 61).

⊠ Prostitution and Commercialized Vice

No other crime has been subjected to more shifts of attitudes and opinions across the centuries and across different cultures than **prostitution.** Throughout much of American history, it was regarded as a "necessary evil." During the American Revolution, camp followers who serviced the sexual needs of the troops were tolerated, and the term *hooker* is apparently derived from the women who serviced the Union troops commanded by General Joseph Hooker during the Civil War (F. Hagan, 2008). Prostitutes were particularly active in the Old West towns, where women were a rare commodity. However, the scourge of venereal disease and the grip of Victorian morality marked a change in American attitudes in the latter part of the 19th century, at which time many jurisdictions criminalized prostitution. In 1910, the federal government got into the act when Congress passed the White-Slave Traffic Act (the Mann Act) prohibiting prostitution and made it a felony to transport females across state lines for immoral purposes.

The FBI defines prostitution and commercialized vice in such a way as to cover people who sell their sexual services (prostitutes), those who recruit (procure) them, those who solicit clients (pander) for them, and those who house them. The common term for a procurer and panderer is a *pimp,* and for the keeper of a bawdy house (a brothel) is a *madam.* There were 56,560 arrests for prostitution and commercialized vice in 2009, down from 74,004 the previous year (FBI, 2010a). This is obviously only the tiniest fraction of all such offenses that actually take place, and the approximately 31% decrease doubtless reflects more lenient police practices rather than an actual reduction of prostitution.

Exchanging sexual favors for some other valued resource is as old as the species, and prostitution has long been referred to as the world's oldest profession. It has not always had the same sordid reputation that is attached to it today, however. Many ancient societies employed prostitutes in temples of worship with whom worshippers "communed," after which they deposited a sum of money into the temple coffers according to their estimation of the worth of the communion. In ancient Greece, many women of high birth who had fallen on hard times became high-class courtesans called *hetaerae,* who supplied their wealthy clients with stimulating conversation and other cultured activities as well as sexual

▲ **Photo 13.3**   A female prostitute parleys with a prospective john.

services. The lower classes had to content themselves with the brothel-based *pornae* or the prettier and more entertaining *auletrides* who would make house calls (Bullough & Bullough, 1994).

This ancient Greek hierarchy of sex workers (as most prostitutes prefer to be called) is mirrored in modern American society. The modern American *hetaerae* belong to the elite escort services and call houses and tend to be much better educated, more sophisticated, and better looking than other sex workers because they cater to a wealthy clientele who want to be made to feel special as well as sexually satisfied. These women (and sometimes men who cater to a gay clientele) can earn six-figure incomes annually and are able to sell their "date books" upon retiring for thousands of dollars (Kornblum & Julian, 1995, p. 109).

Brothel prostitutes are the modern *auletrides.* The only legal brothels in the United States are in certain counties of Nevada, but illegal brothels probably exist in every town of significant size in the United States, although they are not as prominent a part of community life as they used to be. Brothel prostitutes must accept whatever client comes along, but may make from $50 to $100 from each client. The streetwalker is the lowest member of the sex worker hierarchy. These prostitutes solicit customers on the streets and may charge only about $20 a trick (typically a quick act of oral sex).

Becoming a Prostitute

It has been estimated that prostitution is the primary source of income for over 1 million women in the United States, many of whom view sex work as the most financially lucrative option open to them (Bartol, 2002, p. 369). Many brothel and streetwalker prostitutes typically progressed from casual promiscuity at an early age to reasoning that they could sell what they were giving away, under the influence of peer pressure from more experienced girls and from pimps (Kornblum & Julian, 1995). Pimps exploit the strong need for love and acceptance among vulnerable girls. The pimp frequently takes on the roles of father, protector, employer, lover, husband, and often drug supplier, thus making the girl totally dependent on him (Tutty & Nixon, 2003). The girls most vulnerable to pimps and other pressures to enter prostitution are those who have experienced high rates of physical, sexual, and emotional abuse at home and who are drug abusers (Bartol, 2002). Hwang and Bedford (2004), however, show that unlike many Western prostitutes, very few Taiwanese prostitutes cite their own economic motives for entering the profession. As is the case in the United States, many Taiwanese prostitutes ran away from home to escape abuse and were befriended by pimps who supplied them with drugs and a certain amount of affection. However, a certain number of the prostitutes in Taiwan are indentured to a brothel by their parents who were in desperate need of money.

Should Prostitution Be Legalized/Decriminalized?

What are the harms of prostitution, and are they sufficient to warrant state intervention? Most of the harms are obvious, ranging from the spread of sexually transmitted infections to concerns about the exploitation of women. The seamy world of prostitution is also closely related to the drug market and other forms of serious criminality, and to neighborhood blight. The worst exploitation is that of trafficking women and children from poor countries to work in the brothels of rich countries. Some of these women come with their eyes open, but most others are duped, coerced, or forced. Even "voluntary migrants" are forced to sell themselves into prostitution by the poverty and lack of opportunities in their countries of origin (Raymond, 2003).

Prostitution is one of those things that we can never really prevent, although the AIDS epidemic and fear of infection have greatly reduced it. A 1989 study found that about 40% of streetwalkers and 20% of call girls were HIV positive (Kornblum & Julian, 1995). If we can't stop it, should we legalize it and therefore make it safer? When the ancient Greek lawmaker Solon (638–559 B.C.) legalized and taxed prostitution, he was widely praised:

> Hail to you, Solon! You bought public women [prostitutes] for the benefit of the city, for the benefit of the morality of a city that is full of vigorous young men who, in the absence of your wise institution, would give themselves over to the disturbing annoyance of better women. (Durant, 1939, p. 116)

Taxes on prostitution enabled Athens to build the temple to Aphrodite (the Goddess of Love) and provided its "vigorous young men" safe outlets for their urges. To borrow a term from sociology, the citizens of Athens found prostitution to be "functional," meaning that it had a socially useful role to play. Such an attitude, however, ignores the important functional role of morality to society, and the issue of legalization becomes how much morality we are willing to sacrifice for the sake of expediency.

Legalizing prostitution means that it becomes a legitimate occupation and that the state can regulate it by licensing brothels and prostitutes, determining where they can be located, and requiring regular health checkups. Holland is a country that has legalized prostitution. All parties—the prostitutes, pimps, procurers, and customers—are legally sanctioned as long as they remain in prescribed areas. The unfortunate downside of this is that about 70% of Dutch prostitutes are trafficked in from poorer countries because legalization has greatly expanded the demand for a variety of "exotic" foreign females (Raymond, 2003). Thus, legalization has increased demand rather than decreased it.

Decriminalization simply means the removal of laws against prostitution without imposing regulatory controls on it. Decriminalization is the stance favored by the American Civil Liberties Union (ACLU) and prostitution rights groups. These groups oppose legalization since legalization requires regulation and further stigmatizes the "profession" (Weitzer, 1999). The police tolerance (and evidently the public's) of massage parlors in our cities constitutes de facto decriminalization in the United States. Decriminalization basically means business as usual for prostitutes, and saves many millions in taxpayer dollars not expended arresting, prosecuting, and punishing women, most of whom are in the business because they perceive few alternatives or who may have been forced into it by pimps upon whom they rely for love, protection, and drugs (Tutty & Nixon, 2003).

The United Kingdom is an example of a country that has decriminalized prostitution. In the UK, women are not penalized for selling sexual favors, but all third-party activities such as pimping, procuring, or in any way living off "immoral earnings" are criminalized. Whether prostitution is legalized or decriminalized, however, one inevitable upshot has always been an increase in demand as more men come to see it as no big deal ("It's legal, so it must be okay") or to engage in it because there is no longer risk of arrest. Sweden has decriminalized prostitution but also criminalizes purchasing prostitutes' services. If a prostitute and her john (customer) are caught engaging in sex, the man is arrested and the prostitute sent on her way. Since many customers are married or otherwise respectable members of society, an arrest experience has a very large deterrent effect. An American study of arrested johns found that only 18 out of 2,200 (0.8%) men arrested for soliciting were arrested again for soliciting over a 4-year period, a truly remarkably low recidivism rate (Weitzer, 1999).

SUMMARY

- Public order offenses are sometimes dismissed as minor nuisance offenses, but they can be quite serious. Criminologists now use this term rather than victimless with the realization that there are always secondary victims.
- Alcohol is humankind's favorite way of drugging itself and has always been associated with criminal and antisocial behavior. It reduces the inhibiting neurotransmitters and thus reduces impulse control. Contextual factors also play their part in producing the kinds of obnoxious behavior associated with drinking too much alcohol.
- Driving under the influence is the most serious Part II offense because of its sometimes deadly consequences. It was pointed out that more people are killed by drunken drivers in a typical year than are murdered by other means. Activism and legislation since the 1980s have succeeded in significantly reducing drunk driving.
- There are two types of alcoholism: Type I and Type II. Type I is associated with mild abuse, minimal violence, moderate heritability, and character disorders that result from alcoholism. Type II is characterized by early onset, violence, criminality, high heritability, and character disorders that precede alcoholism. Type II alcoholics may have inherited disorders and problems that impact both their alcoholism and their criminality.
- Illicit drug use is also a major problem. Like delinquency, drug usage increases at puberty and drops off in early adulthood to almost zero by the age of 65, as seen in Figure 13.3. Drug addiction is fairly similar to alcoholism in terms of brain mechanisms. Drugs hijack the pleasure centers in the brain and make addicts crave drugs to gain any sort of pleasure at all. Most people who try drugs do not become addicted.
- Drugs are associated with violence in these ways: pharmacological, economic-compulsive, and systemic, with the latter having the strongest association. This is the case because violence is part of "doing business" in the lucrative illicit drug business. The economic-compulsive link with violence is the result of addicts' efforts to gain money to purchase drugs, and the rarest link, pharmacological, is violence induced by ingested drugs.

- Most people arrested for a crime test positive for drugs, but this does not mean that drugs cause crime. Drug abuse is part of a broader propensity of some individuals to engage in all kinds of antisocial behavior, and such behavior is usually initiated before drug abuse behavior. Drug abuse does exacerbate criminal behavior, however.
- Prostitution is as old as the species. While many individuals are coerced into prostitution, others become prostitutes because it is a lucrative business. There are many arguments for and against the legalization or decriminalization of prostitution.

DISCUSSION QUESTIONS

1. Discuss with classmates how each of you act—silly, aggressive, lusty, maudlin, and so on—when you have "gone over the limit" drinking alcohol. Why do you think that the same substance "makes" different people react differently?

2. The traffic fatality rate attributable to drunk driving in 2008 was less than half of what it was in 1982 (see Figure 13.1). Does this tell you anything at all about the deterrent effect of punishment in general, or just about its effect on "ordinary folk"?

3. Given what you know about the history of drug laws in the United States and the link between drug abuse and violence (and crime in general), would legalizing drugs be the lesser of two evils? Give reasons why or why not.

4. Should prostitution be legalized in the United States? Why or why not?

USEFUL WEBSITES

Crimes Against Public Order and Morals. http://faculty.ncwc.edu/mstevens/293/293lect13.htm

Maintaining Public Order. http://en.wikipedia.org/wiki/Public_order_crime

Mothers Against Drunk Driving. www.madd.org/

Prostitution Research and Education. www.prostitutionresearch.com/

CHAPTER TERMS

Alcoholism	Pharmacological violence	Tolerance
Binge drinkers	Physical dependence	Type I alcoholism
Drug addiction	Prostitution	Type II alcoholism
Economic-compulsive violence	Psychological dependence	Withdrawal
Harrison Narcotic Act	Systemic violence	

CHAPTER

14

White-Collar and Organized Crime

On August 10, 1978, teenage sisters Judy and Lyn Ulrich and their cousin Donna were on a 20-mile journey to Goshen, Indiana, in their Ford Pinto when they were rear-ended by another car. As a result, gas spilled onto the highway, caught fire, and all three trapped girls died horrible fiery deaths. Pintos had been fitted with gas tanks that easily ruptured and burst into flames in rear-end collisions of over 25 miles per hour. The problem could have been fixed at a cost of $11 per vehicle, but with 11 million Pintos and 1.5 million light trucks with the problem, Ford accountants calculated that it would cost $137 million to fix. It was calculated that not fixing it would result in 180 burn deaths, 180 serious burn injuries, and 2,100 burned vehicles, which they estimated would cost about $49.5 million dollars in lawsuits and other claims. Comparing those two figures, it was determined that in light of the $87.5 million it would save by not fixing the gas tank problem, to fix it would be unprofitable and irrational.

The consciences of Ford executives evidently did not bother them because they openly used these figures to lobby against federal fuel leakage standards to show how unprofitable such standards would be! According to a Ford engineer, 95 percent of the 700 to 2,500 people who died in Pinto crashes would have survived if the problem had been fixed. If this is an accurate estimate of deaths caused by the defect, the executives who conspired to ignore it may be the worst multiple murderers in U.S. history. Yet no Ford executive was ever imprisoned, and many went on to bigger things. Lee Iacocca, whose personal maxim, "Safety doesn't sell," was still in evidence in 1986 when he opposed mandatory airbags for automobiles, went on to become president of Chrysler Corporation and to chair the committee for the centennial celebrations for the Statue of Liberty.

⊠ The Concept of White-Collar Crime

What images pop into your head when you hear the word *crime*? Whatever images you conjured up, I wager they did not include a well-dressed, middle-aged person sitting in a leather recliner, dictating a memo authorizing the marketing of defective automobiles or the dumping of toxic waste. We seldom think that the chain of events set into motion by a business memo may do more harm than the activities of any "street punk." Yet more money is stolen and more people die every year as the result of scams and willful illegal corporate activity than as the result of the activities of street criminals. Kappeler, Blumberg, and Potter (2000) estimate that crimes committed by corporations result in economic losses whose total is between 17 and 31 times greater than losses resulting from street crimes, and the losses due to non-corporate white-collar crime are roughly the same. There is a huge grubby ring around the white collar that no amount of scrubbing will erase, and most criminologists agree that by virtually all criteria, our most serious crime problem is white-collar crime.

The phrase "white-collar crime" was coined in the 1930s by Edwin Sutherland (1940) who defined it as crime "committed by a person of respectability and high social status in the course of his occupation" (p. 9). In its Administration Improvement Act (AIA) of 1979, the U. S. Congress defined **white-collar crime** as "an illegal act or series of illegal acts committed by non-physical means and by concealment or guile, to obtain money or property, or to obtain business or personal advantage" (Weisburd, Wheeler, Waring, & Bode, 1991, p. 6). This definition focuses on characteristics of the offense as opposed to Sutherland's focus on the offender as a high-status person, because most white-collar crime is not committed by "high-status persons." The AIA definition, however, fails to differentiate between persons who commit crimes for personal gain and those who do so primarily on behalf of an employer. Our analysis of white-collar crime will differentiate among individuals who steal, defraud, and cheat both in and outside of an occupational context, and those who commit the variety of offenses attributed to business corporations. We will follow Rosoff et al. (1998) in using the term *occupational* crime for the former and *corporate* crime for the latter.

⊠ Occupational Crime

Occupational crime is crime committed by individuals in the course of their employment. Such crimes might range from the draining of company funds by sophisticated computer techniques to stealing pens and paper clips, or vandalizing company property by scrambling computer data or scratching graffiti on newly painted walls. Although such activities may seem relatively mundane to most of us, according to the business insurance industry, employee activities such as theft, fraud, and vandalism cost American businesses a total of $660 billion in 2003 (Parekh, 2004). Just as we all pay for street crime through taxes that support the criminal justice system, we all pay for employee crime because companies merely pass on their losses to their customers. Employee crime may also lead to businesses going bankrupt and employees losing the jobs these businesses provided.

Occupational crimes are crimes committed by professionals such as physicians and lawyers in the course of their practices. Frauds committed by physicians include practices such as filing insurance claims for tests or procedures not performed, performing unnecessary operations, steering patients to laboratories or pharmacies in which the doctor has financial stakes, and referrals to other doctors in return for kickbacks. Anyone who has been charged $8 an aspirin, $500 for a nursing bra, or $200 for a pair of crutches on their hospital bill knows how hospitals can rip patients off. In many respects, Medicare and Medicaid programs are welfare programs for physicians. Fraud within these programs has been estimated to cost between $50 and $80 billion per year, and overall medical fraud is estimated to be about $260 billion, or about 10% of the total U.S. health care bill (FBI, 2009b).

Most lawyers also work on a fee-for-service basis, thus generating the same temptations to increase their incomes by fraudulent means. Frauds perpetrated by lawyers can include embezzlement of clients' funds, bribery of witnesses and judges, persuading clients to pursue fraudulent or frivolous lawsuits, billing clients for hours not worked, filing unnecessary motions, and complicating a simple legal matter to keep clients on the hook: "I will defend you all the way to your last dollar." It seems that every occupational category generates a considerable number of criminals, and the higher the prestige of the occupation, the more their criminal activities cost the general public.

Causes of Occupational White-Collar Crime: Are They Different?

According to Hirschi and Gottfredson (1987), occupational crime differs from common street crime only in that it is committed by people in a position to do so—Medicaid fraud can typically only be committed by health care professionals, and bank embezzlement by bank employees in positions of trust. The motives of occupational criminals are the same as those of street criminals—to obtain benefits quickly with minimal effort—and the age, sex, and race profiles of occupational criminals are not that much different from those of street criminals. Hirschi and Gottfredson concluded that "[w]hen opportunity is taken into account, demographic differences in white collar crime are the same as demographic differences in ordinary crime" (p. 967). Walters and Geyer (2004) examined this assertion using indicators of criminal thinking patterns and attitudes. They found that white-collar criminals with prior arrests for non–white-collar crimes were not significantly different from street criminals in their demographics, lifestyle, and criminal thinking patterns, but were significantly different from white-collar criminals with no history of arrest for non–white-collar crimes.

✖ **Corporate Crime**

In *The Wealth of Nations,* economist and founder of modern capitalism Adam Smith (1776/1953) wrote, "Seldom do members of a profession meet . . . that it does not end up in some conspiracy against the public or some contrivance to raise prices" (p.137). Smith was referring to what we call corporate crime today. **Corporate crime** is criminal activity on *behalf* of a business organization committed during the course of fulfilling the legitimate role of the corporation, and in the name of corporate profit and growth. During much of American history, the primary legal stance relating to the activities of business was *laissez-faire* (leave it alone to do as it will). American courts traditionally adopted the view that government should not interfere with business, so for a very long period in U.S. history, victims of defective and dangerous products could not sue corporations for damages because the guiding principle was *caveat emptor* (let the buyer beware). Unhealthy and dangerous working conditions in mines, mills, and factories were excused under the freedom of contract clause of the U.S. Constitution (Walsh & Hemmens, 2008).

Although attitudes in the 20th century changed considerably, the public continued to be victimized by crimes of fraud and misrepresentation committed by businesses. For instance, the savings and loan (S&L) scandal of the 1980s amounted to one of the most costly crime sprees in American history and is likely to cost the U.S. taxpayer (because the Federal Deposit Insurance Corporation insures all bank savings up to $100,000) up to $473 billion. This staggering amount is many times greater than losses from all the "regular" bank robberies in American history put together (Schmalleger, 2004, p. 375).

▲ **Photo 14.1** Corporate crime cases are typically not handled by local or even county prosecutors. Typically, these cases are moved up to the state attorney general's office, which is better equipped with investigators to prepare these cases for prosecution.

Most of the looting took the form of extravagant salaries and bonuses that executives awarded themselves as their banks sank ever further into debt. Other methods involved selling land back and forth ("land flipping") within a few days until its paper value far exceeded its real value, and then finding "sucker" institutions to buy it at the inflated price, and loans made back and forth between employees of different banks with the knowledge that the loans would never be called in (Calavita & Pontell, 1994).

The first major corporate scandal of the first decade of the 21st century was that of Enron Corporation. This crime has been called "one of the most intricate pieces of financial chicanery in history," and for investors in its stock and its employees, "the financial disaster of a lifetime, a harrowing, nerve-racking disaster from which they may never recover" (English, 2004, p. 1). The Enron scandal (and other similar scandals in the first 2 years of the 21st century) did tremendous damage to the economy and "created a crisis of investor confidence the likes of which hasn't been seen since the Great Depression" (Gutman, 2002, p. 1).

Theories About the Causes of Corporate Crime

Rational choice theorists, with their classical assumption that human behavior can be understood in terms of striving to maximize pleasure and minimize pain, have no difficulty appreciating why individuals choose to engage in corporate crime. Opportunities abound in corporate America to gain wealth beyond what individuals could earn legitimately. The weakness of formal and informal controls over business activities and the relatively lenient penalties imposed make corporate crime, immoral as it is, also very rational. We may expect business executives to have assimilated the acquisitiveness and competitiveness fostered by capitalism more completely than most.

The great Greek philosopher Plato knew what the classical theorists knew: Humans are designed to attempt to manipulate the world to the extent that they are able to maximize their pleasure and minimize their pain. His allegory of the Ring of Gyges is particularly useful in helping us to understand white-collar crime. Gyges was a likeable and well socialized young shepherd in the service of the king of Lydia. The gist of the story is that Gyges came upon a gold ring in a cave that allowed him to become invisible whenever he turned it on his finger. When Gyges realized the possibilities this opened for him, he went to the court, seduced the queen, murdered the king, and took over the kingdom. Plato's story makes the point that with the gift of invisibility, no one could resist all the temptations to act unjustly in the name of self-interest:

No man can be imagined to be of such an iron nature that he would stand fast in justice. No man would keep his hands off what is not his own when he could safely take what he liked out of the market, or go into houses and lie with anyone at his pleasure, to kill or release from prison whom he would, and in all respects be like a God among men. (Plato, 1960, p. 44)

All companies and their executives are exposed to the capitalist ethos, the strain of seeking their elevated versions of the American Dream, and the widespread opportunities for illegal gains. We might say that corporate criminals are "high-class innovators," to borrow a phrase from strain theory. But "strain" can only be invoked as a *motive* for corporate crime, not for the *choice* to engage in it. So what differentiates "innovators" from the other modes of adaptation available in the corporate world?

Just as there are criminogenic neighborhoods, there are criminogenic corporations with a "tradition" of wrongdoing. One of the first systematic examinations of corporate recidivism found that 98% of the nation's 70 largest corporations were recidivists with an average of 14 regulatory or criminal decisions against it (Sutherland, 1956). A study of 477 major U.S. corporations found that 60% of them were known to have violated the law, and that 13% of the violator companies accounted for 52% of all violations, with an average of 23.5 violations per company (Clinard & Yeager, 1980). And a study of brokerage firms found that many of the biggest names in the business such as Prudential, Paine Webber, and Merrill Lynch have had an average of two or more serious violations *per year* since 1981 (Wells, 1995). "Three strikes and you're out" laws evidently do not apply in the world of pinstripe suits.

Newcomers entering such corporate environments are socialized into the prevailing way of doing things. If the newcomer does not fit in and conform to the company ways, he or she is not likely to remain employed there very long. This process can produce a sort of moral apathy in well-socialized executives striving to do their jobs in their own and their company's best interests. If each tiny bending of the rules that bring profit to the company brings the rule bender appreciation and bonuses, the individual's behavior will be almost imperceptibly molded in the direction of ever-greater wrongdoing. If he or she is rewarded through the usual stock-based nature of executive compensation, there is further incentive to engage in illegal behavior. Stock-based compensation is designed to increase executives' focus on stockholder's profits, but it is a situation ripe for finagling of stocks through insider trading, falsifying accounts, fraudulent trading, and an emphasis on short-term earnings rather than the long-term success of the corporation (Thorburn, 2004).

In terms of the characteristics of individuals linked to corporate crime, it is obvious that qualities such as low IQ and low self-control do not apply to people who have spent many years of disciplined effort to achieve their positions. But there are many other factors linked to crime that could apply to corporate crooks as easily as to street criminals. For instance, it appears that people who choose business careers tend to have lower ethical and moral standards than people who choose other legitimate careers. A number of studies have concluded that business students are, on average, less ethical than students in other majors, even law (Tang, Chen, & Sutarso, 2008)!

Other studies have focused on locus of control, moral development, and Machiavellianism. People with an internal *locus of control* believe that they can influence life outcomes and are relatively resistant to coercion from others. People with an external locus feel that circumstances have more influence over situations than they themselves do. Those who engage in corporate crime tend to have an external locus of control, and whistleblowers tend to have an internal locus of control (Travino & Youngblood, 1990). Furthermore, people with an internal locus of control operate at higher stages of *moral development* and tend to behave according to their own beliefs about right and wrong. People at lower stages of moral development tend to emphasize conformity to group norms (external locus of control) (Weber, 1990).

Machiavellianism is the manipulation of others for personal gain. People who are high on this trait are shallow individuals who exploit superiors and equals by deceit and ingratiation, and subordinates by bullying. David Simon (2002) describes those who make it to the top of bureaucratic organizations. Such an individual "exudes charisma via a superficial sense of warmth and charm," and he or she exhibits "free floating hostility, competitiveness, a high need for socially approved success, unbridled ambition, aggressiveness [and] impatience" (p. 277). This description sounds very much like that of a psychopath and reminds us of Gordon Gekko in the 1987 movie *Wall Street* and its 2010 sequel. Indeed, a study of 203 corporate professionals participating in management development programs assessed the subjects using the Psychopathy Checklist-Revised (PCL-R; see Chapter 7) in conjunction with copious other data supplied by their parent companies (Babiak, Neumann, & Hare, 2010). It was found that 3% of participants had PCL-R scores at or above the 30-point cut-point used for defining psychopathy compared to 0.2% individuals in a large community sample, and 5.9% had scores indicative of "potential" or "possible" psychopathy compared to 1.2% in the community sample.

Law Enforcement Response to Corporate Crime

Taken together with the NCVS, the UCR provides us with a fairly reliable picture of the extent of street crime, but except for embezzlement, forgery/counterfeiting, and fraud, white-collar crimes are consigned to the relatively unknown Financial Crimes Report (see Chapter 2). Other tallies of corporate wrongdoings are collected and distributed each year by state and federal regulatory agencies such as the FTD (Federal Trade Commission) and the SEC (Securities and Exchange Commission). However, these agencies do not come under the jurisdiction of the Justice Department and thus lack the law enforcement powers of the FBI's Financial Crimes Section. Further complicating things, these agencies cite organizations rather than individuals for wrongdoings, and any one citation may include multiple crimes committed by multiple individuals.

There are a number of difficulties in recognizing and reporting major white-collar crimes. Street crimes are easily defined, discrete events that quickly come to the attention of the police (a body in the street, a house burgled, or a car stolen). With many white-collar crimes, victims often do not even know they have been victimized, and the sequence of events is often quite the opposite; that is, investigators may start with suspected wrongdoing and then try to prove it took place. Many of the most serious white-collar crimes are incredibly complex and require thousands of hours and millions of dollars to unravel.

Conflict theorists have no trouble explaining why white-collar crimes are not represented in the UCR: White-collar crime has never been considered a problem worth pursuing with the same vigor. The people who define the seriousness of criminal acts are the same people who own or have large interests in many of the institutions and corporations that commit the white-collar acts in question, and may also play an "active part in crafting the laws and regulatory standards that circumscribe their conduct" (Shover & Hochstetler, 2006, p. 264). Such people are thus hardly likely to want to define their shady acts as "serious" or "criminal."

Corporate wrongdoing is monitored and responded to by a variety of criminal, administrative, and regulatory agencies, but very few corporate crooks in the past received truly meaningful sanctions. Of the 1,098 defendants charged in the S&L cases, only 451 were sentenced to prison, with the majority (79%) sentenced to less than 5 years, and with the average sentence being 36.4 months (Calavita, Pontell, & Tillman, 1999). The Golden Rule ("Those with the gold make the rules") may explain the differences in punishment between street and corporate criminals because great wealth does confer a certain degree of immunity from prosecution or conviction. This power may be gauged by the 1990 U.S. Justice Department's withdrawal of its support for proposed tougher sentences for corporate offenders in response to heavy lobbying by many prominent industries, the targets of the proposal (F. Hagan, 1994). It is doubtful that street criminals would get very far lobbying against proposals for stricter penalties for them.

The cascade of corporate scandals and failures occurring in the first decade of the 21st century may have finally awakened American law enforcement to the realities of the harm done by elite criminals. As a result of congressional hearings and public outcry, Congress passed the **Sarbanes-Oxley Act (SOA)** of 2002. The SOA increased penalties for corporate criminals, increased the budget of agencies charged with investigating corporate crime, and made prosecution easier. Because of the SOA, it has been said that "prosecutors are driven to go after corporate fraud with an almost evangelical zeal" (Burr, 2004, p. 10). Perhaps the days of leniency for white-collar criminals are over. In 2005, Bernard Ebbers, ex-WorldCom CEO, was sentenced to 25 years in prison for his role in the $11 billion WorldCom fraud; John Rigas, founder of Adelphia Communications Corp, was sentenced to 15 years and his son Timothy to 20 years for their roles in yet another massive fraud; and Tyco executives, Dennis Koslowski and Dennis Swartz, were sentenced to 8 and 25 years, respectively. Enron executive Jeffery Skilling was sentenced to 24 years in 2006, and Bernard Madoff, the biggest swindler in American history, received 150 years in 2008.

▲ **Photo 14.2**   Financier Bernard Madoff arrives at Manhattan Federal Court on March 12, 2009, in New York City. Madoff pleaded guilty to 11 federal crimes and admitted to turning his wealth management business into a massive Ponzi scheme that defrauded thousands of investors of billions of dollars, which under federal law can result in a sentence of about 150 years.

🗒 Organized Crime

What Is Organized Crime?

Criminologists have had a difficult time deciding what organized crime is and how it differs (if it does) from corporate crime. Some argue that "any distinction between organized and white-collar crime may be artificial inasmuch as both involve the important elements of organization and the use of corruption and/or violence to maintain immunity" (Albanese, 2000, p. 412). Corporate crime is "organized crime" in some senses (e.g., corporations are organized, and when their members commit illegal acts they are engaging in crime), but there are major differences. Corporate criminals are created from the opportunities available to them in companies organized around doing legitimate business, whereas members of organized crime must be accomplished criminals before they enter groups that are organized around creating criminal opportunities. Thus, the former make a crime out of business and the latter make a business out of crime.

There is no doubt that high levels of business and political corruption walk hand in hand with organized crime. Dutch criminologist Jan Van Dijk (2008) carried out a book-length research project on this issue using data from 163 nations around the world. He examined the *rule of law* (the extent to which a country has an independent judiciary and the general population's respect for the law), the World Bank's *corruption index* (surveys of the perceived level of corruption in a country), *wealth* (gross domestic product), and the extent of *organized crime*. The results are summarized in Figure 14.1.

Interpreting the model in Figure 14.1, we see that (1) the lower the rule of law in a country, the higher the level of organized crime; (2) the greater the organized crime, the greater the level of corruption;

Figure 14.1   Statistical Model of the Relationship Between Prevalence of Organized Crime and Corruption, the Rule of Law, and Country Wealth

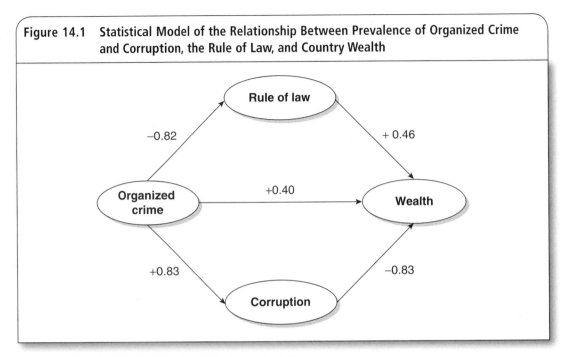

SOURCE: Van Dijk (2008). *The World of Crime.* Sage Publications. Reprinted with permission.

(3) the greater the corruption, the lower the wealth; (4) the greater the rule of law, the greater the wealth; and (5) the greater the wealth, the greater the level of organized crime. The direction of the arrows is arbitrary in a causal sense; for instance, does greater organized crime cause greater corruption, or is it the other way around? It could be either in different countries, but these things really have a reciprocal relationship with one another; that is, each has a causal feedback impact on the other.

The United States was seen as "medium" on the composite organized crime index with a score of 36.4 out of a possible 100, with 100 being the country with the highest level of organized crime (Haiti). The country with the lowest score was Finland (10.4). The United States was ranked 1st in wealth and 17th in the world on the rule of law (Iceland was first; the Democratic Republic of Congo was last). A later publication by Transparency International (2009), a global coalition against corruption, ranked the United States 19th lowest in level of corruption with New Zealand being first (least corrupt) and Somalia being last (the most corrupt). The United States is ranked very favorably with regard to low levels of organized crime and of corruption and high levels of the rule of law. Only the countries of Western Europe and Australasia seem to do better.

According to the President's Commission on Organized Crime (PCOC, 1986), **organized crime** is crime committed by structured criminal enterprises that maintain their activities over time by fear and corruption. PCOC concentrated on **La Cosa Nostra (LCN)** (literally, "our thing"), also commonly referred to as the Mafia, but it should not be inferred from this that La Cosa Nostra and organized crime are synonymous. There are many other organized crime groups in the United States and around the world.

LCN groups are structured in hierarchical fashion reflecting various levels of power and specialization. There are 24 LCN families with a national ruling body known as the **Commission,** which is a kind of "board of directors" and consists of the bosses of the five New York families and four bosses from other important families located in other cities. The Commission functions to arbitrate disputes among the various families and facilitate joint ventures, approve of new members, and authorize the executions of errant members (Lyman & Potter, 2004). Members respect the hierarchy of authority in their organization just as corporate executives respect ordered ranks of authority in their corporations. Although there are occasional family squabbles and coups that remove individuals from the hierarchy, the structure remains intact. The formal structure of an LCN family can be diagramed as in Figure 14.2.

At the top is the boss (the *don* or *capo*) beneath whom is a counselor (*consigliere*) or advisor, and an underboss (*sotto capo*). The counselor is usually an old family member (often a lawyer), wise in the ways of crime; the underboss is being groomed for succession to the top position. Beneath the underboss are the lieutenants (*caporegimes*) who supervise the day-to-day operation of the family through their soldiers (*soldati*), known as *made men, wise guys,* or *button men.* Although soldiers are the lowest-ranking members of the family, they may each run their own crew of nonmember associates.

LCN members are not employees who earn regular incomes from the family. Membership simply entitles the member to run his own rackets using the family's connections and status. A percentage of a soldier's earnings is paid to his lieutenant, who also has his own enterprises of which he pays a percentage to the underboss, and so on up the line.

This model is known as the **corporate model** because of its similarity to corporate structure, but some academics favor the **feudal model.** The feudal model of LCN views it as a loose collection of criminal groups held together by kinship and patronage. The commission may be seen as the king and his ministers in a feudal system who rarely if ever interfered with their nobles; the family bosses may be seen the lords, the lieutenants as the lesser nobility, and the made men and associates as the peasants. The autonomous operations of each family, the semi-autonomous operations of the soldiers and lieutenants, and the provision of status and protection from the family in exchange for a cut of their earnings provide evidence that LCN bosses are more like feudal lords than corporate CEOs. Other scholars see evidence for both the corporate and the feudal models, i.e., a "highly structured feudal system" (Firestone, 1997, p. 78).

Organized crime is like a corporation (or a feudal system, for that matter) in that it continues to operate beyond the lifetime of its individual members. It does not fall apart when key leaders are arrested, die, or are otherwise absent. The criminal group takes on a life of its own, and members subordinate their personal interests to the group's interests, making organized crime different from gangs that spring up and die with their leaders.

LCN is not an equal opportunity employer; it is restricted to males of Italian descent of proven criminal expertise. Prospective members must be sponsored by made guys (established members of the family), and applicants are screened carefully for their criminal activity and loyalty before being allowed to apply. A lifetime commitment to the family is required from the newcomer, and in return he receives a guaranteed and rather lucrative criminal career as part of an organization of great prestige and respect in the underworld. A promising criminal who is not of Italian descent, but who has qualities useful to the organization, may become an associate member of LCN.

The Origins of Organized Crime in the United States

Organized crime groups existed in the United States long before there was any major Italian presence, and many organized crime scholars believe that it is a "normal" product of the competitive and freewheeling nature of American society (Bynum, 1987). Scholars suggest a variety of dates as marking the beginning of

Figure 14.2   Hierarchical Structure of a Typical Organized Crime Family

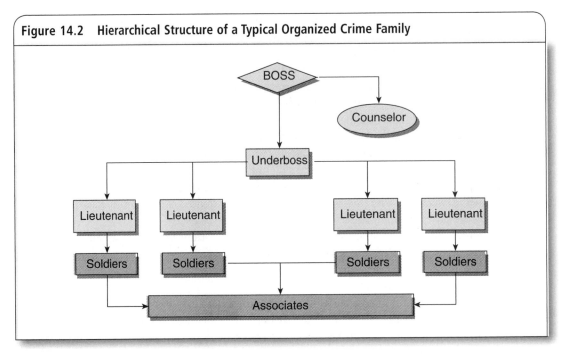

Source: Adapted from President's Commission on Organized Crime (1986, p. 469).

organized crime in America as we have defined it, but the two major candidates are the founding of the Society of Saint Tammany in the late 18th century and Prohibition in the early 20th century.

The **Tammany Society** was a corrupt political machine associated with the Democratic Party, which ran New York City well into the 20th century from the "Hall" (Tammany Hall). The society made use of street gangs to threaten and intimidate political rivals. Prominent among these gangs were the vicious *Whyos* and the *Five Points* gangs. In order for a new member to be accepted by the Whyos, which at its peak had over 500 members, he had to have killed at least once. The Whyos plied their trade among New York's citizenry by passing out price lists on the street for the services they provided (ranging from $2 for punching to $100 for murder) as casually as pizza vendors hawking their wares (Browning & Gerassi, 1980). The Five Points Gang was a confederation of neighborhood gangs, and was said to have over 1,500 members at one time.

In 1920, the United States Congress handed every petty gang in America an invitation to unlimited expansion and wealth with the ratification of the 18th Amendment and the passage of the Volstead Act **(Prohibition),** which prohibited the sale, manufacture, or importation of intoxicating liquors within the United States. Prohibition ushered in a 10-year period of crime, violence, and political corruption as gangsters fought over the right to provide the drinking public with illicit alcohol.

The most successful gangster of the era was Al "Scarface" Capone of Chicago. Capone (a former member of the Whyos) was a ruthless criminal and a flamboyant man who established a criminal empire, which at its height consisted of over 700 gunmen (Abadinsky, 2003). The wealth Capone accumulated from his bootlegging and prostitution enterprises got him into the *Guinness Book of Records* as having the highest gross income ($105 million) of any private citizen in America in 1927. Capone met his demise when the

Supreme Court ruled that unlawful income was subject to taxation. Capone had not paid taxes on his criminal income so was sentenced to 11 years in prison in 1931. He died in 1947 of pneumonia following a stroke at age 48.

With the repeal of Prohibition in 1933, organized crime entered a quieter phase, but the modern face of LCN was already beginning to take form in New York. There were two main factions in Italian organized crime in New York at this time, one headed by Giuseppe Masseria and the other by Salvatore Maranzano, both of whom were struggling for supremacy. This struggle, known as the Castellammarese War, ended with the deaths of both leaders in 1931. The war saw the end of the old Sicilian leaders and the emergence of an Americanized LCN, including five New York LCN families. It also saw Lucky Luciano (another former Whyo) set up the organization's national commission and claim the title of founding father of Italian-American organized crime (Lupsha, 1987).

LCN is today only a shadow of its former self due to the government's prosecution of hundreds of organized crime figures in the so-called Commission trials (a reference to the LCN Commission) of the 1980s. The leaders of 16 of the 24 LCN families were indicted, with the leaders of the Genovese, Lucchese, and Colombo families being sentenced to 100 years each. The leader of the Gambino family, Paul Castellano, avoided prosecution by getting himself murdered. Castellano's successor (and the man who ordered his murder), John Gotti, was sentenced to life in prison in 1992 after being convicted on 13 federal charges.

▲ **Photo 14.3**   Notorious Prohibition Era gangster Al Capone

Other Organized Crime Groups

Russian Mafiya

The Russian "Mafiya" is considered to be the most serious organized crime threat in the world today (Rush & Scarpitti, 2001). Ever since the breakup of the Soviet Union, we have seen an anomic society on steroids; the crime, bribery, and political and police corruption in Russia make the Prohibition period in America look positively benign. As James Finckenauer (2004) put it, "Organized crime has been able to penetrate Russian businesses and state enterprises to a degree inconceivable in most other countries" (p. 62).

The major group in Russian organized crime (ROC) is known as the *vory v zakone (thieves-in-law)*, which began as a large group of political prisoners imprisoned following the Bolshevik Revolution in 1917. The Soviet prison system used this group to maintain order over the general prison population in exchange for many favors. These "elite" prisoners developed their own structural hierarchy and strict code of conduct or "laws" (hence thieves-"in-*law*"). One of their strictest rules was that there is to be absolutely no cooperation with legitimate authority for any reason.

Russian organized crime spread to the United States with the influx of Russian immigrants in the 1970s and 1980s. Unlike the largely uneducated Mafioso who came to the United States in the 1920s and 1930s, many Russian émigré criminals are highly educated individuals driven out by economic hardship. This level of intelligence and expertise should make them a real threat. Indeed, Rush and Scarpitti (2001) write,

> It has been speculated by intelligence agencies such as the IRS, FBI, and CIA that because of their higher level of criminal sophistication Russian organized crime groups will present a greater overall threat to American society than the traditional Italian-American crime families ever have. (p. 538)

The Japanese Yakuza

Japanese organized crime (JOC) groups are probably the oldest and largest in the world, with total membership estimated at 90,000 (Lyman & Potter, 2004). JOC is commonly believed to have evolved from *ronin,* or masterless samurai warriors, who contracted their services out for assassinations and other illegal purposes. They also protected the peasants from other marauding bandits as a sort of vigilante/law enforcement group. The defeat of Japan in WWII and the ensuing chaos (Durkheimian anomie again) provided the impetus for the growth of JOC. This period saw many gang wars erupt over control of lucrative illicit markets. As in the United States, these gang wars led to the elimination of some gangs and to the consolidation and strengthening of others. The Kobe-based *Yamaguchi-gumi,* with an estimated membership of over 10,000, is the largest of these groups (Iwai, 1986).

Members of JOC groups are recruited heavily from the two outcast groups in Japanese society—the *burakumin* (outcasts because their ancestors worked at trades that dealt with dead flesh, such as butchery, tanning, grave digging, etc., which is seen as unclean in the Buddhist religious tradition) and Japanese-born Koreans. Once admitted, a recruit must pledge absolute loyalty to his superiors, and like his LCN counterpart, must generate his own income and contribute part of it to the *ikka* (the family).

JOC enjoys a unique position in Japanese society. Their historical connection with the samurai; their espousal of traditional norms of duty, loyalty, and manliness; their support for nationalistic programs; and their "law enforcement" functions (yakuza neighborhoods are safe from common criminals) endow them with a certain level of respect and admiration among the Japanese. Furthermore, the yakuza are not shadowy underworld figures, their affiliations are proudly displayed on insignia worn on their clothes and on their offices and buildings, and they publish their own newsletter. The headquarters of one crime group, complete with the gang emblem hanging proudly outside, is only three doors away from the local police station (E. Johnson, 1990).

Outlaw Motorcycle Gangs

There are some 900 motorcycle gangs that exist in the United States, with the "big four"—the *Hells Angels, Outlaws, Bandidos,* and *Pagans*—having evolved into a serious organized crime problem (Quinn, 2001). Outlaw motorcycle gangs (OMGs) are of American heritage, but they have been copied around the world. The California-based Hells Angels is the largest of these gangs, with an estimated membership in the United States of 1,000 (Lyman & Potter, 2004).

The organizational structure of OMGs typically consists of a president, vice president, secretary-treasurer, sergeant-at-arms, road captain, and enforcer. Each chapter or club belongs to a regional entity, and to a national organization headed by a mother club, which is usually the founder club. As with LCN "wannabes," prospective OMG members must be sponsored by an active club member, and be willing to demonstrate their worth by committing a serious criminal act in front of "made" members, which enables the club to weed out the weak as well as possible police infiltrators.

Women can join the gangs, but their status is little more than sexual playthings. Wives or steady girl-friends of gang members are referred to as *Old Ladies,* are the exclusive "property" of their men, and wear denim jackets with that designation embroidered on them. Other women associates are referred to as *mammas* or *sheep,* and are the sexual property of any gang member desiring to use them or to prostitute them to nonmembers (Abadinsky, 2003).

Bikers revel in and flaunt their antisocial attitudes, wearing patches on their club colors such as FTW ("fuck the world"), 666 (the sign of Satan), and 1%, which owes its origin to a statement that outlaw bikers were only 1% of motorcyclists (Abadinsky, 2003, p. 11). Bikers are also intensely concerned with maintaining a personal aura of boldness, strength, and toughness, and have a deep fascination with power. Borrowing from Merton's anomie theory, James Quinn (2001) views bikers' social adaptations as "extremes of retreatism, rebellion, and innovation in combinations that vary across groups, regions, and time periods" (p. 382). Certainly all OMGs are involved in many crimes. The manufacturing and distribution of drugs is their main source of income, with extortion, contract murder, gunrunning, prostitution, pornography, and massage parlors being other sources (Lyman & Potter, 2004). The Hells Angels and the Outlaws, who are often in bitter conflict with each other, are known to have ties to LCN and have performed contract killings for it. An Alcohol, Tobacco, and Firearms (ATF) agent stated that the big four OMGs are "priority ATF targets," and are "the largest—and best armed—criminal organizations in the country" (quoted in Quinn, 2001, p. 381).

African American Organized Crime

Powerful African American organized crime groups have existed in the United States since at least the 1920s, meeting the illicit needs for drugs, alcohol, gambling, and prostitution in black communities (Schatzberg & Kelly, 1996). African American organized crime began with the policy syndicates in Harlem in the 1920s, but they were often either muscled out of business or functioned as local franchise operators for white gangsters (Schatzberg & Kelly, 1996). The Vietnam War enabled African American organized crime to gain its independence from white organized crime by establishing its own connections with Asian drug dealers. The Vietnam War was to African American organized crime what Prohibition was to white organized crime, giving birth to such gangs as the Crips and Bloods, and to the kind of inter- and intragang warfare that characterized Prohibition era white gangs (Adamson, 2000).

The *Gangster Disciples,* formed in Chicago in 1969 by the merger of two former rival gangs—the Disciples and the Supreme Gangsters—became an organization that comes closest to matching LCN in scope and sophistication. The number of members in the Gangster Disciples has been estimated to be from 6,000 to 50,000, ranging across 35 states (Abadinsky, 2003). The group is a formal, hierarchical, and authoritarian organization, and it has alliances with other black groups such as the Crips (Knox & Fuller, 1995). Despite this level of organization, Adamson (2000) disputes their status as a "black mafia" because they lack sophistication, stating that "[a]t best, they are a proto-mafia" (p. 289). Yet it is precisely the lack of sophistication that makes such groups more violent and dangerous than traditional organized crime groups (Cureton, 2009).

Latino Groups

After the 1959 revolution that saw Fidel Castro assume power in Cuba, many Cubans fled their homeland and some of them took to crime. The so-called *Marielitos* originated with the Mariel Harbor boatlift from Cuba in 1980 in which Castro rid himself of about 125,000 Cuban dissidents by shipping them to Florida. Among these refugees were a number of common criminals (estimates range from 2,500 to 8,750). The Marielitos teamed up with other Cuban criminal elements as enforcers and executioners, but soon branched off on their own. Today, they are said to exist in a number of states and are active primarily in drug trafficking (Lyman & Potter, 2004).

The Marielitos have connections with the South American drug trafficking groups. Organized crime thrives where political corruption comes easy, and it comes very easy in Mexico. There are a large number of drug cartels operating in Mexico and sending their wares to the United States. These cartels are known to have hundreds of politicians and law enforcement officials on their payroll. Howard Abadinsky (2003) quotes the director of the U.S. Drug Enforcement Administration (DEA) with regard to the danger of Mexican organized crime to the United States:

> Organized crime figures in Mexico have at their disposal an army of personnel, an arsenal of weapons and the finest technology money can buy. They run transportation and financial empires and an insight into how they conduct their day-to-day business leads . . . to the conclusion that the United States is facing a threat of unprecedented proportions and gravity. (p. 186)

⊠ Theories of Organized Crime

Criminologists ask the same questions about organized crime that they do about white-collar crime: What causes it, are these causes unique to it, and are the external social and economic causes of it more important than individual-level causes? Some argue that we create our own organized crime problem by creating laws that prevent members of the public from acquiring goods and services (alcohol, drugs, gambling, pornography, and prostitution) they desire and demand. When such demands are not met legally, there are always those who are willing to supply them illegally.

Early theories of organized crime relied on the anomie/strain tradition to explain it, describing the gangster as "a man with a gun, acquiring by personal merit what was denied him by complex orderings of stratified society," and saw each successive wave of immigrants ascending a "queer ladder of social mobility" in American society (Bell, 1962). According to this **ethnic succession theory,** upon arrival in the United States, each ethnic group was faced with discriminatory attitudes that denied them legitimate means to success. The Irish, Jews, and Italians were each prominent in organized crime before they became assimilated into American culture and gained access to legitimate means of social mobility. According to this view, African Americans, Russians, and other groups may have to climb their own "queer ladder" until they gain full acceptance in American society, or at least until they feel fully American.

The idea of opportunity denial implies that those allegedly denied opportunities would have gladly taken advantage of them if they existed and would have spurned crime. In this view, a criminal career is simply the default option undertaken by the downtrodden, with the unspoken corollary being that no one would actively seek criminal opportunities if legitimate options were available to them. However, memoirs of a number of LCN figures show that they had received good educations, came from involved and intact families, and had many opportunities to enter legitimate careers but saw crime as a more lucrative career than any legitimate alternative (Firestone, 1997).

Differential association theory may provide an alternative explanation. We know that criminal acts arise from the interaction of environmental opportunities and individual risk factors. The environmental risk factors are particularly powerful in some neighborhoods, and therefore the threshold for engaging individual risk factors is lowered for those living in them. That is, when environments are criminogenic, a person with fewer or less potent risk factors will be drawn into antisocial behavior. If you grew up in a neighborhood where organized crime was established and flourishing, you would stand a good chance of at least having a shot at joining the mob if you were so inclined. Such neighborhoods are fertile ground for the constant cultivation of new batches of criminals because they provided their young inhabitants with exposure to an excess of definitions favorable to law violation. Almost all mob members lived in neighborhoods where they were constantly surrounded by criminals and criminal values. They grew up hero-worshiping the neighborhood made men, emulated their dress and mannerisms, and dreamed of becoming one of them. It was the mobster who had the beautiful women, the sleek cars, the fancy clothes, and the respect, not the legitimate "working stiff" (Firestone, 1997, p. 73). The joy and enthusiasm with which made men describe their acceptance into the gang makes nonsense of the idea that gangsters are deprived individuals making the best of a bad deal: "Getting made is the greatest thing that could ever happen to me. . . . I've been looking forward to this day ever since I was a kid." And a made man in the Colombo family gushes, "Since I got made I got a million fuckin' worshipers hanging around" (Abadinsky, 2003, pp. 23–24).

SUMMARY

- White-collar crime is our costliest and most deadly form of crime. White-collar crime is divided into occupational and corporate crime. Occupational crime is crime committed against an employer or the general public in the course of an individual's employment.
- Except for committing crimes requiring high-status occupation for their performance, most white-collar criminals are not all that different from street criminals, and they occupy a middle position between street criminals and "respectable" people in terms of criminal attitudes and behaviors.
- Corporate crime is criminal activity on behalf of an organization. Corporate crimes involve multiple individuals both as perpetrators and victims, the loss of billions of dollars, and—in some cases—the loss of hundreds of lives.
- Corporate crime is explained by a variety of factors, including the juxtaposition of lucrative opportunities and lenient penalties. Personal characteristics associated with corporate criminality include an external locus of control, a low level of cognitive moral development, and a high level of Machiavellianism.
- Corporate wrongdoing is usually investigated and punished by administrative agencies and has typically been treated leniently. New weapons in the fight against white-collar crime, such as the Sarbanes-Oxley Act, have resulted in meaningful penalties imposed upon individual executives of corporations involved in corporate crime.
- Organized crime is defined by its formal structure, its continuity, and restricted membership. La Cosa Nostra (LCN) is a confederation of families, the leaders of which are of Italian heritage, and which restricts membership to ethnic Italians.
- American organized crime grew out of the corrupt political machine and its supporting street gangs known as Tammany Hall, and received its biggest boost from Prohibition. Many of the gangsters who rose to national prominence during this period got their start in the variety of gangs that supported Tammany Hall.
- The repeal of Prohibition ushered in a quiet period in organized crime's history, particularly after the founding of the "Commission" by Lucky Luciano as a judicial body to settle interfamily disputes without resorting to war.
- The Russian "Mafiya" is considered to be the biggest organized crime threat in the world today. The widespread chaos and corruption following the breakup of the U.S.S.R. allowed Russian organized crime (ROC) to come out of the closet and proliferate. The special danger of ROC is that many of its members are highly intelligent and educated men who held professional jobs in the old Soviet Union.

- The Japanese yakuza is the oldest organized crime group in the world. Having evolved from masterless samurai warriors, it received a major boost by the chaos existing in Japan after its defeat in WWII. The yakuza has many characteristics in common with LCN, but it operates openly and proudly, even publishing its own newsletter.
- Other organized crime groups such as outlaw motorcycle gangs and African American and Asian gangs point to the widespread existence of organized crime.
- Theories of organized crime tend to be in the anomie/strain tradition whereby gangsters are presumed to be mainly from disadvantaged groups who are climbing the "queer (illegitimate) path to success."

DISCUSSION QUESTIONS

1. Do you think there is any moral difference between (A) setting off a bomb outside a building for some political reason, knowing that a certain number of people would be killed, and (B) marketing 11 million defective automobiles knowing that a certain proportion of them will explode into flames when rear-ended and burn the occupants alive? Explain your answer.

2. Do you think that white-collar crime (occupational and corporate) can be explained by the same principles as street crime? Read one or two of the relevant cited articles for guidance.

3. Looking back at all the theories presented in this book, make a case for one of them as the best in terms of explaining organized crime.

4. Go to any website dealing with organized crime and choose an organized crime figure. Write a one-page report on the person.

USEFUL WEBSITES

Criminal Justice Resources—Organized Crime. www.fbi.gov/about-us/investigate/organizedcrime

FBI—Organized Crime. www.fbi.gov/hq/cid/orgcrime/ocshome.htm

FBI—White Collar Crime. www.fbi.gov/whitecollarcrime.htm.

National White Collar Crime Center. www.nw3c.org/

RICO Act. www.ricoact.com/

CHAPTER TERMS

Commission	Feudal model	Prohibition
Corporate crime	La Cosa Nostra	Sarbanes-Oxley Act (SOA)
Corporate model	Occupational crime	White-collar crime
Ethnic succession theory	Organized crime	

CHAPTER 15

Victimology

Exploring the Experience of Victimization

John Sutcliff's entire adult life has been devoted to the sexual seduction of teenage boys. At the age of 33, he was arrested and sentenced to prison for sexually assaulting a 13-year-old boy who was a member of his "Big Brother's Club." By his own admission, he had sexually molested over 200 "members" of his club. John's favorite activity with these boys was giving and receiving enemas. John became involved with the fetish while enrolled in a residential boy's school where many of the boys were subjected to enemas administered in front of the entire dormitory.

After his release from prison, John became much more "scientific" in his efforts to procure victims. A "theoretical" paper he wrote indicated that father-absent boys were "ripe" for seduction, and he would entice them with his friendly ways and with a houseful of electronic equipment he would teach the boys to repair and operate. He weeded out boys with a father in the home and would spend at least 6 weeks grooming each victim. He used systematic desensitization techniques, starting with simply getting the boys to agree to type in answers to innocuous questions, and escalating to having them view pornographic homosexual pictures, and giving them "pretend" enemas, actual enemas, and enemas accompanied by homosexual activity. With each successive approximation toward John's goal, the boys were reinforced by material and nonmaterial rewards (friendship, attention, praise) that made the final events seem almost natural.

(Continued)

(Continued)

John's activities came to light when U.S. postal inspectors found a package containing pictures, letters, and tapes John exchanged with like-minded individuals. On the basis of this evidence, the police raided John's home and found neatly catalogued files detailing 475 boys that he had seduced. His methods were so successful that his actions were never reported to the authorities (indeed, some of the boys were recruited for him by earlier victims). Some of his earlier victims still kept in touch with him and were victimizing boys themselves. Only one victim agreed to testify, but John was allowed to plea to one count of lewd and lascivious conduct. He received a sentence of 1 year in prison and was paroled after serving 10 months, and thus served 15.7 hours for each of his 475 known victims. This case illustrates how victims (totally innocent as children) can be turned into victimizers (totally responsible as adults) and how the distinction between victim and perpetrator can sometimes be blurred.

The Emergence of Victimology

Except for minor public order crimes, for every criminal act there is necessarily at least one victim. Criminologists have spent decades trying to determine the factors that contribute to making a person a criminal, but it wasn't until the German criminologist Hans von Hentig's (1941) work that they began seriously thinking about the role of the victim. It turned out that although victimization can be an unfortunate random event where the victim was simply in the wrong place at the wrong time, in many—perhaps even in most—cases of victimization, there is a systematic pattern if one looks closely enough. Just as criminologists want to find out why some people commit crimes and others do not, and why some who commit crimes commit more crimes than others, victimologists want to discover why some people become victims, and why some victims become repeat victims.

Victimology is a subfield of criminology that specializes in studying the victims of crime. Victimologists study the series of events that typically lead to victimization acts of various kinds in attempts to (1) arrive at general theories of victimization and (2) try to arrive at insights relevant to how victimization can be avoided. They also examine the way victims are treated in the criminal justice system in its attempts to compensate crime victims and attend to their practical and emotional needs (Karmen, 2005). Criminologists interested in perpetrators of crime ask what the risk factors for becoming involved in crime are; criminologists interested in victims of crime ask pretty much the same questions, i.e., why are some individuals, households, groups, and other entities targeted and others are not (Doerner & Lab, 2002). The labels "offender" and "victim" are sometimes blurred distinctions that hide the details of the interactions of the offender/victim dyad. Burglars often prey on their own kind, robbers prey on drug dealers, and homicides are frequently the outcome of minor arguments in which the victim was the instigator. As victimologist Andrew Karmen puts it,

> Predators prey on each other as well as upon innocent members of the public. . . . When youth gangs feud with each other by carrying out "drive-by" shootings, the young members who get gunned down are casualties of their own brand of retaliatory street justice. (p. 14)

Of course, we should not think of all victims, or even most victims, this way. There are millions of innocent victims who in no way contribute to their victimization, and even lawbreakers can be genuine victims deserving of protection and redress in the criminal courts.

✍ Who Gets Victimized?

Victimization is not a random process. Becoming a victim is a process encompassing a host of systematic environmental, demographic, and personal characteristics. According to the 2008 NCVS study (Rand, 2009), the individual most likely to be victimized is a young black unmarried male living in poverty in an urban environment. Victimization, like criminal behavior, drops precipitously from 25 years of age onward, it also drops with increasing household income, and being married is a protective factor against victimization as it is against crime.

Victim characteristics also differ according to the type of crime. The study cited above found that females were 4.3 times more likely than males to be victimized by rape/sexual assault, but males were 1.6 times more likely to be victimized by aggravated assault. Females were more likely to be victimized by someone they know, and males tended to be victimized by strangers. Note from Figure 15.1 that male and female rates were nowhere near as far apart in 2008 as they were in 1973. This convergence is mostly the result of much lower rates of male victimization in 2008, although it is clear that victimization is decreasing

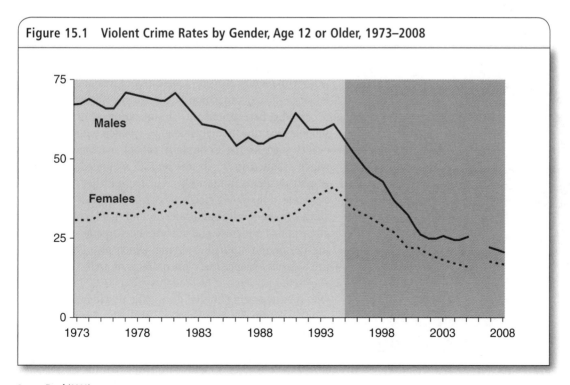

Figure 15.1   Violent Crime Rates by Gender, Age 12 or Older, 1973–2008

Source: Rand (2009).

for both genders. Blacks were 1.7 times more likely than "other races" (Asian, American Indian/Alaska Native) to be victims of aggravated assault, but slightly less likely than whites to be victims of simple assault. Individuals 65 or older were 20 times less likely than individuals 20–24 to be victimized by any type of violent crime, but slightly more likely to be victimized by a personal theft.

▨ Victimization in the Workplace and School

Two important demographic variables not included in the 2008 NCVS study are victimization at work and at school. It is important to consider these variables since most of us spend the majority of our waking hours either on the job or at school. The last systematic effort by the Department of Justice to assess the level of workplace violence in the United States was in 1998. This report dealt with workplace violence taking place from 1992 to 1996 (inclusive) and found that on average, over 2 million incidents take place annually (Warchol, 1998). Table 15.1 provides the average annual level of workplace violence over the 5-year period. Two-thirds of the victims in the survey were male, almost 90% were white, and the age category most likely to be victimized was the 35–49 group. The three occupations most at risk were police officers (at a rate of 306 per 1,000 workers), corrections officers (217.8), and taxi drivers (183.8); taxi drivers were the most likely to be killed, however. Taxi drivers had a homicide victimization rate of 26.9 per 100,000 workers; all protective services workers had a rate more than 5 times lower than taxi drivers (5.0 per 100,000, which is approximately the same as the general population in 2009). The most dangerous jobs are those in which the workers must deal with the public in a protective/supervisory capacity or who work alone and are relatively isolated from others, work at night, or work with money. The safest job category was university professor (with a victimization rate of 2.5 per 100,000).

Public perceptions of victimization at the nation's schools are unfortunately fueled by isolated but spectacular events such as the Columbine school massacre in 1999, which resulted in 13 deaths and 25 wounded, and other similar incidents in the 1990s. The truth is that our schools are some of the safest places we can be. DeVoe et al.'s (2003) study of school crime and safety found that less than 1% of all juvenile homicides and suicides occurred at school during the period studied. Bullying, which also gets a lot of press, seems surprisingly uncommon as well. Figure 15.2 shows the percentage of students from 12 to 18 years old who reported being bullied during the previous school year (Dinkes, Kemp, Baum, & Snyder, 2010). The vast majority of the bullying reported consists of being the subject of rumors, being made fun of, or being excluded from activities. Only 11% of respondents said they were pushed, tripped, or spit on, and 6% said they were threatened with physical harm.

▲ **Photo 15.1** Where does normal male rough-housing end and bullying begin?

Table 15.1	Average Annual Number of Violent Victimizations in the Workplace, 1992–1996	
	Annual average	**Percent**
Homicide	1,023	0.05
Rape/Sexual Assault	50,500	2.50
Robbery	83,700	4.20
Aggravated Assault	395,500	19.70
Simple Assault	1,480,000	73.60
Total	2,010,723	100.00

Source: Warchol (1998).

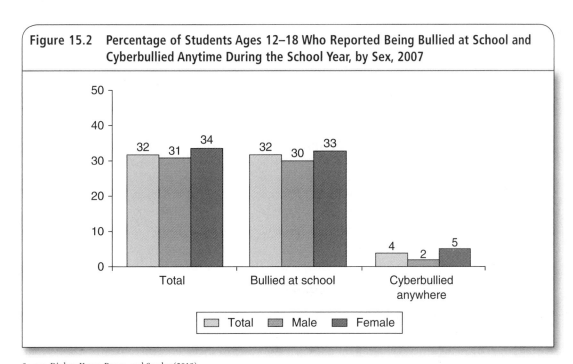

Figure 15.2   Percentage of Students Ages 12–18 Who Reported Being Bullied at School and Cyberbullied Anytime During the School Year, by Sex, 2007

Source: Dinkes, Kemp, Baum, and Snyder (2010).

Child Molestation: Who Gets Victimized?

Child molestation is perhaps the most prevalent crime against the person in the United States, with approximately two-thirds of incarcerated sex offenders having offended against children (Talbot, Gilligan, Carter, & Matson, 2002). It is problematic to accurately gauge the prevalence of child molesting, as rates depend on how broadly or how narrowly molesting is defined. A "best guess" arrived at from a variety of sources is that the percentage of children in the United States experiencing sexual abuse sometime during their childhood is 25% for girls and 10% for boys (Knudson, 1991). Girls are more likely to be abused within the family, and boys are more likely to be victimized by acquaintances and by strangers (Walsh, 1994). The strongest single predictor of victimization for girls is having a stepfather. Stepfathers are about 5 times more likely to sexually abuse their daughters than are biological fathers (Glaser & Frosh, 1993). The strongest predictor for boys is growing up in a father-absent home (Walsh, 1988). There are many other factors predictive of child sexual abuse, and the more that are present, the more likely abuse is to occur.

Finkelhor (1984) developed a risk factor checklist for the likelihood of girls' victimization containing the following predictors:

1. Living with a stepfather

2. Living without a biological mother

3. Not close to mother

4. Mother never finished high school

5. Sex-punitive mother

6. No physical affection from (biological) father

7. Family income under $10,000 (in 1980 dollars; $25,750 in 2010 dollars)

8. Two friends or fewer in childhood

Finkelhor found that the probability of victimization was virtually zero among girls with none of the predictors in their background and rose steadily to 66% among girls with five. Given the large number of divorces, births to single mothers, and reconstituted families we are seeing in the United States, these risk factors for sexual abuse will be experienced by an increasing number of children. For instance, a large nationwide study of children conducted by Turner, Finkelhor, and Ormrod (2006) found that child molestation is over 4 times more likely to occur when a stepfather or live-in boyfriend is in the home as opposed to the child's biological father married to the mother.

✂ **Domestic Violence Victimization**

Domestic violence victimization encompasses a variety of acts, and refers to any abusive act (physical, sexual, or psychological) that occurs within a domestic setting. Family violence is the most prevalent form of violence in the United States today, and most of it is intimate partner (spouse or lover) violence (Tolan, Gorman-Smith, & Henry, 2006). Except for minor forms of abuse, intimate partner violence is overwhelmingly committed by males against females (see Figure 15.3), although when females commit such violence they are more likely to use a weapon to equalize the size and strength difference between the sexes (E. Smith & Farole, 2009). However, while just over 33% of all murders of females in the United States are committed by intimate partners, less than 4% of males are killed by intimate partners (Rennison, 2003).

Violent victimization of spouses or lovers perpetrated by males is primarily driven by male sexual ownership, jealousy, and suspicion of infidelity. Evidence from around the world indicates that the single most important cause of domestic violence (including homicide) is male jealousy and suspicion of infidelity

Figure 15.3   Highlights of the 2003 Report on Intimate Partner Violence

Demographic characteristic	Percent of intimate partner violence—	
	Victims	Defendants
Total	100%	100%
Gender		
Male	140%	86.3%
Female	86.0	13.7
Race/Hispanic origin		
White non-Hispanic	37.1%	33.6%
Black non-Hispanic	26.4	33.5
Hispanic	33.6	30.8
Other non-Hispanic	2.8	2.0
Age at offense		
17 or younger	2.7%	0.2%
18–24	26.1	24.2
25–34	34.9	34.8
35–54	34.0	38.2
55 or older	2.3	2.6

Source: Smith and Farole (2009).

(Lepowsky, 1994). DNA data indicate that between 1 and 30% (depending on the culture or subculture) of children are fathered by someone other than the presumed father (Birkenhead & Moller, 1992). The threat of cuckoldry (being fooled into raising someone else's child) is thus real, which suggests that male violence against spouses and lovers should be most common in environments where the threat of infidelity is most real. Such environments would be those in which marriages are most precarious, where moral restrictions on pre- and extramarital sexual relationships are weakest, and where out-of-wedlock birthrates are highest. This is what we see in Figure 15.4.

Although by no means limited to the lower classes, domestic violence is most often committed by competitively disadvantaged (CD) males (Graham-Kevan & Archer, 2009). CD males have low mate value because they have less to offer in terms of resources or prospects of acquiring them, which tends to make their mates less desirous of maintaining the relationship with them, and more likely to seek other partners. Lacking alternative means of controlling their partner's behavior, CD males may turn to violently coercive tactics to intimidate them. This economic disadvantage may be one of the reasons that intimate personal violence is 2 to 3 times more prevalent and more deadly among African American males than among

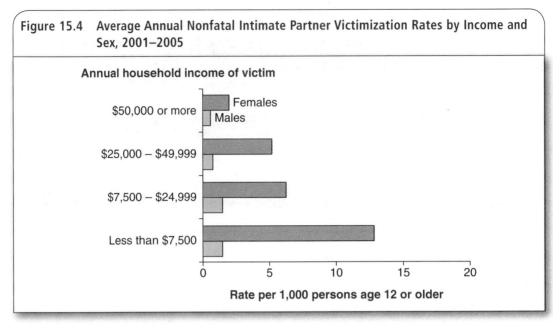

Figure 15.4 Average Annual Nonfatal Intimate Partner Victimization Rates by Income and Sex, 2001–2005

SOURCE: Catalano (2007).

males of other races (Hampton, Oliver, & Magarian, 2003). Hampton and his colleagues also list the anger and frustration born of poverty and unemployment, the reluctance of black females to report incidents, and the generally fractious and antagonistic relationship that allegedly exists between black men and women as reasons.

The victimization of children in domestic situations is particularly heinous, and we need to know under what circumstances it is most prevalent. It seems that every research program examining this problem finds that it is most likely to occur outside of homes in which children reside with two biological parents. A national representative sample of over 2,000 children ages 2 through 9 found that children of single parents were 6.7 times more likely to witness family violence, 3.9 times more likely to be maltreated, and 2.7 times more likely to be sexually assaulted than children with both biological parents present. The figures for stepparent families were even worse at 9.2, 4.6, and 4.3, respectively (Turner et al., 2006). This same study found that stepchildren were 9.2 times more likely to witness family violence, 4.6 times more likely to be maltreated, and 4.3 times more likely to be sexually assaulted than children living with two biological parents. A child living with a stepfather or live-in boyfriend is approximately 65 times more likely to be fatally abused than a child living with both biological parents (M. Daly & Wilson, 1996). The vast majority of stepparents do not abuse their stepchildren, of course, but the risk is greatly elevated.

⬚ Victimization Theories

Victimization can occur at any time or place without warning. Who could have predicted someone putting gas in her car at the filling station would be gunned down by the Washington, D.C., snipers in 2002, or the typist at his desk in the World Trade Center would be obliterated seconds later by a passenger jet on

September 11, 2001? There is no systematic way to evaluate events such as these from a victimology perspective. But as previously noted, most victimizing events are not random or unpredictable. Criminologists no longer view victims as simply passive players in crime who were unfortunate enough to be in the wrong place at the wrong time (as of course were the victims of 9/11 and the D.C. snipers). In the majority of cases, victims are now seen as individuals who in some way, knowingly or unknowingly, passively or actively, influenced their victimization. Obviously, the role of the victim, however provocative it may be, is never a necessary and sufficient cause of his or her victimization, and therefore cannot fully explain the actions of the person committing the criminal act.

Victim Precipitation Theory

Victim precipitation theory was first presented by von Hentig (1941) and applies only to violent victimization. Its basic premise is that by acting in certain provocative ways, some individuals initiate a chain of events that lead to their victimization. Most murders of spouses and boyfriends by women, for example, are victim precipitated in that the "perpetrator" is defending herself from the victim (Mann, 1990). Likewise, serious delinquent and criminal behavior and serious victimization are inextricably linked. Shaffer and Ruback (2002), for instance, found that violent offenders, when all other things are statistically controlled, are about twice as likely as nonviolent offenders to be victimized themselves. Furthermore, past victimization is the best predictor of future victimization (odds ratio = 5.7, which means that if you were victimized in Year 1, the odds of your being victimized in Year 2 are 5.7 times greater than they are if you were not victimized in Year 1). Another study using

▲ **Photo 15.2**   Gary Ridgway became known as the Green River Killer for his habit of depositing victims' bodies along this waterway. Serial killers frequently victimize marginalized groups, such as prostitutes. Some of his victims' bodies were only discovered years after their untimely deaths, by searchers such as these, revisiting kill sites.

data from the longitudinal Pittsburgh and Denver studies of delinquency risk factors (e.g., low SES, single-parent household, hyperactivity, impulsiveness, drug usage, etc.) showed that the same factors predicted victimization as well (Loeber, Kalb, & Huizinga, 2001). As seen in Figure 15.5, overall, 50% of seriously violent delinquents were themselves violently victimized, compared with 10% of non-delinquents from the same neighborhoods.

Victim precipitation theory has been most contentious when it is applied to rape victims ever since Menachem Amir's (1971) study of police records found that 19% of forcible rapes were supposedly victim precipitated. Amir defined victim-precipitated rape as a case of rape in which the victim initially agreed to sexual relations and then reneged. We saw in Chapter 10 that a number of surveys of high school and college students have shown that a majority of males and a significant minority of females believe that it is justifiable

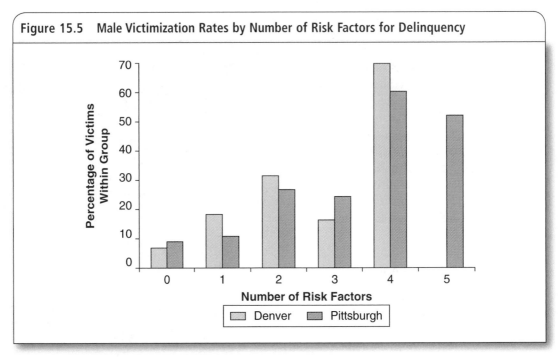

Figure 15.5   Male Victimization Rates by Number of Risk Factors for Delinquency

Source: Loeber, Kalb, and Huizinga (2001).

for a man to use a measure of violence to obtain sex if the victim had "led him on" (Herman, 1990). It is for this reason that many criminologists have disparaged victim precipitation theory as victim blaming, although it was never meant to be that. Hopefully, the attitudes revealed in these 1980s surveys have diminished with the greater awareness of the horrible nature of this crime, and a greater awareness that women have every right to change their minds even if they have "excited" their partners physically.

Figure 15.6 provides four scenarios illustrating various levels of victim/offender responsibility from the victim precipitation perspective. In the first scenario, the woman who stabbed her husband after suffering years of abuse is judged blameless, although some lacking a little in empathy and understanding of the psychology of domestic abuse may argue that she must take some responsibility for remaining in the relationship. In the second scenario, both the offender and the victim were engaging in a minor vice crime and both are judged equally responsible for the crime (morally he should not have been there, and he was careless with his wallet). In the third scenario, the victim facilitated the crime by carelessly leaving his keys in the car. In the last scenario, the child is totally innocent of any responsibility for what happened to her. I want to strongly emphasize that whatever the degree of responsibility, "responsibility" does not mean "guilt."

Routine Activities/Lifestyle Theory

Routine activities and **lifestyle theories** are separate entities, but in victimology they are similar enough to warrant being merged into one (Doerner & Lab, 2002). As we saw in Chapter 3, routine activities theory stresses that criminal behavior takes place via the interaction of three variables that reflect individuals' everyday routine activities: (1) the presence of motivated offenders, (2) the availability of suitable targets,

Figure 15.6   Four Scenarios Illustrating the Degree of Victim/Offender Responsibility According to Victim Precipitation Theory

Degree of Criminal Intent of the Perpetrator			
None →	Some →	More →	Much
Victim Provocation A woman who has suffered years of abuse stabs and kills her husband in self-defense as he is beating her again.	*Equal Responsibility* Victim using the services of a prostitute leaves his wallet on the bed stand and leaves. She decides to keep the money in his wallet.	*Victim Facilitation* Victim leaves keys in his car while he runs into a store. A teenager impulsively steals the car and wrecks it.	*Victim Innocent* A sex offender kidnaps a screaming young girl from a playground and molests her.
Much	← More	← Some	← None
Degree of Victim Facilitation or Provocation/Precipitation			

and (3) the absence of capable guardians. The basic idea of lifestyle theory is that there are certain lifestyles (routine activities) that disproportionately expose some people to high risk for victimization. Lifestyles are the routine patterned activities that people engage in on a daily basis, both obligatory (e.g., work-related) and optional (e.g., recreational). A high-risk lifestyle may mean getting involved with deviant peer groups or drugs, just "hanging out," or frequenting bars until late into the night and drinking heavily. Routine activities/lifestyle theory explains some of the data relating to demographic profiles and risk presented by Loeber et al. (2001) discussed earlier. Males, the young, the unmarried, and the poor are more at risk for victimization than females, older people, married people, and more affluent people because they have riskier lifestyles. On average, the lifestyles of the former are more active and action-oriented than those of the latter.

These lifestyles sometimes lead to repeat victimization. Prior victimization has been called "arguably the best readily available predictor of future victimization" and one that "appears a robust finding across crime types and data sources" (Tseloni & Pease, 2003, p. 196). Lisa Bostaph (2004) reviews the literature on what she calls "career victims" and among the various interesting research findings on this phenomenon, she lists the following as attributable to lifestyle patterns:

- A British crime survey that found that 20.2% of the respondents were victims of 81.2% of all offenses
- A study that found 24% of rape victims had been raped before
- A study of assault victims in the Netherlands that found 11.3% of victims accounted for 25.3% of hospital admissions for assault over 25 years
- A study reporting that 67% of sexual assault victims had experienced prior sexual assaults

Why we see so much repeat victimization is an important question on victimology's research agenda. The explanations offered by various theorists almost inevitably revolve around routine activities theory—a motivated offender taking advantage of suitable victims lacking capable guardians. The repeat victimization of domestic partners or of children occurs because the perpetrator is typically the person who is supposed to be the capable guardian, and the "suitable targets" are often trapped in the same household as the motivated offenders. Over time, such victims may come to accept victimization as normal and inevitable.

Repeat victims of violent assault are often found to be individuals who tend to frequent places known to have violent reputations such as certain bars where people come with the stated intention of getting drunk and involved in fights (Farrell, Phillips, & Pease, 1995). These individuals can be repeat offenders, repeat victims, or both.

Most of the research on routine activities/lifestyle theory has been done on rape victimization. Bonnie Fisher and her colleagues' (Fisher, Cullen, & Turner, 2001) national sample of college women found that 2.8% had been raped, although 46.5 % of this 2.8% whom Fisher et al. defined as rape victims said that they did not *experience* the event as rape. Fisher and colleagues report that four lifestyle factors are consistently found to increase the risk of sexual assault: (1) frequently drinking enough to get drunk, (2) being unmarried, (3) having previously been a victim of sexual assault, and (4) living on campus (for on-campus victimization only). A later study by Daigle, Fisher, and Cullen (2008) among 4,432 college females found that a mere 3.3% of them experienced 45.2% of all the forcible sexual incidents reported in the study, indicating many repeat victimizations. The major factor that appears to distinguish one-time from repeat victims is that one-time victims take steps to avoid circumstances that led to their victimization whereas repeat victims tend not to. It is also possible that repeat victims suffer from what has been called *traumatic sexualization,* which is the result of prior victimization by someone the victim trusted. This can lead to risk taking and provocative behavior. Such victimization can also undermine trust in others and result in feelings of powerlessness. All of these symptoms increase the risk of revictimization (J. Reid & Sullivan, 2009).

Is Victimology "Blaming the Victim"?

Some victim advocates reject victimology theories as "victim blaming." Victimologists counter by emphasizing that they do not "blame" victims; rather, they simply explore the process of victimization with the goal of understanding it and *preventing* it. Although victimology research is used to develop crime-prevention strategies, not to berate victims, some victim advocates even reject "as ideologically tainted" crime-prevention tips endorsed by victimologists (Karmen, 2005, p. 129). Victimologists maintain, however, that crime-prevention tips and strategies are ignored at our peril. We all agree that we *should* be able to leave our cars unlocked, sleep with the windows open in summer, leave the doors to our homes unlocked, frequent any bar we choose, and walk down any alley in any neighborhood at any time we please, but we cannot. Common sense demands that we take what steps we can to safeguard ourselves and our property in this imperfect world. Crime-prevention tips are really no different from tips we get all the time about staying healthy: Eat right, exercise, and quit smoking if you want to avoid health problems. Similarly, avoid certain places, dress sensibly, don't provoke, take reasonable precautionary measures, and don't drink too much if you want to avoid victimization. Victims deserve our sympathy even if they somehow provoked or facilitated their own victimization. Victimologists do not "blame"; they simply remind us that complete innocence and full responsibility lie on a continuum.

The Consequences of Victimization

Some crime victims suffer lifelong pain from wounds and some suffer permanent disability, but for the majority of victims, the worst consequences are psychological. We all like to think that we live in a safe, predictable, and lawful world in which people treat one another decently. When we are victimized, this comfortable "just world" view is shattered. With victimization come stressful feelings of shock, personal vulnerability, anger, fear of further victimization, and suspicion of others.

Victimization also produces feelings of depression, guilt, self-blame, and lowered self-esteem and self-efficacy. Rape in particular has these consequences for its victims ("Did I contribute to it?" "Could I have done more to prevent it?"). The shock, anger, and depression that typically afflict a rape victim are collectively known as **rape trauma syndrome,** which is similar to post-traumatic stress syndrome (re-experiencing the event via "flashbacks," avoiding anything at all associated with the event, and a general numbness of affect) often suffered by those who have

▲ **Photo 15.3** Efforts to better recognize victims and their rights have become more common over the past 20 years. This photo memorializes the victims of Columbine High School, including the killers themselves, who committed suicide.

experienced the horrors of war (van Berlo & Ensink, 2000). Victimization "also changes one's perceptions of and beliefs about others in society. It does so by indicating others as sources of threat and harm rather than sources of support" (Macmillan, 2001, p. 12).

Victims of property crimes, particularly burglary, also have the foundations of their world shaken. The home is supposed to be a personal sanctuary of safety and security, and when it is "touched" by an intruder, some victims describe is as the "rape" of their home (Bartol, 2002, p. 336). A British study of burglary victims found that 65% reacted with anger, 30% experienced fear of revictimization, and 29% suffered insomnia as a consequence. The type and severity of these reactions were structured by the victims' place in the social structure, with those most likely to be affected being women, older, and poorer individuals, and residents of single-parent households (Mawby, 2001).

As mentioned in Chapter 14, with the advent of the computer age we are all "victims-in-waiting." One of the most terrifying fictional depictions of cyber victimization is provided in the movie *The Net*. In this movie, Angela Bennett, played by Sandra Bullock, is a computer expert whose life turns into a nightmare when her records are wiped clean and she is given a new identity by people who have it in for her. Her new identity comes complete with a police record, and the rest of the movie is about her struggle to find out who has done this to her and why.

Michelle Brown, a 29-year-old white female, is one of a number of real-life Angela Bennetts. Her nightmare began in January of 1998. While Michelle's records were not erased, they were "cloned" by a woman who had gained access to her personal information. Her identity clone was Heddi Larae Ille, a 33-year-old white female. With a line of credit established with Michelle's social security number and driver's license number, Heddi racked up $1,443 in phone bills, bought a $32,000 automobile, had $4,800 worth of liposuction, and bought numerous other items. Worse yet, Heddi was arrested under Michelle Brown's name for smuggling 3,000 pounds of marijuana into Texas from Mexico. Michelle was thus named in the indictment and listed as a DEA informant. Returning from a trip overseas, Michelle (the real one) was detained for over an hour at Los Angeles airport because the DEA had posted a lookout for her. Only a phone call from a police detective aware of Michelle's predicament got her released. Heddi was arrested and booked again under Michelle's identity for grand theft and possession of stolen property in 1999. Her true identity finally came to light and she was sentenced to 73 months in federal prison and 24 months in state prison.

Michelle presented her story before a U.S. Senate committee hearing on identity theft (Kyl & Feinstein, 2000). She also informed the committee of the traumatic effect her victimization had had on her life. She said that she spent over 500 hours (the equivalent of 12-and-a-half workweeks) trying to unravel the mess, lost countless hours of sleep, lost her appetite, and lost a valued 3-year relationship with her boyfriend. She added that she also "lost identification with the person I really was inside and shut myself out of social functions because of the negativity this caused in my life." She indicated that she was afraid to leave the country again in case her name is still on some country's computer listing her as "wanted." Michelle's victimization is just one of the many thousands of such cases that occur annually in the United States.

In summing up the consequences of victimization, it is important to note that just as offending behavior shapes the life course trajectories of offenders, violent victimization affects the life course trajectories of victims. Scott Menard's (2002) study of the National Survey of Youth samples, a longitudinal study involving individuals from age 11 to 33, found that violent victimization during adolescence has pervasive effects on problem outcomes as adults. Figure 15.7 shows that the expected probability of a variety of negative outcomes in adulthood is much greater for victims of violence during adolescence than for nonvictims during the same period.

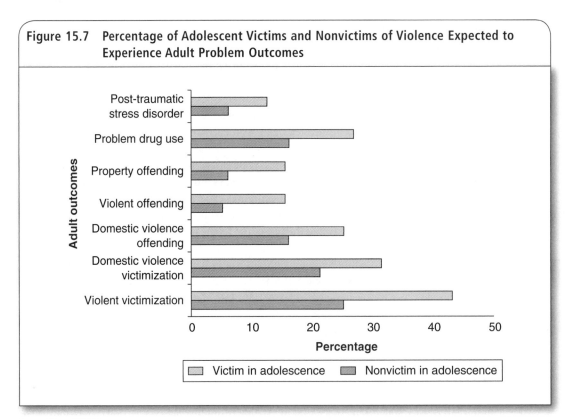

Figure 15.7   Percentage of Adolescent Victims and Nonvictims of Violence Expected to Experience Adult Problem Outcomes

Source: Menard (2002).

Victimization and the Criminal Justice System

Until fairly recently, the victim had been the forgotten party in the criminal justice system. In the United States, crime is considered as an act against the state rather than against the individual who was actually victimized. In 2004, the U.S. Senate passed a crime victims' bill of rights (see below) that has gone some considerable way toward recognizing the previously discounted victim. Although these rights apply only to victims of federal crimes, all 50 states have implemented constitutional amendments or promulgated bills guaranteeing similar rights. We owe much of this increased attention to victim issues to the women's movement and to feminist criminologists.

CRIME VICTIMS' BILL OF RIGHTS

(1) The right to be reasonably protected from the accused

(2) The right to reasonable, accurate, and timely notice of any public proceeding involving the crime or of any release or escape of the accused

(3) The right not to be excluded from any such public proceeding

(4) The right to be reasonably heard at any public proceeding involving release, plea, or sentencing

(5) The right to confer with the attorney for the Government in the case

(6) The right to full and timely restitution as provided in law

(7) The right to proceedings free from unreasonable delay

(8) The right to be treated with fairness and with respect for the victim's dignity and privacy.

Source: Senate Bill S2329, April 21, 2004.

Crime victims are eligible for partial compensation from the state to cover medical and living expenses incurred as a result of their victimization. All 50 states and all U.S. protectorates have established programs that typically cover what private insurance does not, assuming the state has sufficient funds. According to the National Association of Crime Victim Compensation Board (NACVCB), in 2004, victims of violent crime nationwide received a total of $426 million in compensation, with the majority (51%) going for medical expenses (NACVB, 2005).

Victim–Offender Reconciliation Programs

Victim–offender reconciliation programs (VORPs) are an integral component of restorative justice philosophy, which was briefly discussed under the heading of "Peacemaking Criminology" in Chapter 6. Many crime victims are seeking fairness, justice, and restitution *as defined by them* (restorative justice) as opposed to revenge and punishment. Central to the VORP process is the bringing together of victim and offender in face-to-face meetings mediated by a person trained in mediation theory and practice (Price, 2005). Meetings are voluntary for both offender and victim and are designed to iron out ways in which the offender can make amends for the hurt and damage caused to the victim.

Victims participating in VORPs gain the opportunity to make offenders aware of their feelings of personal violation and loss and to lay out their proposals of how offenders can restore the situation. Offenders are afforded the opportunity to see firsthand the pain they have caused their victims, and perhaps even to express remorse. The mediator assists the parties in developing a contract agreeable to both. The mediator monitors the terms of the contract and may schedule further face-to-face meetings.

VORPs are used most often in the juvenile system but rarely used for victims and offenders in personal violent crimes in either juvenile or adult systems. Where they are used, about 60% of victims invited to participate actually become involved, and a high percentage (mid- to high 90s) result in signed contracts (Coates, 1990). Mark Umbreit (1994) sums up the various satisfactions expressed by victims who participate in VORPs:

1. Meeting offenders helped reduce their fear of being revictimized.

2. They appreciated the opportunity to tell offenders how they felt.

3. Being personally involved in the justice process was satisfying to them.

4. They gained insight into the crime and into the offender's situation.

5. They received restitution.

However, VORPs do not suit all victims, especially those who feel that the wrong done to them cannot so easily be "put right" and want the offender punished (Olson & Dzur, 2004).

SUMMARY

- Victimology is the study of the risk factors for and consequences of victimization, and criminal justice approaches dealing with victims and victimization. The risk factors for victimization are basically the same as the risk factors for victimizing in terms of gender, race, age, SES, personal characteristics, and neighborhood.
- Domestic violence (mostly intimate partner violence) is the most prevalent form of violence in the United States today. Much of it is driven by jealousy and real or imagined infidelity and is most likely to be committed by competitively disadvantaged males.
- Theories of victimization such as victim precipitation theory and routine activities/lifestyle theory examine the victim's role in facilitating or precipitating his or her victimization. This is not "victim blaming," but rather an effort to understand and prevent victimization. Victimologists apportion responsibility within the victim/offender dyad on a continuum from complete victim innocence to victim precipitation.
- The consequences of victimization can be devastating both physically and psychologically. Although the severity of the psychological consequences of the same sort of victimization can vary widely according to the characteristics of the victim, consequences can range from short-lived anger to post-traumatic stress syndrome, especially for victims of rape.
- Until fairly recently, victims were the forgotten party in a criminal justice system that tended to think of them only as "evidence" or witnesses. Things have changed over the last 25 years with the passage of victims' rights bills by the federal government and all 50 states. There are also various victim-centered programs designed to ease the pains of victimization such as through victim compensation.

DISCUSSION QUESTIONS

1. Interview a willing classmate or friend who has been victimized by a serious crime and ask about his or her feelings both shortly after victimization and now. Did it change his or her attitudes about crime and punishment?

2. Is it a surprise to you that perpetrators of crimes are more likely to also be victims of crime than people in general? Why or why not?

3. Go to your state's official website and determine funding levels and what services are available to crime victims.

4. Discuss what learning about victimology helps you to further understand about offending behavior.

5. Domestic violence has been falling dramatically over the last 10 to 20 years. In your view, is this attributable to mandatory arrest policies or to some other factor? Explain your answer.

6. In your opinion, does the criminal justice system do enough to guarantee victims' rights, and what other steps can be taken to lessen the harm suffered by victims in the aftermath of being victimized?

USEFUL WEBSITES

American Society of Victimology. www.american-society-victimology.us/

International Victimology Institute. www.victimology.nl/

National Crime Victimization Survey Resource Guide. www.icpsr.umich.edu/NACJD/NCVS/

National Incident-Based Reporting System Resource Guide. www.icpsr.umich.edu/NACJD/NIBRS/

The World Society of Victimology. www.worldsocietyofvictimology.org/

CHAPTER TERMS

Lifestyle theory	Routine activities/lifestyle theory	Victim precipitation theory
Rape trauma syndrome	Routine activities theory	Victimology

Glossary

Actus reus: Literally, *guilty act.* Refers to the principle that a person must commit some forbidden act or neglect some mandatory act before he or she can be subjected to criminal sanctions.

Adaptations: The products of the process of natural selection. Adaptations may be anatomical, physiological, or behavioral.

Age-graded theory: Sampson and Laub's theory stressing the power of informal social controls to explain onset, continuance, and desistance from crime. Emphasizes the concepts of social capital, turning points in life, and human agency.

Aggravated assault: An unlawful attack by one person upon another for the purpose of inflicting severe or aggravated bodily injury.

Aggravated murder or first-degree murder: The most serious kind of murder, it requires that the act be committed with malice aforethought, deliberation, and premeditation.

Agreeableness: The tendency to be friendly, considerate, courteous, helpful, and cooperative with others.

Alcoholism: A chronic disease marked by progressive incapacity to control alcohol consumption despite psychological, social, or physiological disruptions.

Alienation: A condition that describes the estrangement or distancing of individuals from something, such as another person or from society in general.

Altruism: The action component of empathy; i.e., an *active* concern for the well-being of others.

Anomie: A term meaning "lacking in rules" or "normlessness" used by Durkheim to describe a condition of normative deregulation in society.

Antisocial personality disorder: A psychiatric label described as "a pervasive pattern of disregard for, and violation of, the rights of others that begins in childhood or early adolescence and continues into adulthood."

Arraignment: A court proceeding in which the defendant answers to the charges against him or her by pleading guilty, not guilty, or no contest (nolo contendere).

Arrest: The act of being legally detained to answer criminal charges on the basis of an arrest warrant or the belief of a law enforcement officer that there is probable cause to think that the person arrested has committed a felony crime.

Arson: Any willful or malicious burning or attempting to burn, with or without intent to defraud, a dwelling house, public building, motor vehicle or aircraft, or personal property of another.

Atavism: Cesare Lombroso's term for his "born criminals," meaning that they are evolutionary "throwbacks" to an earlier form of life.

Attachment: One of the elements of the social bonds in social control theory; the emotional component of conformity referring to one's attachment to others and to social institutions.

Attention-deficit/hyperactivity disorder (ADHD): A chronic neurological condition that is manifested as constant restlessness, impulsiveness, difficulty with peers, disruptive behavior, short attention span, academic underachievement, risk-taking behavior, and extreme boredom.

Autonomic nervous system (ANS): Part of the peripheral nervous system that carries out the body's basic housekeeping functions by funneling messages from the environment to the various internal organs; the physiological basis of the conscience.

Behavior genetics: A branch of genetics that studies the relative contributions of heredity and environment to behavioral and personality characteristics.

Behavioral activating (or approach) system (BAS): A reward system associated chemically with the neurotransmitter dopamine, and anatomically with pleasure areas in the limbic system.

Behavioral inhibition system (BIS): A system that inhibits or modulates behavior; associated with the neurotransmitter serotonin.

Belief: One of the elements of the social bond; it refers to the acceptance of the social norms regulating conduct.

Binge drinkers: People who frequently consume anywhere between 5 and 10 drinks in few hours' time (go on a binge).

Bourgeoisie: In Marxism, the owners of the means of production.

Burglary: The unlawful entry of a structure to commit a felony or theft.

Carjacking: The theft or attempted theft of a motor vehicle from its occupant by force or threat of force.

Cartographic criminologists: Criminologists who employ maps and other geographic information in their research to study where and when crime is most prevalent.

Causation: A legal principle stating that there must be an established proximate causal link between the criminal act and the harm suffered.

Cheats: Individuals in a population of cooperators who gain resources from others by signaling their cooperation and then defaulting.

Chicago Area Project (CAP): A project designed by Clifford Shaw to "treat" communities from which most delinquents came.

Choice structuring: A concept in rational choice theory referring to how people decide to offend, and defined as "the constellation of opportunities, costs, and benefits attaching to particular kinds of crime."

Class struggle: Marxist concept stating that all history is the history of class struggle.

Classical conditioning: A mostly passive visceral form of learning that depends on autonomic nervous system arousal, which forms an association between two paired stimuli.

Classical school: The classical school of criminology was a non-empirical mode of inquiry similar to the philosophy practiced by the classical Greek philosophers.

Cleared offenses: A crime is cleared by the arrest of a suspect or by exceptional means (cases in which a suspect has been identified but he or she is not immediately available for arrest).

Collective efficacy: The shared power of a group of connected and engaged individuals to influence an outcome that the collective deems desirable.

Commission: A national ruling body of La Cosa Nostra consisting of the bosses of the five New York families and four bosses from other important families.

Commitment: One of the four social bonds in social bonding theory; the rational component of conformity referring to a lifestyle in which one has invested considerable time and energy in the pursuit of a lawful career.

Concurrence: The legal principle stating that the act (actus reus) and the mental state (mens rea) concur in the sense that the criminal intention actuates the criminal act.

Conduct disorder (CD): The persistent display of serious antisocial actions that are extreme given the child's developmental level and have a significant impact on the rights of others.

Conscience: A complex mix of emotional and cognitive mechanisms that we acquire by internalizing the moral rules of our social group in the ongoing socialization process.

Conscientiousness: A personality trait composed of several secondary traits such as well-organized, disciplined, scrupulous, responsible, and reliable at one pole, and disorganized, careless, unreliable, irresponsible, and unscrupulous at the other.

Consensus or functionalist perspective: A view of society as a system of mutually sustaining parts and one that is characterized by broad normative consensus.

Constrained vision: One of the two so-called ideological visions of the world. The constrained vision views human activities as constrained by an innate human nature that is self-centered and largely unalterable.

Contrast effect: The effect of punishment on future behavior depends on how much the punishment and the usual life experience of the person being punished differ or contrast.

Corporate crime: Criminal activity on behalf of a business organization.

Corporate model: A model of La Cosa Nostra that sees it as similar to corporate structure, i.e., as a formal hierarchy in which the day-to-day activities of the organization are planned and coordinated at the top and carried out by subordinates.

Corpus delicti: A Latin term meaning "body of the crime." Refers to the elements of a given act that must be present in order to legally define it as a crime.

Correlates: Correlated factors that are linked, associated, or related (co-related) with the phenomenon we are trying to explain.

Counterfeiting: The illegal creation or altering of currency.

Crime: An intentional act in violation of the criminal law, committed without defense or excuse and penalized by the state.

Crime rate: The rate of a given crime is the actual number of reported crimes standardized by some unit of the population.

Criminality: A continuously distributed trait composed of a combination of other continuously distributed traits that signals the willingness to use force, fraud, or guile to deprive others of their lives, limbs, or property for personal gain.

Criminaloid: One of Lombroso's criminal types. They had none of the physical peculiarities of the born or insane criminal and were considered less dangerous.

Criminology: An interdisciplinary science that gathers and analyzes data on crime and criminal behavior.

Cybercrime: A wide variety of crimes committed using computer technology.

Dark figure of crime: The dark (or hidden) figure of crime refers to all of the crimes committed that never come to official attention.

Decommodification: The process of freeing social relationships from economic considerations.

Defensible space: A model for residential environments that is supposed to inhibit crime by creating the physical expression of a social fabric that defends itself.

Definitions: Term used by Edwin Sutherland to refer to meanings our experiences have for us, our attitudes, values, and habitual ways of viewing the world.

Delinquency: A legal term that distinguishes between youthful (juvenile) offenders and adult offenders. Acts forbidden by law are called delinquent acts when committed by juveniles.

Determinism: The position that events have causes that preceded them.

Deterrence: The prevention of criminal acts by the use or threat of punishment; deterrence may be either specific or general.

Developmental theories: A group of theories that typically integrate biological, psychological, and sociological variables and that follow the same individuals over an extended period of time.

Differential association theory (DAT): Criminological theory devised by Edwin Sutherland asserting that criminal behavior is behavior learned through association with others who communicate their values and attitudes.

Differential reinforcement: The balance of anticipated or actual rewards and punishments that follow or are consequences of behavior.

Discrimination: A term applied to stimuli that provide clues that signal whether a particular behavior is likely to be followed by reward or punishment.

Domestic violence: Any abusive act (physical, sexual, or psychological) that occurs within the family setting. Intimate partner violence is the most common form.

Drug addiction: Compulsive drug-seeking behavior where acquiring and using a drug becomes the most important activity in the user's life.

Dual pathway developmental theory: Moffitt's theory, based on the notion that there are two main pathways to offending: One pathway is followed by individuals with neurological and temperamental difficulties that are exacerbated by inept parenting, and the other by "normal" individuals temporarily derailed during adolescence.

Ecological fallacy: Making inferences about individuals and groups on the basis of information derived from a larger population of which they are a part.

Economic marginalization hypothesis: This hypothesis argues that much of female crime is related to economic need.

Economic-compulsive violence: Violence associated with efforts to obtain money to finance the cost of illicit drugs.

Emancipation hypothesis: Rita Simon's view that increased participation in the workforce affords women greater opportunities to commit job-related crime without undergoing "masculinization."

Embezzlement: The misappropriation or misapplication of money or property entrusted to the embezzler's care, custody, or control.

Empathy: The emotional and cognitive ability to understand the feelings and distress of others as if they were one's own—to be able to "walk in another's shoes."

Enlightenment: A major intellectual shift in the way people viewed the world and their place in it, questioning traditional religious and political values, and substituting humanism, rationalism, and naturalism for supernaturalism.

Ethnic succession theory: Theory about the causes of organized crime that posits that upon arrival in the United States, each ethnic group was faced with prejudicial and discriminatory attitudes that denied them legitimate means to success in America.

Evolutionary psychology: A way of thinking about human behavior using a Darwinian evolutionary theoretical framework.

Felony murder: Murder committed during the commission of some other felony crime.

Fence: A person who regularly buys stolen property for resale and who often has a legitimate business to cover his or her activities.

Fetal alcohol syndrome (FAS): A chronic condition affecting the brain resulting from an individual's prenatal alcohol exposure.

Feudal model: A model of La Cosa Nostra that sees it as similar to the old European feudal system based on patronage, oaths of loyalty, and semi-autonomy.

Fight or flight system: An autonomic nervous system mechanism that mobilizes the body for action in response to threats by pumping out epinephrine.

Flynn effect: The upward creep in average IQ scores that has been taking place over the last three or four generations in all countries examined.

Focal concerns: Walter Miller's description of the value system and lifestyle of the lowest classes; characteristics include trouble, toughness, excitement, smartness, fate, and autonomy.

Forcible rape: The carnal knowledge of a female forcibly and against her will.

Forgery: The creation or alteration of documents to give them the appearance of legality and validity with the intention of gaining some fraudulent benefit from doing so.

Fraud: Obtaining the money or property of another through deceptive practices such as false advertising, impersonation, and other misrepresentations.

Free will: That which enables human beings to purposely and deliberately choose to follow a calculated course of action.

Gender ratio problem: An issue in feminist criminology asking why always and everywhere females commit far less crime than males.

Gene–environment correlation (rGE): The notion that genotypes and the environments they find themselves in are related because parents provide children with both.

Gene–environment interaction (GxE): The interaction of a genotype with its environment. People are differentially sensitive to identical environmental influences because of their genes and will thus respond in different ways to them.

General deterrence: The assumed preventive effect of the threat of punishment on the general population, i.e., *potential* offenders.

General strain theory (GST): Agnew's extension of anomie theory into the realm of social psychology stressing multiple sources of strain and how people cope with it.

Generalizability problem: An issue in feminist criminology asking if traditional theories based on male offender samples apply to women offenders.

Genetic polymorphisms: Variations in the same gene allele (alternate form of a gene) such as SNPs and VNTRs.

Genotype: A person's genetic makeup.

Grand jury: An investigatory jury composed of 7 to 23 citizens before which the prosecutor presents evidence that sufficient grounds exist to try the suspect for a crime. If the prosecutor is successful, he or she obtains an indictment from the grand jury listing the charges a person is accused of.

Hacker: A person who illicitly accesses someone else's computer system.

Harm: The legal principle that states that a crime must have a negative impact either on the victim or on the general values of the community to be a crime.

Harrison Narcotic Act: A 1914 congressional act that criminalized the sale and use of narcotics.

Hedonism: A doctrine assuming that the achievement of pleasure or happiness is the main goal of life.

Hedonistic calculus: Combining hedonism and rationality to logically weigh the anticipated benefits of a given course of action against its possible costs.

Hedonistic serial killer: A killer that kills for the pure thrill and joy of it.

Hegemonic masculinity: Concept in structured action theory that refers to the cultural ideal of masculinity that men are expected to live up to, i.e., "work in the paid-labor market, the subordination of women, heterosexism, and the driven uncontrollable sexuality of men."

Heritability: A concept defined by a number ranging between 0 and 1 indicating the extent to which variance in a phenotypic trait in a population is due to genetic factors.

Hierarchy rule: A rule requiring the police to report only the most serious offense committed in a multiple-offense single incident to the FBI and to ignore the others.

Honor subcultures: Communities in which young men are hypersensitive to insult, rushing to defend their reputation in dominance contests.

Human agency: A concept that maintains humans have the capacity to make choices and the moral responsibility to make moral ones, regardless of the internal or external constraints on one's ability to do so.

Hypotheses: Statements about relationships between and among factors we expect to find, based on the logic of our theories.

Identity theft: The use of someone else's personal information without their permission to perpetrate an illegal act.

Ideology: A way of looking at the world; a general emotional picture of "how things should be" that forms, shapes, and colors our concepts of the phenomena we study.

Impulsiveness: A personality trait reflecting a person's varying tendencies to act on matters without giving much thought to the possible consequences (not looking before one leaps).

Incarceration: The act of being confined to a secure institution as punishment for a crime.

Insane criminal: One of Lombroso's criminal types. Insane criminals bear some stigmata but are not born criminals. Among their ranks are alcoholics, kleptomaniacs, nymphomaniacs, and child molesters.

Institutional anomie theory (IAT): Messner and Rosenfeld's extension of anomie theory, which avers that high crime rates are intrinsic to the structural and cultural arrangements of American society.

Integrated cognitive antisocial potential (ICAP) theory: Farrington's theory, which is based on the notion that people have varying levels of antisocial propensity due to a variety of environmental and biological factors.

Intellectual imbalance: A significant difference between a person's verbal and performance IQ scores.

Intelligence: The capacity of individuals to act purposefully, to think rationally, and to deal effectively with their environment.

Involuntary manslaughter: A criminal homicide where an unintentional killing results from a reckless act.

Involvement: One of the elements of the social bond, it is a direct consequence of commitment; it is part of an overall conventional pattern of existence.

Italian School: Positivist school of criminology associated with Cesare Lombroso, Raffaele Garofalo, and Enrico Ferri.

Labeling theory: A theory stating that the act of being caught and labeled delinquent or criminal serves as a self-fulfilling prophesy leading to further delinquent/criminal acts.

La Cosa Nostra (LCN): Literally, "our thing"; also commonly referred to as the Mafia; an organized crime group of Italian/Sicilian origins.

Larceny-theft: The unlawful taking, leading, or riding away of property from the possession or constructive possession of another.

Left realists: Group of Marxist criminologists who want to work within the system to make things better for the working classes.

Level of analysis: That segment of the phenomenon of interest that is measured and analyzed, i.e., individuals, families, neighborhoods, states, etc.

Lifestyle theory: A theory stressing that crime is not just a behavior but a general pattern of life.

Lumpenproletariat: The lowest class in Marxist theory; the "criminal class."

Madrasas: Islamic religious schools that stress the immorality and materialism of Western life and the need to convert all infidels to Islam.

Mala in se: Universally condemned crimes that are "inherently bad."

Mala prohibita: Crimes that are "bad" simply because they are prohibited.

Masculinization hypothesis: Freda Adler's idea that as females increasingly adopt male roles, they will increasingly masculinize their attitudes and behavior and become as crime-prone as men.

Mass murder: The killing of several people at one location within minutes or hours.

Mating effort: The proportion of total reproductive effort allotted to acquiring sexual partners; traits facilitating mating effort are associated with antisocial behavior.

Maturity gap: In Moffitt's theory, the gap between the average age of puberty and the acquisition of socially responsible adult roles.

Mechanical solidarity: A form of social solidarity existing in small, isolated, pre-state societies in which individuals sharing common experiences and circumstances share common values and develop strong emotional ties to the collectivity.

Mens rea: Literally, *guilty mind.* Refers to whether or not the suspect had a wrongful purpose in mind when carrying out the *actus reus* (guilty act).

Middle-class measuring rods: According to Albert Cohen, because low-class youths cannot measure up to middle-class standards, they experience status frustration and this frustration spawns an oppositional culture.

Minority power groups: Groups whose interests are sufficiently on the margins of mainstream society that just about all their activities are criminalized.

Mission-oriented serial killer: A killer that feels it to be his mission in life to kill certain kinds of people.

Modes of adaptation: Robert Merton's concept of how people adapt to the alleged disjunction between cultural goals and structural barriers to the means of obtaining them. These modes are conformity, ritualism, retreatism, innovation, and rebellion.

Motor vehicle (MV) theft: The theft or attempted theft of a motor vehicle.

Murder: The willful (non-negligent) killing of one human being by another.

National Crime Victimization Survey (NCVS). A biannual survey of a large number of people and households requesting information on crimes committed against individuals and households (whether reported to the police or not) and circumstances of the offense (time and place it occurred, perpetrator's use of a weapon, any injuries incurred, and financial loss).

National Incident-Based Reporting System (NIBRS): A comprehensive crime statistic collection system, which is currently a component of the UCR program and is eventually expected to replace it entirely.

Natural selection: The evolutionary process that selects genetic variants that best fit organisms in their present environments and preserves them in later generations.

Negative emotionality: A personality trait that refers to the tendency to experience many situations as aversive, and to react to them with irritation and anger more readily than with positive affective states.

Negligent manslaughter: An unintentional homicide that is charged when a death or deaths arise from some negligent act that carries a substantial risk of death to others.

Neurons: Brain cells consisting of the cell body, an axon, and a number of dendrites.

Neurotransmitters: Brain chemicals that carry messages from neuron to neuron across the synaptic gap.

Neutralization theory: A theory maintaining that delinquents and criminals tend to "neutralize" their responsibility for antisocial acts through various excuses and rationalizations.

Occupational crime: Crime committed by individuals in the course of their employment.

Operant psychology: A perspective on learning that asserts that behavior is governed and shaped by its consequences (reward or punishment).

Opportunity: In self-control theory, a situation that presents itself to those with low self-control by which they can immediately satisfy their needs with minimal effort.

Opportunity structure theory: An extension of anomie theory claiming that lower-class youth join gangs as a path to monetary success.

Organic solidarity: A form of social solidarity characteristic of modern societies in which there is a high degree of occupational specialization and a weak normative consensus.

Organized crime: A continuing criminal enterprise that works rationally to profit from illicit activities that are often in great public demand. Its continuing existence is maintained through the use of force, threats, and/or corruption of public officials.

Parenting effort: The proportion of total reproductive effort invested in rearing offspring; traits facilitating parenting effort are associated with prosocial behavior.

Parole: A conditional release from prison granted to certain inmates sometime prior to the completion of their sentences.

Part I offenses (or **index crimes**): The four violent (homicide, assault, forcible rape, and robbery) and four property offenses (larceny-theft, burglary, motor vehicle theft, and arson) reported in the Uniform Crime Reports.

Part II offenses: The less serious offenses reported in the Uniform Crime Reports; they are recorded based on arrests made rather than cases reported to the police.

Patriarchy: Any social system that is male dominated at all levels, from the family to the highest reaches of government, and supported by the belief of male superiority.

Peacemaking criminology: Theory based on the postmodernist tradition that rejects the notion that the scientific view is any better than any other view, and that any method of understanding can be objective.

Personality: The relatively enduring, distinctive, integrated, and functional set of psychological characteristics that results from people's temperaments interacting with their cultural and developmental experiences.

Pharmacological violence: Violence induced by the pharmacological properties of a drug.

Physical dependence: The state in which a person is physically dependant on a drug because of changes to the body that have occurred after repeated use of it, and that necessitates its continued administration to avoid withdrawal symptoms.

Policy: A course of action designed to solve some problem that has been selected by appropriate authorities from among alternative courses of action.

Positivism: An extension of the scientific method to social life, from which more *positive* knowledge can be obtained.

Power/control serial killer: A killer that gains the most satisfaction from exercising complete power over his victims.

Power-control theory: A feminist theory that views gender differences in criminal and delinquent behavior as a function of power differentials in the family.

Prefrontal cortex (PFC): Part of the brain that occupies about one third of the front part of the cerebrum. It has many connections with other brain structures and plays the major integrative and supervisory roles in the brain.

Prefrontal dysfunction theory: Theory maintaining that disinhibited and impulsive behavior is often the result of some dysfunction of the brain's "command and control" system—the prefrontal cortex (PFC).

Preliminary arraignment: The presenting of suspects in court before a magistrate or municipal judge to advise them of their constitutional rights and of the tentative charges against them, and to set bail.

Preliminary Hearing: A proceeding before a magistrate or municipal judge in which three major matters must be decided: (1) whether or not a crime has actually been committed, (2) whether or not there are reasonable grounds to believe that the person before the bench committed it, and (3) whether or not the crime was committed in the jurisdiction of the court.

Primary deviance: In labeling theory, the initial nonconforming act that comes to the attention of the authorities, resulting in the application of a criminal label.

Primitive rebellion hypothesis: Marxist idea that crime is simply the product of people rebelling against unjust and alienating social conditions.

Principle of utility: A principle that posits that human action should be judged moral or immoral by its effect on the happiness of the community and that the proper function of the legislature is to promulgate laws aimed at maximizing the pleasure and minimizing the pain of the largest number in society—"the greatest good for the greatest number."

Probation: A probation sentence is a suspended commitment to prison that is conditional on the offender's good behavior.

Prohibition: Common term for the Volstead Act, which prohibited the sale, manufacture, or importation of intoxicating liquors within the United States.

Proletariat: The working class in Marxist theory.

Prostitution: The provision of sexual services in exchange for money or other tangible reward as the primary source of income.

Psychological dependence: The deep craving for a drug and the feeling that one cannot function without it; psychological dependence is synonymous with addiction.

Psychopathy: A syndrome characterized by the inability to tie the social emotions with cognition. Psychopaths come from all social classes and may or may not be criminals.

Psychopathy Checklist–Revised (PCL-R): An instrument designed by Robert Hare to assess psychopathy.

Punishment: A process that leads to the weakening or eliminating of the behavior preceding it.

Rape trauma syndrome: A syndrome sometimes suffered by rape victims that is similar to post-traumatic stress syndrome (re-experiencing the event via "flashbacks," avoiding anything at all associated with the event, and a general numbness of affect).

Rational: Rational behavior is behavior consistent with logic; a logical "fit" between the goals people strive for and means they use to achieve them.

Rational choice theory: A neoclassical theory asserting that offenders are free actors responsible for their own actions. Rational choice theorists view criminal acts as specific examples of the general principle that all human behavior reflects the rational pursuit of benefits and advantages. People are conscious social actors free to choose crime, and they will do so if they perceive that its utility exceeds the pains they might conceivably expect if discovered.

Recidivism: Refers to "falling back" into criminal behavior after having being punished.

Reinforcement: A process that leads to the repetition and strengthening of behavior.

Restorative justice: A system of mediation and conflict resolution oriented toward justice by repairing the harm that has been caused by the crime using a face-to-face confrontation between victim and perpetrator.

Reticular activating system (RAS): A finger-sized bundle of brain cells situated at the top of the spinal cord that can be thought of as the brain's filter system, determining what incoming stimuli the higher brain centers will pay attention to.

Reward dominance theory: A neurological theory based on the proposition that behavior is regulated by two opposing mechanisms, the behavioral activating system (BAS) and the behavioral inhibition system (BIS).

Risk factor: Something in individuals' personal characteristics or their environment that increases the probability of offending.

Robbery: The taking or attempted taking of anything of value from the care, custody, or control of a person or persons by force or threat of force or violence and/or putting the victim in fear.

Routine activities theory: A neoclassical theory pointing to the routine activities in that society or neighborhood that invite or prevent crime. Routine activities are defined as "recurrent and prevalent activities which provide for basic population and individual needs." Crime is the result of (a) *motivated offenders* meeting (b) *suitable targets* that lack (c) *capable guardians.*

Routine activities/lifestyle theory: A victimization theory that states that there are certain lifestyles (routine activities) that disproportionately expose some people to a high risk for victimization.

Sarbanes-Oxley Act (SOA): An act passed in 2002 in response to numerous corporate scandals. The provisions of this act include increased funding for the Securities Exchange Commission, penalty enhancement for white-collar crimes, and the relaxing of some legal impediments to gaining convictions.

Secondary deviance: Deviance that results from society's reaction to offenders' primary deviance.

Self-control: The extent to which different people are vulnerable to the temptations of the moment.

Self-control theory: Theory developed by Gottfredson and Hirschi that maintains all crime is attributable to an individual's lack of self-control.

Self-report surveys: The collecting of data by criminologists themselves asking people to disclose their delinquent and criminal involvement on anonymous questionnaires.

Sensation seeking: The active desire for novel, varied, and extreme sensations and experiences, often to the point of taking physical and social risks to obtain them.

Serial murder: The killing of three or more victims over an extended period of time.

Short-run hedonism: The seeking of immediate gratification of desires without regard for any long-term consequences.

Social bond theory: A theory focusing on a person's bonds to others. The four elements of the social bond are attachment, commitment, involvement, and belief. The absence of these bonds in criminals does not cause crime; it permits it.

Social capital: The store of positive relationships in social networks built on norms of reciprocity and trust developed over time upon which the individual can draw for support.

Social control: Any action on the part of others, deliberate or not, that facilitates conformity to social rules.

Social defense: A theory of punishment asserting that its purpose is not to deter or to rehabilitate but to defend society against criminals.

Social disorganization: The central concept of the Chicago School of Social Ecology. It refers to the breakdown or serious dilution of the power of informal community rules to regulate conduct in poor neighborhoods.

Social ecology: Term used by the Chicago School to describe the interrelations of human beings and the communities in which they live.

Social learning theory (SLT): A theory designed to explain how people learn criminal behavior using the psychological principles of operant conditioning.

Social push hypothesis: The idea that if an individual lacks environmental risk factors that predispose him or her toward antisocial behavior yet still engages in antisocial behavior, then the causes of this behavior are more likely to be biological than social.

Social sentiments: Willem Bonger's proposition that individuals vary in their risk for crime because they vary in the innate social sentiments of altruism and egoism.

Social structure: How society is organized by social institutions—the family, and educational, religious, economic, and political institutions—and stratified on the basis of various roles and statuses.

Sociopaths: All sociopaths are criminals by definition. The development of sociopathy is not as closely tied to genetics as it is in psychopaths, but is developed primarily through inadequate socialization and hostile childhood experiences.

Software piracy: Illegally copying and distributing computer software.

Specific deterrence: The effect of punishment on the future behavior of the person who experiences the punishment.

Spree murder: The killing of several people at different locations over several days.

Status frustration: A form of frustration experienced by lower-class youth who desire approval and status but who cannot meet middle-class criteria and thus seek status via alternative means.

Staying alive hypothesis: Campbell's thesis that staying alive is more crucial to a mother's reproductive success than to a father's and that greater fear in females is a result.

Strain theory: Robert Merton's original extension of Durkheim's anomie theory in which he put forth his five modes of adaptation to a society that exhorts all to strive for monetary success but denies certain groups access to legitimate means of achieving it.

Structured action theory: A feminist theory formulated by James Messerschmidt that focuses on how individuals "do gender."

Subculture of violence: A part of a larger culture in which the norms, attitudes, and values of its people legitimizes the use of violence.

Super traits theory: A developmental theory that asserts that five life domains interact over the life course once individuals are set on a particular developmental trajectory by their degree of low self-control and irritability.

Symbolic interactionism: A perspective in sociology that focuses on how people interpret and define their social reality and the meanings they attach to it in the process of interacting with one another via language (symbols).

Systemic violence: Violence associated with aggressive patterns of interaction within the system of drug distribution and use.

Tammany Society: A corrupt political machine that ran New York into the early 20th century, associated with the Democratic Party and with organized crime.

Techniques of neutralization: Techniques by which offenders justify their behavior as "acceptable" on a number of grounds.

Temperament: An individual characteristic identifiable as early as infancy that constitutes a habitual mode of emotionally responding to stimuli.

Terrorism: The FBI defines terrorism as "the unlawful use of force or violence against persons or property to intimidate or coerce a government, the civilian population, or any segment thereof, in furtherance of political or social goals."

Theory: A set of logically interconnected propositions explaining how phenomena are related, and from which a number of hypotheses can be derived and tested.

Thinking errors: Criminals' typical patterns of faulty thoughts and beliefs.

Traits: Components or features of a person's overall personality that all humans share to varying degrees, such as self-control, shyness, and conscientiousness.

Transition Zone: An area or neighborhood in the process of being "invaded" by members of "alien" racial or ethnic groups bringing with them values and practices that conflicted with those established by the "natural" inhabitants of the area.

Turning points: Transition events in life (getting married, finding a job, moving to a new neighborhood) that may change a person's life trajectory in prosocial directions.

Type I alcoholism: A form of alcoholism characterized by mild abuse, minimal criminality, and passive-dependent personality.

Type II alcoholism: A form of alcoholism characterized by early onset, violence, and criminality, and that is largely limited to males.

Unconstrained vision: One of the two so-called ideological visions of the world. The unconstrained vision denies an innate human nature, viewing it as formed anew in each different culture.

Uniform Crime Reports (UCR): Annual report compiled by the Federal Bureau of Investigation (FBI) containing crimes known to the nation's police and sheriff's departments, the number of arrests made by these agencies, and other crime-related information.

Victim precipitation theory: A theory in victimology that examines how violent victimization may have been precipitated by the victim by acting in certain provocative ways.

Victimology: A subfield of criminology that specializes in studying the victims of crime.

Violent crime: Crime in which the use of force is exercised without excuse or justification to achieve a goal at the expense of a victim.

Visionary serial killer: A killer that feels impelled to commit murder by visions or "voices in my head."

Voluntary manslaughter: The intentional killing of another human being without malice aforethought, often in response to the mistaken belief that self-defense required the use of deadly force, or in response to adequate provocation while in the heat of passion.

White-collar crime: An illegal act or series of illegal acts committed by nonphysical means and by concealment or guile, to obtain money or property or to obtain business or personal advantage.

Withdrawal: A process involving a number of adverse physical reactions that occur when the body of a drug abuser is deprived of his or her drugs.

References

Abadinsky, H. (2003). *Organized crime* (7th ed.). Belmont, CA: Wadsworth.

Adams, J. (1971). *In defense of the Constitution of the United States* (Vol. 1). New York: De Capo Press. (Original work published 1778)

Adamson, C. (2000). Defensive localism in black and white: A comparative history of European-American and African-American youth gangs. *Ethnic and Racial Studies, 23,* 272–298.

Adler, F. (1975). *Sisters in crime: The rise of the new female criminal.* New York: McGraw-Hill.

Adler, F., Mueller, G., & Laufer, W. (2001). *Criminology and the criminal justice system.* Boston: McGraw-Hill.

Agnew, R. (1994). The techniques of neutralization and violence. *Criminology, 32,* 555–580.

Agnew, R. (2001). *Juvenile delinquency: Causes and control.* Los Angeles: Roxbury.

Agnew, R. (2002). Foundation for a general strain theory of crime. In S. Cote (Ed.), *Criminological theories: Bridging the past to the future* (pp. 113–124). Thousand Oaks, CA: Sage.

Agnew, R. (2005). *Why do criminals offend? A general theory of crime and delinquency.* Los Angeles: Roxbury.

Akers, R. (1994). *Criminological theories: Introduction and evaluation.* Los Angeles: Roxbury.

Akers, R. (1999). Social learning and social structure: Reply to Sampson, Morash, and Krohn. *Theoretical Criminology, 3,* 477–493.

Akers, R. (2002). A social learning theory of crime. In S. Cote (Ed.), *Criminological theories: Bridging the past to the future* (pp. 135–143). Thousand Oaks, CA: Sage.

Akers, R. (2009). *Social learning and social structure: A general theory of crime and deviance.* New Brunswick, NJ: Transaction.

Albanese, J. (2000). The causes of organized crime. *Journal of Contemporary Criminal Justice, 16,* 409–432.

Albanese, J., & Pursley, R. (1993). *Crime in America: Some existing and emerging issues.* Englewood Cliffs, NJ: Regents/ Prentice Hall.

Amateau, S., & McCarthy, M. (2004). Induction of PGE2 by estradiol mediates developmental masculinization of sex behavior. *Nature Neuroscience, 7,* 643–650.

American Psychiatric Association. (1994). *Diagnostic and statistical manual of mental disorders* (4th ed.). Washington, DC: Author.

Amir, M. (1971). *Patterns of forcible rape.* Chicago: University of Chicago Press.

Anderson, D. (1999). The aggregate burden of crime. *Journal of Law and Economics, 42,* 611–642.

Anderson, E. (1994). The code of the streets. *The Atlantic Monthly, 5,* 81–94.

Anderson, E. (1999). *Code of the street: Decency, violence, and the moral life of the inner city.* New York: Norton.

Andrews, D., & Bonta, J. (1998). *The psychology of criminal conduct.* Cincinnati, OH: Anderson.

Ardila, R. (2002). The psychology of the terrorist: Behavioral perspectives. In C. Stout (Ed.), *The psychology of terrorism* (Vol. 1, pp. 9–15). Westport, CT: Praeger.

Armanios, F. (2003). *Islamic religious schools, madrasas: Background.* Congressional Research Service. Available at http://fpc.state.gov/documents/organization/26014.pdf

Associated Press. (1996, August 27). U.S. fugitive convicted in Cuba. *The Idaho Statesman,* p. 6a.

Babiak, P., Neumann, C., & Hare, R. (2010). Corporate psychopathy: Talking the walk. *Behavioral Sciences and the Law, 28,* 174–193.

Badcock, C. (2000). *Evolutionary psychology: A critical introduction.* Cambridge, UK: Polity Press.

Baines, H. (1996). *The Nigerian scam masters: An exposé of a modern international gang.* Hauppauge, NY: Nova Science.

Baker, L., Bezdjian, S., & Raine, A. (2006). Behavior genetics: The science of antisocial behavior. *Law and Contemporary Problems, 69,* 7–46.

Barak, G. (1998). *Integrating criminologies.* Boston: Allyn & Bacon.

Barash, D., & Lipton, J. (2001). Making sense of sex. In D. Barash (Ed.), *Understanding violence* (pp. 20–30). Boston: Allyn & Bacon.

Barber, N. (2004). Single parenthood as a predictor of cross-national variation in violent crime. *Cross Cultural Research, 38,* 343–358.

Barkow, J. (Ed.). (2006). *Missing the revolution: Darwinism for social scientists.* Oxford, UK: Oxford University Press.

Bartol, C. (2002). *Criminal behavior: A psychosocial approach* (6th ed.). Englewood Cliffs, NJ: Prentice Hall.

Bartol, C., & Bartol, A. (1989). *Juvenile delinquency: A systems approach.* Englewood Cliffs, NJ: Prentice Hall.

Bartollas, C. (2005). *Juvenile delinquency* (7th ed.). Boston: Allyn & Bacon.

Baumeister, R., Smart, L., & Boden, J. (1996). Relation of threatened egoism to violence and aggression: The dark side of self-esteem. *Psychological Review, 103,* 5–33.

Beaver, K., Wright, J., & Walsh, A. (2008). A gene-based evolutionary explanation for the association between criminal involvement and number of sex partners. *Biodemography and Social Biology, 54,* 47–55.

Beccaria, C. (1963). *On crimes and punishment* (H. Paulucci, Trans.). Indianapolis, IN: Bobbs-Merrill. (Original work published 1764)

Bell, D. (1962). *The end of ideology.* New York: Collier Books.

Bellinger, D. (2008). Neurological and behavioral consequences of childhood lead exposure. *PLoS Medicine, 5,* 690–692.

Bennett, S., Farrington, D., & Huesmann, L. (2005). Explaining gender differences in crime and violence: The importance of social cognitive skills. *Aggression and Violent Behavior, 10,* 263–288.

Bentham, J. (1948). *A fragment on government and an introduction to the principles of morals and legislation* (W. Harrison, Ed.). Oxford, UK: Basil Blackwell. (Original work published 1789)

Berger, P., & Hoffman, B. (2010). *Assessing the terrorist threat: Report to the Bipartisan Policy Center's National Security Preparedness Center.* Available at http://www.bipartisan policy.org/library/report/assessing-terrorist-threat

Bernard, T., Snipes, J., & Gerould, A. (2010). *Vold's theoretical criminology.* New York: Oxford University Press.

Bing, L. (1991). *Do or die.* New York: HarperCollins.

Birkenhead, T., & Moller, A. (1992, July). Faithless females seek better genes. *New Scientist,* 34–38.

Blonigen, D. (2010). Explaining the relationship between age and crime: Contributions from the developmental literature on personality. *Clinical Psychology Review, 30,* 89–100.

Bohm, R. (2001). *A primer on crime and delinquency* (2nd ed.). Belmont, CA: Wadsworth.

Bonger, W. (1969). *Criminality and economic conditions.* Bloomington: Indiana University Press. (Original work published 1905)

Bostaph, L. (2004). Race and repeat victimization: Does the repetitive nature of police motor vehicle stops impact racially biased policing? Unpublished doctoral dissertation, University of Cincinnati.

Brennan, P., Grekin, E., & Sarnoff, M. (1999). Maternal smoking during pregnancy and adult male criminal outcomes. *Archives of General Psychiatry, 56,* 215–219.

Brennan, P., Raine, A., Schulsinger, F., Kirkegaard-Sorenen, L., Knop, J., Hutchings, B., et al. (1997). Psychophysiological protective factors for male subjects at high risk for criminal behavior. *American Journal of Psychiatry, 154,* 853–855.

Brett, A. (2004). "Kindling theory" in arson: How dangerous are firesetters? *Australian and New Zealand Journal of Psychiatry, 38,* 419–425.

Browning, F., & Gerassi, J. (1980). *The American way of crime.* New York: G.P. Putnam.

Brownmiller, S. (1975). *Against our will: Men, women, and rape.* New York: Simon & Schuster.

Buck, K., & Finn, D. (2000). Genetic factors in addiction: QTL mapping and candidate gene studies implicate GABAergic genes in alcohol and barbiturate withdrawal in mice. *Addiction, 96,* 139–149.

Bullough, B., & Bullough, V. (1994). Prostitution. In V. Bullough & B. Bullough (Eds.), *Human sexuality: An encyclopedia* (pp. 494–499). New York: Garland Press.

Burgess, R., & Akers, R. (1966). A differential association-reinforcement theory of criminal behavior. *Social Problems, 14,* 363–383.

Burns, E. (2004). *The spirits of America: A social history of alcohol.* Philadelphia: Temple University Press.

Burns, R. (2004, June 6). Rumsfeld fearful of losing broader battle against extremists. *Idaho Statesman,* p. 5a.

Burr, M. (2004, December). SEC gains power, prestige in post Enron era. *Corporate Legal Times,* 10–13.

Business Journal. (2005). *Survey: Shoplifting losses mount.* Available at http://orlando.bizjournals.com/orlando/stories/2005/12/05/daily.html

Business Software Alliance. (2005). *2005 Piracy Study.* Available at http://www.bsa.org/globalstudy

Bynum, T. (1987). Controversies in the study of organized crime. In T. Bynum (Ed.), *Organized crime in America: Concepts and controversies* (pp. 3–11). Monsey, NY: Willow Tree Press.

Calavita, K., & Pontell, H. (1994). "Heads I win, tails you lose": Deregulation, crime, and crisis in the savings and loan industry. In D. Curran & C. Renzetti (Eds.), *Contemporary societies: Problems and prospects* (pp. 460–480). Englewood Cliffs, NJ: Prentice Hall.

Calavita, K., Pontell, H., & Tillman, R. (1999). *Big money game: Fraud and politics in the savings and loan crisis.* Berkeley: University of California Press.

Campbell, A. (1999). Staying alive: Evolution, culture, and women's intrasexual aggression. *Behavioral and Brain Sciences, 22,* 203–214.

Campbell, A. (2006). Sex differences in direct aggression: What are the psychological mediators? *Aggression and Violent Behavior, 6,* 481–497.

Campbell, A. (2008). Attachment, aggression, and affiliation: The role of oxytocin in female social behavior. *Biological Psychology, 77,* 1–10.

Campbell, A. (2009). Gender and crime: An evolutionary perspective. In A. Walsh & K. Beaver (Eds.), *Criminology and biology: New directions in theory and research* (pp. 117–136). New York: Routledge.

Canter, D. (2004). Offender profiling and investigative psychology. *Journal of Investigative Psychology and Offender Profiling, 1,* 1–15.

Cao, L. (2004). *Major criminological theories: Concepts and measurement.* Belmont, CA: Wadsworth.

Cao, L., Adams, A., & Jensen, V. (1997). A test of the black subculture of violence thesis: A research note. *Criminology, 35,* 367–369.

Carey, G. (2003). *Human genetics for the social sciences.* Thousand Oaks, CA: Sage.

Carlisle, A. L. (1993). The divided self: Toward an understanding of the dark side of the serial killer. *American Journal of Criminal Justice, 17,* 23–26

Cartier, J., Farabee, D., & Prendergast, M. (2006). Methamphetamine use, self-reported violent crime, and recidivism among offenders in California who abuse substances. *Journal of Interpersonal Violence, 21,* 435–445.

Cartwright, J. (2000). *Evolution and human behavior.* Cambridge: MIT Press.

Casey, E. (1978). *History of drug use and drug users in the United States.* Schaffer Library of Drug Policy. Available at http://www.druglibrary.org/schaffer/History/CASEY1.htm

Caspi, A. (2000). The child is the father of the man: Personality continuities from childhood to adulthood. *Journal of Personality and Social Psychology, 78,* 158–172.

Caspi, A., McClay, J., Moffitt, T., Mill, J., Martin, J., Craig, I., et al. (2002). Evidence that the cycle of violence in maltreated children depends on genotype. *Science, 297,* 851–854.

Caspi, A., Moffitt, T., Silva, P., Stouthamer-Loeber, M., Krueger, R., & Schmutte, P. (1994). Are some people crime-prone? Replications of the personality–crime relationship across countries, genders, races, and methods. *Criminology, 32,* 163–194.

Catalano, S. (2006). *Criminal victimization, 2005.* Washington, DC: Bureau of Justice Statistics.

Catalano, S. (2007). *Intimate partner violence in the United States.* Washington, DC: Bureau of Justice Statistics.

Cauffman, E., Steinberg, L., & Piquero, A. (2005). Psychological, neuropsychological, and physiological correlates of serious antisocial behavior in adolescence: The role of self-control. *Criminology, 43,* 133–175.

Cecil, K., Brubaker, C., Adler, C., Dietrich, K., Altaye, M., Egelhoff, J., et al. (2008). Decreased brain volume in adults with childhood lead exposure. *PLoS Medicine, 5,* 742–750.

Cernkovich, S., Giordano, P., & Rudolph, J. (2000). Race, crime, and the American dream. *Journal of Research in Crime and Delinquency, 37,* 131–170.

Chambliss, W. (1976). *Criminal law in action.* Santa Barbara, CA: Hamilton.

Chamorro-Premuzic, T., & Furnham, A. (2005). Intellectual competence. *Psychologist, 18,* 352–354.

Champion, D. (2005). *Probation, parole, and community corrections* (5th ed.). Upper Saddle River, NJ: Prentice Hall.

Chapple, C., & Johnson, K. (2007). Gender differences in impulsivity. *Youth Violence and Juvenile Justice, 5,* 221–234.

Chesney-Lind, M. (1995). Girls, delinquency, and juvenile justice: Toward a feminist theory of young women's crime. In B. Price & N. Sokoloff (Eds.), *The criminal justice system and women: Offenders, victims, and workers* (pp. 71–88). New York: McGraw-Hill.

Clark, R., & Cornish, D. (1985). Modeling offenders' decisions: A framework for research and policy. In M. Tonry & N. Morris (Eds.), *Crime and Justice annual review of research* (Vol. 6, pp. 147–185). Chicago: University of Chicago Press.

Cleveland, H., Wiebe, R., van den Oord, E., & Rowe, D. (2000). Behavior problems among children from different family structures: The influence of genetic self-selection. *Child Development, 71,* 733–751.

Clinnard, M., & Yeager, P. (1980). *Corporate crime.* New York: The Free Press.

Cloward, R., & Ohlin, L. (1960). *Delinquency and opportunity.* New York: The Free Press.

Coates, R. B. (1990). Victim–offender reconciliation programs in North America: An assessment. In B. Galaway & J. Hudson (Eds.), *Criminal justice, restitution, and reconciliation* (pp. 125–134). Monsey, NY: Criminal Justice Press.

Cohen, A. (1955). *Delinquent boys.* New York: The Free Press.

Cohen, L., & Felson, M. (1979). Social change and crime rate trends: A routine activities approach. *American Sociological Review, 44,* 588–608.

Coleman, J. (1986). *The criminal elite: The sociology of white-collar crime.* New York: St. Martin's Press.

Collins, R. (2004). Onset and desistence in criminal careers: Neurobiology and the age–crime relationship. *Journal of Offender Rehabilitation, 39,* 1–19.

Collins, R. (2010). The micro-sociology of violence. *British Journal of Sociology, 60,* 566–575.

Connell, R., & Messerschmidt, J. (2005). Hegemonic masculinity: Rethinking the concept. *Gender and Society, 19,* 829–859.

Consumer Freedom. (2005). *America's number one terrorists.* Available at http://consumerfreedom.com

Coolidge, F., Thede, L., & Young, S. (2000). Heritability and the comorbidity of attention deficit hyperactivity disorder with behavioral disorders and executive function deficits: A preliminary investigation. *Developmental Neuropsychology, 17,* 273–287.

Cooper, J., Walsh, A., & Ellis, L. (2010). Is criminology ripe for a paradigm shift? Evidence from a survey of American criminologists. *Journal of Criminal Justice Education, 2,* 332–347.

Copes, H. (2003). Streetlife and the rewards of auto theft. *Deviant Behavior, 24,* 309–332.

Cornish, D., & Clarke, R. (Eds.). (1986). *The reasoning criminal.* New York: Springer-Verlag.

Council on Foreign Relations. (2004). *American militant extremists.* Available at http://cfrterrorism.org/groups/American_print.html

Covell, C., & Scalora, M. (2002). Empathetic deficits in sexual offenders: An integration of affective, social, and cognitive constructs. *Aggression and Violent Behavior, 37,* 251–270.

Crabbe, J. (2002). Genetic contributions to addiction. *Annual Review of Psychology, 53,* 435–462.

Crockett, M., Clark, L., Lieberman, M., Tabinia, G., & Robbins, T. (2010). Impulsive choice and altruistic punishment are correlated and increase in tandem with serotonin depletion. *Emotion, 10,* 855–862.

Cullen, F. (2005). Challenging individualistic theories of crime. In S. Guarino-Ghezzi & J. Trevino (Eds.), *Understanding crime: A multidisciplinary approach* (pp. 55–60). Cincinnati, OH: Anderson.

Cullen, F., & Agnew, R. (2011). *Criminological theory: Past to present.* New York: Oxford University Press.

Cureton, S. (2009). Something wicked this way comes: A historical account of black gangsterism offers wisdom and warning for African American leadership. *Journal of Black Studies, 40,* 347–361.

Curran, D., & Renzetti, C. (2001). *Theories of crime.* Boston: Allyn & Bacon.

Currie, E. (1989). Confronting crime: Looking toward the twenty-first century. *Justice Quarterly, 6,* 5–25.

Daigle, L., Fisher, B., & Cullen, F. (2008). The violent and sexual victimization of college women: Is repeat victimization a problem? *Journal of Interpersonal Violence, 23,* 1296–1313.

D'Alessio, S., & Stolzenberg, L. (2003). Race and the probability of arrest. *Social Forces, 81,* 1381–1397.

Daly, K., & Chesney-Lind, M. (2002). Feminism and criminology. In S. Cote (Ed.), *Criminological theories: Bridging the past to the future* (pp. 267–284). Thousand Oaks, CA: Sage.

Daly, M. (1996). Evolutionary adaptationism: Another biological approach to criminal and antisocial behavior. In G. Bock & J. Goode (Eds.), *Genetics of criminal and antisocial behaviour* (pp. 183–195). Chichester, UK: Wiley.

Daly, M., & Wilson, M. (1996). Violence against stepchildren. *Current Directions in Psychological Science, 5,* 77–81.

Daly, M., & Wilson, M. (2000). Risk-taking, intersexual competition, and homicide. *Nebraska Symposium on Motivation, 47,* 1–36.

Davenport-Hines, R. (2002). *The pursuit of oblivion: A global history of narcotics.* New York: Norton.

Day, J., & Carelli, R. (2007). The nucleus accumbens and Pavlovian reward learning. *Neuroscientist, 13,* 148–159.

De Haan, W., & Vos, J. (2003). A crying shame: The overrationalized conception of man in the rational choice perspective. *Theoretical Criminology, 7,* 29–54.

DeLisi, M. (2009). Psychopathy is the unified theory of crime. *Youth Violence and Juvenile Justice, 7,* 257–273.

DeLisi, M., Beaver, K., Vaughn, M., & Wright, J. (2009). All in the family: Gene x environment interaction between DRD2 and criminal father is associated with five antisocial phenotypes. *Criminal Justice and Behavior, 36,* 1187–1197.

Demir, B., Ucar, G., Ulug, B., Ulosoy, S., Sevinc, I., & Batur, S. (2002). Platelet monoamine oxidase activity in alcoholism subtypes: Relationship to personality traits and executive functions. *Alcohol and Alcoholism, 37,* 597–602.

Depue, R., & Collins, P. (1999). Neurobiology of the structure of personality: Dopamine, facilitation of incentive motivation, and extraversion. *Behavioral and Brain Sciences, 22,* 491–569.

DeVoe, J., Peter, K., Kaufman, P., Ruddy, S., Miller, A., Planty, M., et al. (2003). *Indicators of school crime and safety.* Washington, DC: U.S. Department of Education and U.S. Department of Justice.

Dickens, W., & Flynn, J. (2001). Heritability estimates versus large environmental effects: The IQ paradox resolved. *Psychological Review, 108,* 346–349.

Dinkes, R., Kemp, J., Baum, K., & Snyder, T. (2010). *Indicators of school crime and safety: 2009.* Washington, DC: Bureau of Justice Statistics.

Dishman, C. (2001). Terrorism, crime, and transformation. *Studies in Conflict & Terrorism, 24,* 43–58.

Doerner, W., & Lab, S. (2002). *Victimology* (3rd ed.). Cincinnati, OH: Anderson.

Drug Enforcement Administration. (2003). *Drugs of abuse.* Arlington, VA: U.S. Department of Justice.

Drummond, J. (2002). From the northwest imperative to global jihad: Social psychological aspects of the construction of the enemy, political violence, and terror. In E. Stout (Ed.), *The psychology of terrorism* (Vol. 1, pp. 49–95). Westport, CT: Praeger.

Dunworth, T. (2001). Criminal justice and the IT revolution. *Criminal Justice, 3,* 371–426. Available at http://www.ncjrs.gov/criminal_justice2000/vol_3/03h.pdf

DuPont, R. (1997). *The selfish brain: Learning from addiction.* Washington, DC: American Psychiatric Press.

Durant, W. (1939). *The life of Greece.* New York: Simon & Schuster.

Durant, W., & Durant, A. (1968). *The lessons of history.* New York: Simon & Schuster.

Durkheim, E. (1951). *The division of labor in society.* Glencoe, IL: The Free Press.

Durkheim, E. (1982). *Rules of sociological method.* New York: The Free Press.

Durston, S. (2003). A review of the biological bases of ADHD: What have we learned from imaging studies? *Mental Retardation and Developmental Disabilities, 9,* 184–195.

Eisener, M. (2001). Modernization, self-control, and lethal violence: The long-term dynamics of European homicide rates in theoretical perspective. *British Journal of Criminology, 41,* 618–638.

Elliot, D., Huizinga, D., & Menard, S. (1989). *Multiple problem youth: Delinquency, substance abuse, and mental health problems.* New York: Springer-Verlag.

Ellis, L. (2003). Genes, criminality, and the evolutionary neuroandrogenic theory. In A. Walsh & L. Ellis (Eds.), *Biosocial criminology: Challenging environmentalism's supremacy* (pp. 12–34). Hauppauge, NY: Nova Science.

Ellis, L., & Hoffman, H. (1990). Views of contemporary criminologists on causes and theories of crime. In L. Ellis & H. Hoffman (Eds.), *Crime in biological, social, and moral contexts* (pp. 50–58). New York: Praeger.

Ellis, L., & Walsh, A. (1997). Gene-based evolutionary theories in criminology. *Criminology, 35,* 229–276.

Ellis, L., & Walsh, A. (2000). *Criminology: A global perspective.* Boston: Allyn & Bacon.

Ellis, L., & Walsh, A. (2003). Crime, delinquency and intelligence: A review of the worldwide literature. In H. Nyborg (Ed.), *The scientific study of general intelligence: A tribute to Arthur Jensen* (pp. 343–365). Amsterdam: Pergamon.

Ellis, L., Widmayer, A., & Palmer, C. (2009). Perpetrators of sexual assault continuing to have sex with their victims following the initial assault: Evidence for evolved reproductive strategies. *International Journal of Offender Therapy and Comparative Criminology, 53,* 454–463.

Ember, M., & Ember, C. (1998, October). Facts of violence. *Anthropology Newsletter,* 14–15.

English, S. (2004). Enron legal bills will cost $780m. *Business Telegraph.* http://www.telegraph.co.uk/money.jhtml?xml=/money/2004/1

Farrell, G., Phillips, C., & Pease, K. (1995). Like taking candy from a baby: Why does repeat victimization occur? *British Journal of Criminology, 35,* 384–399.

Farrington, D. (1982). Longitudinal analyses of criminal violence. In M. Wolfgang & N. Weiner (Eds.), *Criminal violence* (pp. 171–200). Beverly Hills, CA: Sage.

Farrington, D. (2003). Developmental and life-course criminology: Key theoretical and empirical issues—the 2002 Sutherland Award address. *Criminology, 41,* 221–255.

Fast Track Project. (2005). *Fast track project overview.* Available at http://www.fasttrackproject.org/

Federal Bureau of Investigation. (2005). *Uniform Crime Reports handbook.* Washington, DC: U.S. Government Printing Office.

Federal Bureau of Investigation. (2008). *Serial murder—multi-disciplinary perspectives for investigators.* Available at http://www.asiaing.com/serial-murder-multi-disciplinary-perspectives-for-investigators.html

Federal Bureau of Investigation. (2009a). *Crime in the United States: 2008.* Washington, DC: U.S. Government Printing Office.

Federal Bureau of Investigation. (2009b). *Financial crimes report to the public, fiscal year 2008.* Washington, DC: U.S. Government Printing Office.

Federal Bureau of Investigation. (2010a). *Crime in the United States, 2009: Uniform Crime Reports.* Washington, DC: U.S. Government Printing Office.

Federal Bureau of Investigation. (2010b). *Financial Crimes Report, 2009.* Available at http://www.fbi.gov/stats-services/publications/financial-crimes-report-2009

Federal Trade Commission. (2010). *Consumer sentinel network data book.* Washington, DC: Author.

Feeney, D. (2002). Enhancement in Islamic fundamentalism. In C. Stout (Ed.), *The psychology of terrorism* (Vol. 3, pp. 192–209). Westport, CT: Praeger

Ferguson, C. (2010). Genetic contributions to antisocial personality and behavior: A meta-analytic review from an evolutionary perspective. *Journal of Social psychology, 150,* 160–180.

Fergusson, D., Swain-Campbell, N., & Horwood, J. (2004). How does childhood economic disadvantage lead to crime? *Journal of Child Psychology and Psychiatry, 45,* 956–966.

Ferri, E. (1917). *Criminal sociology.* Boston: Little, Brown. (Original work published 1897)

Finckenauer, J. (2004, July/August). The Russian "Mafia." *Society,* 61–64.

Finkelhor, D. (1984). *Child sexual abuse: New theory and research.* New York: The Free Press.

Firestone, T. (1997). Mafia memoirs: What they tell us about organized crime. In P. Ryan & G. Rush (Eds.), *Understanding organized crime in global perspective* (pp. 71–86). Thousand Oaks, CA: Sage.

Fishbein, D. (1992). The psychobiology of female aggression. *Criminal Justice and Behavior, 19,* 99–126.

Fishbein, D. (2001). *Biobehavioral perspectives in criminology.* Belmont, CA: Wadsworth.

Fishbein, D. (2003). Neuropsychological and emotional regulatory processes in antisocial behavior. In A. Walsh & L. Ellis (Eds.), *Biosocial criminology: Challenging environmentalism's supremacy* (pp. 185–208). Hauppauge, NY: Nova Science.

Fisher, B., Cullen, F., & Turner, M. (2001). *The sexual victimization of college women.* Washington, DC: National Institute of Justice.

Flynn, J. (2007). *What is intelligence? Beyond the Flynn effect.* Cambridge, UK: Cambridge University Press.

Fox, J., & Levin, J. (2001). *The will to kill: Making sense of senseless murder.* Boston: Allyn & Bacon.

Freeman, N. (2007). Predictors of rearrest for rapists and child molesters on probation. *Criminal Justice and Behavior, 34,* 752–758.

Freud, S. (1923). The ego and the id. (J. Strachey, Ed. & Trans.). *The complete psychological works of Sigmund Freud* (Vol. 19). New York: Norton. (Original work published 1923)

Friedman, L. (2005). *A history of American law.* New York: Simon & Schuster.

Garofalo, R. (1968). *Criminology.* Montclair: NJ: Patterson Smith. (Original work published 1885)

Garrison, A. (2004). Defining terrorism: Philosophy of the bomb, propaganda by deed and change through fear and violence. *Criminal Justice Studies, 17,* 259–279.

Gatzke-Kopp, L., Raine, A., Loeber, R., Stouthamer-Loeber, M., & Steinhauer, S. (2002). Serious delinquent behavior, sensation seeking, and electrodermal arousal. *Journal of Abnormal Child Psychology, 30,* 477–486.

Gaulin, S., & McBurney, D. (2001). *Psychology: An evolutionary approach.* Upper Saddle River, NJ: Prentice Hall.

Geary, D. (2000). Evolution and proximate expression of human paternal investment. *Psychological Bulletin, 126,* 55–77.

Giannangelo, S. (1996). *The psychopathology of serial murder: A theory of violence.* Westport, CT: Praeger.

Gibson, M. (2002). *Born to crime: Cesare Lombroso and the origins of biological criminology.* Westport, CT: Praeger.

Gilsinan, J. (1990). *Criminology and public policy: An introduction.* Englewood Cliffs, NJ: Prentice Hall.

Given, J. (1977). *Society and homicide in thirteenth-century England.* Stanford, CA: Stanford University Press.

Glaser, D., & Frosh, S. (1993). *Child sex abuse.* Toronto, Ont., Canada: University of Toronto Press.

Glueck, S. (1956). Theory and fact in criminology: A criticism of differential association theory. *British Journal of Criminology, 7,* 92–109.

Glueck, S., & Glueck, E. (1950). *Unraveling juvenile delinquency.* New York: Commonwealth Fund.

Goldberg, E. (2001). *The executive brain: Frontal lobes and the civilized mind.* New York: Oxford University Press.

Goldstein, P. (1985). The drugs/violence nexus: A tripartite conceptual framework. *Journal of Drug Issues, 15,* 493–506.

Goldstein, P., Brownstein, H., Ryan, P., & Belluci, P. (1989). Crack and homicide in New York City: A conceptually-based event analysis. *Contemporary Drug Problems, 16,* 651–687.

Goode, E. (2005). *Drugs in American society* (6th ed.). Boston: McGraw-Hill.

Goodlett, C., Horn, K., & Zhou, F. (2005). Alcohol teratogenesis: Mechanisms of damage and strategies for intervention. *Developmental Biology and Medicine, 230,* 394–406.

Gottfredson, M. (2006). The empirical status of control theory in criminology. In F. Cullen, J. Wright, & K. Blevins (Eds.), *Taking stock: The status of criminological theory* (pp. 77–100). New Brunswick, NJ: Transaction.

Gottfredson, M., & Hirschi, T. (1990). *A general theory of crime.* Stanford, CA: Stanford University Press.

Gottfredson, M., & Hirschi, T. (1997). National crime control policies. In M. Fisch (Ed.), *Criminology 97/98* (pp. 27–33). Guilford, CT: Dushkin.

Gove, W., & Wilmoth, C. (2003). The neurophysiology of motivation and habitual criminal behavior. In A. Walsh & L. Ellis (Eds.), *Biosocial criminology: Challenging environmentalism's supremacy* (pp. 227–24). Hauppauge, NY: Nova Science.

Graham-Kevan, N., & Archer, J. (2009). Control tactics and partner violence in heterosexual relationships. *Evolution and Human Behavior, 30,* 445–452.

Grana, S. (2002). *Women and (in) justice: The criminal and civil effects of the common law on women's lives.* Boston: Allyn & Bacon.

Grasmick, H., Tittle, C., Bursik, R., & Arneklev, B. (1993). Testing the core empirical implication of Gottfredson and Hirschi's general theory of crime. *Journal of Research in Crime and Delinquency, 30,* 5–29.

Greenberg, D. (1981). *Crime and capitalism: Readings in Marxist criminology.* Palo Alto, CA: Mayfield.

Griffin, S. (2002). Actors or activities? On the social construction of "white-collar crime" in the United States. *Crime, Law, and Social Change, 37,* 245–276.

Gudjonsson, G., Sigurddsson, J., Young, S., Newton, A., & Peersen, M. (2009). Attention deficit hyperactivity disorder (ADHD): How do ADHD symptoms relate to personality among prisoners? *Personality and Individual Differences, 47,* 64–68.

Guo, G., Tong, Y., & Cai, T. (2008). Gene by social context interactions for number of sexual partners among white male youths: Genetics-informed sociology. *American Journal of Sociology, 114,* S36–66.

Gutman, H. (2002). Dishonesty, greed, and hypocrisy in corporate America. *Statesman* (Kolkata, India). Available at http://www.commondreams.org/views02/0712-02.htm

Hacker, F. (1977). *Crusaders, criminals, crazies: Terror and terrorism in our time.* New York: Norton.

Hagan, F. (1994). *Introduction to criminology.* Chicago: Nelson-Hall.

Hagan, F. (2008). *Introduction to criminology* (6th ed.) Thousand Oaks, CA: Sage.

Hagan, J. (1985). *Modern criminology: Crime, criminal behavior, and its control.* New York: McGraw-Hill.

Hagan, J. (1989). *Structural criminology.* New Brunswick, NJ: Rutgers University Press.

Hampton, R., Oliver, W., & Magarian, L. (2003). Domestic violence in the African American community. *Violence Against Women, 9,* 533–557.

Hare, R. (1993). *Without conscience: The disturbing world of the psychopaths among us.* New York: Pocket Books.

Hare, R. (1996). Psychopathy: A clinical construct whose time has come. *Criminal Justice and Behavior, 23,* 25–54.

Harris, A., Thomas, S., Fisher, G., & Hirsch, D. (2002). Murder and medicine: The lethality of criminal assault 1960–1999. *Homicide Studies, 6,* 128–166.

Harris, J. (1998). *The nurture assumption: Why children turn out the way they do.* New York: The Free Press.

Harris, K. (1991). Moving into the new millennium: Toward a feminist view of justice. In H. Pepinsky & R. Quinney (Eds.), *Criminology as peacemaking* (pp. 83–97). Bloomington: Indiana University Press.

Hazelwood, R., Ressler, R., Depue, K., & Douglas, J. (1987). Criminal personality profiling: An overview. In R. Hazelwood & A. Burgess (Eds.), *Practical aspects of rape investigation: A multidisciplinary approach* (pp. 137–149). New York: Elsevier.

Henry, B., Caspi, A., Moffitt, T., & Silva, P. (1996). Temperament and familial predictors of violent and non-violent criminal convictions: From age 3 to age 8. *Developmental Psychology, 32,* 614–623.

Herman, J. (1990). Sex offenders: A feminist perspective. In W. Marshall, D. Laws, & H. Barbaree (Eds.), *Handbook of sexual assault: Issues, theories, and treatment of the offender* (pp. 177–193). New York: Plenum.

Hermans, E., Putman, P., & van Honk, J. (2006). Testosterone reduces empathetic mimicking in healthy young women. *Psychoneuroendocrinology, 31,* 859–866.

Herrnstein, R., & Murray, C. (1994). *The bell curve: Intelligence and class structure in American Society.* New York: The Free Press.

Hickey, E. (2006). *Serial murderers and their victims* (4th ed.). Belmont, CA: Wadsworth.

Hill, P. (2003). *The Japanese Mafia: Yakuza, law, and the state.* Oxford, UK: Oxford University Press.

Hindelang, M., Hirschi, T., & Weis, J. (1981). *Measuring delinquency.* Beverly Hills, CA: Sage.

Hirschi, T. (1969). *The causes of delinquency.* Berkeley: University of California Press.

Hirschi, T. (2004). Self-control and crime. In R. Baumeister & K. Vohs (Eds.), *Handbook of self-regulation research, theory, and applications* (pp. 537–552). New York: Guilford.

Hirschi, T., & Gottfredson, M. (1987). Causes of white-collar crime. *Criminology, 25,* 949–974.

Hoffman, B. (2010, Fall). The evolving nature of terrorism— Nine years after the 9/11 attacks *The Social Contract, 21*(1), 33–40.

Hoffman, C. (2002). Rethinking terrorism and counterterrorism since 9/11. *Studies in Conflict & Terrorism, 25,* 303–316.

Holmes, R., & DeBurger, J. (1998). Profiles in terror: The serial murderer. In R. Holmes & A. Holmes (Eds.), *Contemporary perspectives on serial murder* (pp. 1–16). Thousand Oaks, CA: Sage.

Hopcroft, R. (2009). Gender inequality in interaction: An evolutionary account. *Social Forces, 87,* 1845–1872.

Howard, C. (1979). *Zebra: The true account of the 179 days of terror in San Francisco.* New York: Richard Marek.

Huber, J. (2008). Reproductive biology, technology, and gender inequality: An autobiographical essay. *Annual Review of Sociology, 34,* 1–13.

Hudson, R. (1999). *The sociology and psychology of terrorism: Who becomes a terrorist and why?* Washington, DC: Library of Congress, Federal Research Division.

Huizink, A., & Mulder, E. (2006). Maternal smoking, drinking, or cannabis use during pregnancy and neurobehavioral and cognitive functioning in human offspring. *Neuroscience and Biobehavioral Reviews, 30,* 24–41.

Hunnicutt, G., & Broidy, L. (2004). Liberation and economic marginalization: A reformulation and test of (formerly?) competing models. *Journal of Research in Crime and Delinquency, 41,* 130–155

Hurd, M. (2003, December 14). The psychology of junior sniper Lee Malvo. *Capitalism Magazine.* Available at http://www.capitalismmagazine.com/law/crime/3402-the-psychology-of-junior-sniper-lee-malvo.html?print

Hwang, S.-L., & Bedford, O. (2002). Juveniles' motivation for remaining in prostitution. *Psychology of Women Quarterly, 28,* 136–146.

Insurance Information Institute. (2009). *Drunk driving.* Available at http://www.iii.org/media/hottopics/insurance/drunk/

INTERPOL. (1992). *International crime statistics.* Lyons, France: Author.

Irwin, K., & Chesney-Lind, M. (2008). Girls' violence: Beyond dangerous masculinity. *Sociology Compass, 2/3,* 837–855.

Iwai, H. (1986). Organized crime in Japan. In R. Kelly (Ed.), *Organized crime: A global perspective* (pp. 208–233). Totowa, NJ: Rowman & Littlefield.

Jacobs, B., Topalli, V., & Wright, R. (2003). Carjacking, streetlife, and offender motivation. *British Journal of Criminology, 43,* 673–688.

Jacobs, B., & Wright, R. (1999). Stick-up, street culture, and offender motivation. *Criminology, 37,* 149–173.

Jalata, A. (2011). Terrorism from above and below in the age of globalization. *Sociology Mind, 1*(1), 1–15.

Jarboe, J. (2202). The threat of domestic terrorism. Testimony to the Congressional Committee on Forests and Forest Health. Available at http://www.fbi.gov/congress/congress02/jarboe021202.htm

Jenkins, P. (1994). *Using murder: The social construction of serial homicide.* New York: Aldine De Gruyter.

Johnson, E. (1990). Yakuza (criminal gangs) in Japan: Characteristics and management in prison. *Journal of Contemporary Criminal Justice, 6,* 113–126.

Johnson, L., O'Malley, P., & Bachman, J. (2000). *Monitoring the Future: National survey results on drug use, 1975–1999.* Bethesda, MD: National Institute of Drug Abuse.

Johnson, P., & Feldman, T. (1992). Personality types and terrorism: Self-psychology perspectives. *Forensic Reports, 5,* 293–303.

Justice Research and Statistics Association. (2010). *Status of NIBRS in the states.* Available at http://www.jrsa.org/ibrrc/background-status/nibrs_states.shtml

Kanazawa, S. (2003). A general evolutionary psychological theory of criminality and related male-typical behavior. In A. Walsh & L. Ellis (Eds.), *Biosocial criminology: Challenging environmentalism's supremacy* (pp. 37–60). Hauppauge, NY: Nova Science.

Kappeler, V., Blumberg, M., & Potter, G. (2000). *The mythology of crime and criminal justice* (3rd ed.). Prospect Heights, IL: Waveland.

Karmen, A. (2005). *Crime victims: An introduction to victimology* (5th ed.). Belmont, CA: Wadsworth.

Keeney, B., & Heide, K. (1995). Serial murder: A more accurate and inclusive definition. *International Journal of Offender Therapy and Comparative Criminology, 39,* 299–306.

Kesteren, J. van, Mayhew, P., & Nieuwbeerta, P. (2000). Criminal victimization in seventeen in strialised countries: Key findings from the 2000 Internati l Crime Victims Survey. The Hague, the Netherlands: Min ry of Justice.

Kim, J., Fendrich, M., & Wislar, J. (2000). The validity of juvenile arrestees' drug use reporting: A gender comparison, *Journal of Research in Crime and Delinquency, 37,* 429–432.

Kim-Cohen, J., Caspi, A., Taylor, A., Williams, B., Newcombe, R., Craig, I., et al. (2006). MAOA, maltreatment, and gene-environment interaction predicting children's mental health: New evidence and a meta-analysis. *Molecular Psychiatry, 11,* 903–913.

Kimura, D. (1992). Sex differences in the brain. *Scientific American, 267,* 119–125.

Klaus, P. (2004). Carjacking, 1993–2002. *Bureau of Justice Statistics Crime Data Briefs.* Washington, DC: U.S. Department of Justice.

Kleber, H. (2003). Pharmacological treatments for heroin and cocaine dependence. *American Journal on Addictions, 12,* S5–S18.

Knox, G., & Fuller, L. (1995). The Gangster Disciples: A gang profile. *Journal of Gang Research, 3,* 58–76.

Knudsen, D. (1991). Child sexual coercion. In E. Grauerholz & M. Koralewski (Eds.), *Sexual coercion: A sourcebook on its nature, causes, and prevention* (pp. 17–28). Lexington, MA: D.C. Heath.

Kochanska, G., & Aksan, N. (2004). Conscience in childhood: Past, present, and future. *Merrill-Palmer Quarterly, 50,* 299–310.

Koller, K., Brown, T., Spurfeon, A., & Levy, L. (2004). Recent developments in low-level lead exposure and intellectual impairment in children. *Environmental Health Perspectives, 112,* 987–994.

Kornblum, W., & Julian, J. (1995). *Social problems* (8th ed.). Englewood Cliffs, NJ: Prentice Hall.

Kornhauser, R. (1978). *Social sources of delinquency: An appraisal of analytical methods.* Chicago: University of Chicago Press.

Kramer, M. (1990). The moral logic of Hizballah. In W. Reich (Ed.), *Origins of terrorism: Psychologies, ideologies, theologies, states of mind* (pp. 131–157). New York: Cambridge University Press.

Krueisi, M., Leonard, H., Swedo, S., Nadi, S., Hamburger, S., Lui, J. et al. (1994). Endogenous opioids, childhood psychopathology, and Quay's interpretation of Jeffrey Gray. In D. Routh (Ed.), *Disruptive Behavior Disorders in Childhood* (pp. 207–219). New York: Plenum.

Krug, E., Dahlberg, L., Mercy, J., Zwi, A., & Lozano, R. (2002). *World report on violence and health.* Geneva, Switzerland: World Health Organization.

Kshetri, N. (2006, January/February). The simple economics of cybercrimes. *IEEE Security & Privacy,* 33–39.

Kurtz, H. (2002, February 2). America is "doomed" bin Laden says on tape. *International Herald Tribune,* p. 5.

Kyl, J., & Feinstein, D. (2000). *Written testimony of Michelle Brown.* Posted by Privacy Rights Clearing House. Available at http://www.freerepublic.com/focus/f-news/1241208/posts

LaFree, G., & Ackerman, G. (2009). The empirical study of terrorism: Social and legal research. *Annual Review of Law and Social Science, 5,* 347–374.

LaFree, G., Drass, K., & O'Day, P. (1992). Race and crime in postwar America: Determinants of African-American and white rates, 1957–1988. *Criminology, 30,* 157–185.

Lamontagne, Y., Boyer, R., Hetu, C., & Lacerte-Lamontagne, C. (2000). Anxiety, significant losses, depression, and irrational beliefs in first-offense shoplifters. *Canadian Journal of Psychiatry, 45,* 63–66.

Lanier, M., & Henry, S. (2010). *Essential criminology* (3rd ed.). Boulder, CO: Westview.

Laub, J., & Sampson, R. (2003). *Shared beginnings, divergent lives: Delinquent boys at age 70.* Cambridge, MA: Harvard University Press.

Law Enforcement Agency Resource Network. (2004). *Aryan Nations/Church of Jesus Christ Christian.* Available at http://www.adl.org/learn/ext_us/Aryan_Nations.asp?xpicked=3&

Leader, S., & Probst, P. (2006). *The Earth Liberation Front and environmental terrorism.* Available at http://www1.umn.edu/des/earthliberationfront3pub.htm

Lemert, E. (1974). Beyond Mead: The societal reaction to deviance. *Social Problems, 21,* 457–468.

Leonard, E. (1995). Theoretical criminology and gender. In B. Price & N. Sokoloff (Eds.), *The criminal justice system and women: Offenders, victims, and workers* (pp. 54–70). New York: McGraw-Hill.

Lepowsky, M. (1994). Women, men, and aggression in egalitarian societies. *Sex Roles, 30,* 199–211.

Lester, D. (2010). Suicide in mass murderers and serial killers. *Suicidology, 1,* 19–27.

Levin, J., & Fox, J. (1985). *Mass murder: America's growing menace.* New York: Plenum.

Levin, Y., & Lindesmith, A. (1971). English ecology and criminology of the past century. In H. Voss & D. Petersen (Eds.), *Ecology, crime, and delinquency* (pp. 47–76). New York: Appleton-Century-Crofts.

Levine, D. (2006). Neural modeling of the dual motive theory of economics. *Journal of Socio-Economics, 35,* 613–625.

Levy, F., Hay, D., McStephen, M., Wood, C., & Waldman, I. (1997). Attention-deficit hyperactivity disorder: A category or a continuum? Genetic analysis of a large-scale twin study. *Journal of the American Academy of Child and Adolescent Psychiatry, 36,* 737–744.

Levy, S., & Stone, B. (2005, July 4). Grand theft identity. *Newsweek Business Online.* Available at http://www.highbeam.com/doc/1G1-133640543.html

Leyton, E. (1986). *Hunting humans: Inside the minds of mass murderers.* New York: Pocket Books.

Lilly, J., Cullen, F., & Ball, R. (2007). *Criminological theory: Context and consequences* (4th ed.). Thousand Oaks, CA: Sage.

Linden, R., & Chaturvedi, R. (2005). The need for comprehensive crime prevention planning: The case of motor vehicle theft. *Canadian Journal of Criminology and Criminal Justice, 47,* 251–270.

Lo, C., & Stephens, R. (2002). The role of drugs in crime: Insights from a group of incoming prisoners. *Substance Use and Misuse, 37,* 121–131.

Loeber, R., Kalb, L., & Huizinga, D. (2001). Juvenile delinquency and serious injury victimization. *Juvenile Justice Bulletin.* Washington, DC: U.S. Department of Justice.

Lombroso, C. (1876). *Criminal man.* Milan, Italy: Hoepli.

Lombroso, C. (1920). *The female offender.* New York: Appleton.

Lombroso-Ferrero, G. (1972). *Criminal man according to the classification of Cesare Lombroso.* Montclaire, NJ: Patterson Smith. (Original work published 1911)

Lupsha, P. (1987). La Cosa Nostra in drug trafficking. In T. Bynum (Ed.), *Organized crime in America: Concepts and controversies* (pp. 31–41). Monsey, NY: Willow Tree Press.

Lykken, D. (1995). *The antisocial personalities.* Hillsdale, NJ: Lawrence Erlbaum.

Lyman, M., & Potter, G. (2004). *Organized crime* (3rd ed.). Upper Saddle River, NJ: Prentice Hall.

Lynam, D. (1996). Early identification of chronic offenders: Who is the fledgling psychopath? *Psychological Bulletin, 120,* 209–234.

Lynam, D., Moffitt, T., & Stouthamer-Loeber, M. (1993). Explaining the relation between IQ and delinquency: Class, race, test motivation, school failure, or self control? *Journal of Abnormal Psychology, 102,* 187–196.

Lyons, M., True, W., Eusen, S., Goldberg, J., Meyer, J., Faraone, S., et al. (1995). Differential heritability of adult and juvenile antisocial traits. *Archives of General Psychiatry, 53,* 906–915.

Lytton, H., & Romney, D. (1991). Parents' differential socialization of boys and girls: A meta-analysis. *Psychological Bulletin, 109,* 267–296.

MacDonald, K., & MacDonald, T. (2010). The peptide that binds: A systematic review of oxytocin and its prosocial effects in humans. *Harvard Review of Psychiatry, 18,* 1–21.

Machetta, J. (2011). *Washington bomb suspect linked to Kirksville white supremacist.* Available at http://www.missourinet.com/2011/03/17/washington-bomb-suspect-linked-to-kirksville-white-supremacist-website/

Macmillan, R. (2001). Violence and the life course: The consequences of victimization for personal and social development. *Annual Review of Sociology, 27,* 1–22.

Mallon, R. (2007). A field guide to social construction. *Philosophy Compass, 2,* 93–108.

Mann, C. (1990). Black female homicides in the United States. *Journal of Interpersonal Violence, 5,* 176–201.

Martin, S. (2001). The links between alcohol, crime, and the criminal justice system: Explanations, evidence and interventions. *American Journal on Addictions, 10,* 136–158.

Martínez, Jr., R., Rosenfeld, R., & Mares, D. (2008). Social disorganization, drug market activity, and neighborhood violent crime. *Urban Affairs Review, 43,* 846–874.

Maruschak, L. (1999). *DWI offenders under correctional supervision.* Washington, DC: Bureau of Justice Statistics.

Marx, K., & Engels, F. (1948). *The communist manifesto.* New York: International Publishers.

Marx, K., & Engels, F. (1965). *The German ideology.* London: Lawrence and Wishart.

Massey, D. (2004). Segregation and stratification: A biosocial perspective. *Du Bois Review, 1,* 7–25.

Matarazzo, J. (1976). *Weschler's measurement and appraisal of adult intelligence.* Baltimore: Williams & Wilkins.

Mauer, M. (2005). *New incarceration figures: Growth in population continues.* Washington, DC: The Sentencing Project.

Maughan, B. (2005). Developmental trajectory modeling: A view from developmental psychopathology. *Annals of the American Academy of Political and Social Science, 602,* 118–130.

Mawby, R. (2001). *Burglary.* Devon, UK: Willan Publishing.

Maynard, R., & Garry, E. (1997). Adolescent motherhood: Implications for the juvenile justice system. *OJJDP Fact sheet #50.* Washington, DC: U.S. Department of Justice.

Mazur, A., & Booth, A. (1998). Testosterone and dominance in men. *Behavioral and Brain Sciences, 21,* 353–397.

McBride, D., & McCoy, C. (1993). The drugs-crime relationship: An analytical framework. *Prison Journal, 73,* 257–278.

McCabe, B., O'Reilly, C., & Pfeffer, J. (1991). Context, values, and moral dilemmas: Comparing the choices of business and law school students. *Journal of Business Ethics, 10,* 951–960.

McCarthy, B. (2002). New economics of sociological criminology. *Annual Review of Sociology, 28,* 417–442.

McCrae, R., Costa, P., Ostendorf, F., Angleitner, A., Hrebickova, A., Avia, M., et al. (2000). Nature over nurture: Temperament, personality, and life span development. *Journal of Personality and Social Psychology, 78,* 173–186.

McDermott, P. A., Alterman, A. I., Cacciola, J. S., Rutherford, M. J., Newman, J. P., & Mulholland, E. M. (2000). Generality of Psychopathy Checklist–Revised factors over prisoners and substance-dependent patients. *Journal of Consulting and Clinical Psychology, 68*(1), 181–186.

McGoey, C. (2005). *Carjacking facts: Robbery prevention advice.* Available at http://www.crimedoctor.com/carjacking.htm

McGue, M. (1999). The behavioral genetics of alcoholism. *Current Directions in Psychological Science, 8,* 109–115.

McGue, M., Bacon, S., & Lykken, D. (1993). Personality stability and change in early adulthood: A behavioral genetic analysis. *Developmental Psychology, 29,* 96–109.

McMurren, M. (2003). Alcohol and crime. *Criminal Behaviour and Mental Health, 13,* 1–4.

Mealey, L. (1995). The sociobiology of sociopathy: An integrated evolutionary model. *Behavioral and Brain Sciences, 18,* 523–559.

Mealey, L. (2003). Combating rape: Views of an evolutionary psychologist. In R. Bloom & N. Dess (Eds.), *Evolutionary psychology and violence* (pp. 83–113). Westport, CT: Praeger.

Mears, D., Ploeger, M., & Warr, M. (1998). Explaining the gender gap in delinquency: Peer influence and moral evaluations of behavior. *Journal of Research in Crime and Delinquency, 35,* 251–266.

Menard, S. (2002, February). Short- and long-term consequences of adolescent victimization. *Youth Violence Research Bulletin.* Washington, DC: U.S. Department of Justice.

Menard, S., & Mihalic, S. (2001). The tripartite conceptual framework in adolescence and adulthood: Evidence from a national sample. *Journal of Drug Issues, 31,* 905–940.

Menard, S., Mihalic, S., & Huizinga, D. (2001). Drugs and crime revisited. *Justice Quarterly, 18,* 269–299.

Merton, R. (1938). Social structure and anomie. *American Sociological Review, 3,* 672–682.

Messerschmidt, J. (1993). *Masculinities and crime: Critique and reconceptualization of theory.* Lanham, MD: Rowman & Littlefield.

Messerschmidt, J. (2002). On gang girls, gender, and a structured action theory. *Theoretical Criminology, 6,* 461–475.

Messner, S., & Rosenfeld, R. (2001). *Crime and the American Dream* (3rd ed.). Belmont, CA: Wadsworth.

Miller, J. (1998). Up it up: Gender and the accomplishment of street robbery. *Criminology, 36,* 37–65.

Miller, J., & Lynam, D. (2001). Structural models of personality and their relation to antisocial behavior: A meta-analytic review. *Criminology, 39,* 765–798.

Miller, L. (1987). Neuropsychology of the aggressive psychopath: An integrative review. *Aggressive Behavior, 13,* 119–140.

Miller, L. (2011). The terrorist mind. Part 1: A psychological and political analysis. In A. Walsh & C. Hemmens (Eds.), *Introduction to criminology: A text/reader* (pp. 423–432). Thousand Oaks, CA: Sage.

Miller, W. (1958). Lower-class culture as a generating milieu of gang delinquency. *Journal of Social Issues, 14,* 5–19.

Mills, J., Anderson, D., & Kroner, D. (2004). The antisocial attitudes of sex offenders. *Criminal Behavior and Mental Health, 14,* 134–145.

Moffitt, T. (1993). Adolescent-limited and life-course-persistent antisocial behavior: A developmental taxonomy. *Psychological Review, 100,* 674–701.

Moffitt, T., Caspi, A., Rutter, M., & Silva, P. (2001). *Sex differences in antisocial behaviour: Conduct disorder, delinquency and violence in the Dunedin longitudinal study.* Cambridge, UK: Cambridge University Press.

Moffit, T., & the E-Risk Study Team. (2002). Teen-aged mothers in contemporary Britain. *Journal of Child Psychology and Psychiatry, 43,* 1–16.

Moffitt, T., & Walsh, A. (2003). The adolescence-limited/life-course persistent theory and antisocial behavior: What have we learned? In A. Walsh & L. Ellis (Eds.), *Biosocial criminology: Challenging environmentalism's supremacy* (pp. 125–144). Hauppauge, NY: Nova Science.

Moore, J., & Hagedorn, J. (2001). Female gangs: Focus on research. *OJJDP Juvenile Justice Bulletin.* Washington, DC: U.S. Department of Justice.

Moore, R. (1984). Shoplifting in Middle America: Patterns and motivational correlates. *International Journal of Offender Therapy and Comparative Criminology, 23,* 29–40.

Morley, K., & Hall, W. (2003). Is there a genetic susceptibility to engage in criminal acts? *Trends and Issues in Crime and Criminal Justice, 263,* 1–10.

Mowbray, J. (2002). *Justice interrupted.* Available at http://www.nationalreview.com

Mullins, C., & Wright, R. (2003). Gender, social networks, and residential burglary. *Criminology, 41,* 813–839.

Mustaine, E., & Tewksbury, R. (2004). Alcohol and violence. In S. Holmes & R. Holmes (Eds.), *Violence: A contemporary reader* (pp. 9–25). Upper Saddle River, NJ: Prentice Hall.

Nagin, D. (1998). Criminal deterrence research at the onset of the twenty-first century. *Crime and Justice, 23,* 1–42. Chicago: University of Chicago Press.

National Association of Crime Victim Compensation Board. (2005). *FY 2004 compensation to victims continues to increase.* Available at http://www.nacvcb.org/

National Center for Policy Analysis. (1998). *Falsified crime data.* Available at http://www.ncpa.org/sub/dpd/index.php?Article_ID=15747

National Counterterrorism Center. (2009). *NCTC report on terrorism.* Washington, DC: Author.

National Highway Traffic Safety Commission. (2009). *Traffic safety facts.* Washington, DC: U.S. Department of Transportation.

National Institute of Drug Abuse. (1996). The brain's drug reward system. *NIDA Notes.* Washington, DC: National Institutes of Health, U.S. Department of Health and Human Services.

National Institute on Alcohol Abuse and Alcoholism. (1998). *Economic costs of alcohol and drug abuse estimated at $246 billion in the United States.* Available at http://www.nih.gov/news/pr/may98/nida-13.htm

National Insurance Crime Bureau. (2009). *Hot wheels 2008.* Available at https://www.nicb.org/newsroom/nicb_campaigns/hot%E2%80%93wheels

National Insurance Crime Bureau. (2010). *Hot wheels 2009.* Available at https://www.nicb.org/newsroom/news-releases/hot-wheels-2010

National Youth Gang Center. (2009). *National Youth Gang Survey analysis.* Available at http://www.nationalgangcenter.gov/

Neisser, U., Boodoo, G., Bouchard, T., Boykin, A., Brody, N., Ceci, S., et al. (1995). Intelligence: Knowns and unknowns: Report of a task force established by the Board of Scientific Affairs of the American Psychological Association. Washington, DC: American Psychological Association.

Nettler, G. (1984). *Explaining crime* (3rd ed.). New York: McGraw-Hill.

Newman, O. (1972). *Defensible space.* New York: Macmillan.

Newton, M. (2000). *The encyclopedia of serial killers.* New York: Checkmark.

Niehoff, D. (2003). A vicious circle: The neurobiological foundations of violent behavior. *Modern Psychoanalysis, 28,* 235–245.

O'Brien, R. (2001). Crime facts: Victim and offender data. In J. Sheley (Ed.), *Criminology: A contemporary handbook* (pp. 59–83). Belmont: CA, Wadsworth.

Office of Drug Control Policy. (2010). *ADAM II 2009 annual report.* Washington, DC: U.S. Government Printing Office.

Office of the Surgeon General of the United States. (2001). *Youth violence: A report of the Surgeon General.* Washington, DC: U.S. Department of Health and Human Services.

Olds, D., Hill, P., Mihalic, S., & O'Brien, R. (1998). *Blueprints for violence prevention, book seven: Prenatal and infancy home visitation by nurses.* Boulder, CO: Center for the Study and Prevention of Violence.

Olson, S., & Dzur, A. (2004). Revisiting informal justice: Restorative justice and democratic professionalism. *Law and Society Review, 38,* 139–176.

O'Manique, J. (2003). *The origins of justice: The evolution of morality, human rights, and law.* Philadelphia: University of Pennsylvania Press.

Oscar-Berman, M., Valmas, M., Sawyer, K., Kirkley, S., Gansler, D., Merritt, D., et al. (2009). Frontal brain dysfunction in alcoholism with and without antisocial personality disorder. *Neuropsychiatric Disease and Treatment, 5,* 309–326.

Osgood, D., & Chambers, J. (2003, May). Community correlates of rural youth violence. *Juvenile Justice Bulletin.* Washington, DC: U.S. Department of Justice.

Osofsky, J. (1995). The effects of exposure to violence on young children. *American Psychologist, 50,* 782–788.

Palermo, G. (1997). The berserk syndrome: A review of mass murder. *Aggression and Violent Behavior, 2,* 1–8.

Parekh, R. (2004). Fraud by employees on the rise, survey finds. *Business Insurance, 38,* 4–6.

Parker, N., & Auerhahn, K. (1998). Alcohol, drugs, and violence. *Annual Review of Sociology, 24,* 291–311.

Paus, T. (2010). Population neuroscience: Why and how. *Human Brain Mapping, 31,* 891–903.

Perlman, D. (2002). Intersubjective dimensions of terrorism and its transcendence. In C. Stout (Ed.), *The psychology of terrorism* (Vol. 3, pp. 57–81). Westport, CT: Praeger.

Perry, B., & Pollard, R. (1998). Homeostasis, stress, trauma, and adaptation: A neurodevelopmental view of childhood trauma. *Child and Adolescent Psychiatric Clinics of America, 7,* 33–51.

Pinel, J. (2000). *Biopsychology* (4th ed.). Boston: Allyn & Bacon.

Plato. (1960). *The republic and other works.* Garden City, NY: Doubleday.

Pope, C., & Snyder, H. (2003). Race as a factor in juvenile arrests. *Juvenile Justice Bulletin.* Washington, DC: U.S. Department of Justice.

Power, R. (2000). *Tangled web: Tales of digital crime from the shadows of cyberspace.* Indianapolis, IN: Que Books.

President's Commission on Organized Crime. (1986). *The impact: organized crime today.* Washington, DC: U.S. Government Printing Office.

Price, M. (2005). *Can mediation produce restorative justice for victims and offenders?* VORP Information and Resource Center. Available at http:/www.vorp.com/articles/crime

Pridemore, W. (2004). Weekend effects on binge drinking and homicide: The social connection between alcohol and violence in Russia. *Addiction, 99,* 1034–1041.

Primakov, E. (2004). *A world challenged: Fighting terrorism in the twenty-first century.* Washington, DC: The Nixon Center and Brookings Institute Press.

Quartz, S., & Sejnowski, T. (1997). The neural basis of cognitive development: A constructivist manifesto. *Behavioral and Brain Sciences, 20,* 537–596.

Quinn, J. (2001). Angels, bandidos, outlaws, and pagans: The evolution of organized crime among the Big Four 1% motorcycle clubs. *Deviant Behavior, 22,* 379–399.

Quinney, R. (1975). Crime control in capitalist society: A critical philosophy of legal order. In I. Taylor, P. Walton, & J. Young (Eds.), *Critical criminology* (pp. 181–202). Boston: Routledge & Kegan Paul.

Quinsey, V. (2002). Evolutionary theory and criminal behavior. *Legal and Criminological Psychology, 7,* 1–14.

Raine, A. (1997). Antisocial behavior and psychophysiology: A biosocial perspective and a prefrontal dysfunction hypothesis. In D. Stoff, J. Breiling, & J. Maser (Eds.), *Handbook of antisocial behavior* (pp. 289–304). New York: Wiley.

Raine, A., Meloy, J., Bihrle, S., Stoddard, J., LaCasse, L., & Buchsbaum, M. (1998). Reduced prefrontal and increased subcortical brain functioning assessed using positron emission tomography in predatory and affective murderers. *Behavioral Sciences and the Law, 16,* 319–332.

Rand, M. (2009). *Criminal victimization, 2008.* Washington, DC: Bureau of Justice Statistics.

Raymond, J. (2003). Ten reasons for not legalizing prostitution, and a legal response to the demand for prostitution. *Journal of Trauma Practice, 2,* 315–332. Available at http://www.prostitutionresearch.com/laws/000022.html

Raz, A. (2004, August). Brain imaging data of ADHD. *Neuropsychiatry,* 46–50.

Reckdenwald, A., & Parker, K. (2008). The influence of gender inequality and marginalization on types of female offending. *Homicide Studies, 12,* 208–226.

Reich, W. (1990). Understanding terrorist behavior: The limits and opportunities of psychological inquiry. In W. Reich (Ed.), *Origins of terrorism: psychologies, ideologies, theologies, states of mind* (pp. 261–279). New York: Cambridge University Press.

Reid, J., & Sullivan, C. (2009). A model of vulnerability for adult sexual victimization: The impact of attachment, child maltreatment, and scarred sexuality. *Violence and Victims, 24,* 485–501.

Reid, W. (2002). Controlling political terrorism: Practicality, not psychology. In C. Stout (Ed.), *The psychology of terrorism: Public understanding* (pp. 1–8). Westport, CT: Praeger.

Rengert, G., & Wasilchick, J. (2001). *Suburban burglary: A tale of two suburbs.* Springfield, IL: Charles C. Thomas.

Rennison, C. (2003). Intimate partner violence, 1993–2003. *Bureau of Justice Statistics Report.* Washington, DC: U.S. Department of Justice.

Restak, R. (2001). *The secret life of the brain.* New York: Dana Press and Joseph Henry Press.

Rhee, S., & Waldman, I. (2002). Genetic and environmental influences on antisocial behavior: A meta-analysis of twin and adoption studies. *Psychological Bulletin, 128,* 490–529.

Rice, K., & Smith, W. (2002). Socioecological models of automotive theft: Integrating routine activities and social disorganization approaches. *Journal of Research in Crime and Delinquency, 39,* 304–336.

Richardson, A., & Budd, T. (2003). Young adults, crime and disorder. *Criminal Behaviour and Mental Health, 13,* 5–17.

Robertiello, G., & Terry, K. (2007). Can we profile sex offenders? A review of sex offender typologies. *Aggression and Violent Behavior, 12,* 508–518.

Robinson, M. (2004). *Why crime? An integrated systems theory of antisocial behavior.* Upper Saddle River, NJ: Prentice Hall.

Robinson, M. (2005). *Justice blind: Ideals and realities of American criminal justice.* Upper Saddle River, NJ: Prentice Hall.

Robinson, M. (2009). No longer taboo: Crime prevention implications of biosocial criminology. In A. Walsh & K. Beaver (Eds.), *Biosocial criminology: New directions in theory and research* (pp. 243–263). New York: Routledge.

Robinson, T., & Berridge, K. (2003). Addiction. *Annual Review of Psychology, 54,* 25–53.

Rodkin, P., Farmer, T., Pearl, R., & Van Acker, R. (2000). Heterogeneity of popular boys: Antisocial and prosocial configurations. *Developmental Psychology, 36,* 14–24.

Roper v. Simmons, 112 S.W. 3rd 397 (2005).

Rosenbaum, D., Lurigio, A., & Davis, R. (1998). *The prevention of crime: Social and situational strategies.* Belmont, CA: West/Wadsworth.

Rosenfeld, R. (2000). Patterns in adult homicide. In A. Blumstein & J. Wallman (Eds.), *The crime drop in America* (pp. 130–163). Cambridge, UK: Cambridge University Press.

Rösler, M., Retz, W., Retz-Junginger, P., Hengesch, G., Schneider, M., Supprian, T., et al. (2004). Prevalence of attention deficit-hyperactivity disorder and comorbid disorders in young male prison inmates. *European Archives of Psychiatry Clinical Neuroscience, 254,* 365–371.

Rosoff, S., Pontell, H., & Tillman, R. (1998). *Profit without honor: White-collar crime and the looting of America.* Upper Saddle River, NJ: Prentice Hall.

Ross, J. (1999, February 28). Suspect with satanic impulses confesses to burning churches. *Associated Press.* Available at http://www.rickross.com//reference/satanism36.html

Rothbart, M., Ahadi, A., & Evans, D. (2000). Temperament and personality: Origins and outcomes. *Journal of Personality and Social Psychology, 78,* 122–135.

Rowe, D. (1996). An adaptive strategy theory of crime and delinquency. In J. Hawkins (Ed.), *Delinquency and crime: Current theories* (pp. 268–314). Cambridge, UK: Cambridge University Press.

Rowe, D. (2002). *Biology and crime.* Los Angeles: Roxbury.

Ruden, R. (1997). *The craving brain: The biobalance approach to controlling addictions.* New York: HarperCollins.

Rush, R., & Scarpitti, F. (2001). Russian organized crime: The continuation of an American tradition. *Deviant Behavior, 22,* 517–40.

Saffron, I. (1997, February 2). Chance leads to capture in serial killer-cannibal case. *Idaho Statesman,* p. 19a.

Sampson, R. (2000). Whither the sociological study of crime. *Annual Review of Sociology, 26,* 711–714.

Sampson, R. (2004). Neighborhood and community: Collective efficacy and community safety. *New Economy, 11,* 106–113.

Sampson, R., & Laub, J. (1999). Crime and deviance over the life course: The salience of adult social bonds. In F. Scarpitti & A. Nielsen (Eds.), *Crime and criminals: Contemporary and classical readings in criminology* (pp. 238–246). Los Angeles: Roxbury.

Sampson, R., & Laub, J. (2005). A life-course view of the development of crime. *Annals of the American Academy of Political and Social Sciences, 602,* 12–45.

Sampson, R., Raudenbush, S., & Earls, F. 1997. Neighborhoods and crime: A multilevel study of collective efficacy. *Science, 277,* 918–924.

Sanchez-Jankowski, M. (2003). Gangs and social change. *Theoretical Criminology, 7,* 191–216.

Sanjiv, K., & Thaden, E. (2004, January). Examining brain connectivity in ADHD. *Psychiatric Times,* 40–41.

Santtila, P., Hakkanen, H., Alison, L., & Whyte, C. (2003). Juvenile firesetters: Crime scene actions and offender characteristics. *Legal and Criminological Psychology, 8,* 1–20.

Sawhill, I., & Morton, J. (2007). *Economic mobility: Is the American Dream alive and well?* Washington, DC: The Economic Mobility Project/Pew Charitable Trusts.

Scarpa, A., & Raine, A. (2003). The psychophysiology of antisocial behavior: Interactions with environmental experiences. In A. Walsh & L. Ellis (Eds.), *Biosocial criminology: Challenging environmentalism's supremacy* (pp. 209–226). Hauppauge, NY: Nova Science.

Schatzberg, R., & Kelly, R. (1996). *African-American organized crime: A social history.* New Brunswick, NJ: Rutgers University Press.

Schilling, C., Walsh, A., & Yun, I. (2011). ADHD and criminality: A review of the genetic, neurobiological, evolutionary, and treatment literature. *Journal of Criminal Justice, 39,* 3–11.

Schmalleger, F. (2004). *Criminology today* (3rd ed.). Upper Saddle River, NJ: Prentice Hall.

Schwarz, B. (1995). Characteristics and typologies of sex offenders. In B. Schwartz & H. Cellini (Eds.), *The sex offender: Corrections, treatment, and legal practice* (pp. 1-22). Kingston, NJ: Civic Research Institute.

Seale, D., Polakowski, M., & Schneider, S. (1998). It's not really theft! Personal and workplace ethics that enable software piracy. *Behavior and Information Technology, 17,* 27–40.

Sears, D. (1991). *To kill again: The motivation and development of serial murder.* Wilmington, DE: Scholarly Resources.

Sederberg, P. (1989). *Terrorist myths: Illusions, rhetoric, and reality.* Englewood Cliffs, NJ: Prentice Hall.

Segrave, K. (1992). *Women serial and mass murderer: A worldwide reference, 1580 through 1990.* Jefferson, NC: McFarland.

Seligman, D. (1992). *A question of intelligence: The IQ debate in America.* New York: BirchLane.

Shaffer, J., & Ruback, B. (2002). Violent victimization as a risk factor for violent offending among juveniles. *Juvenile Justice Bulletin.* Washington, DC: U.S. Department of Justice.

Shaw, C., & McKay, H. (1972). *Juvenile delinquency and urban areas* (Rev. ed.). Chicago: University of Chicago Press.

Shelden, R., Tracy, S., & Brown, W. (2001). *Youth gangs in American society* (2nd ed.). Belmont, CA: Wadsworth.

Sherman, L., Gottfredson, D., McKenzie, D., Eck, J., Reuter, P., & Bushway, S. (1997). *Preventing crime: What works, what doesn't, what's promising.* Washington, DC: U.S. Department of Justice.

Shore, R. (1997). *Rethinking the brain: New insights into early development.* New York: Families and Work Institute.

Shover, N., & Hochstetler, A. (2006). *Choosing white-collar crime.* New York: Cambridge University Press.

Siegel, L. (1986). *Criminology.* Belmont, CA: Wadsworth.

Siegel, L. (1992). *Criminology* (4th ed.). St. Paul, MN: West Publishing.

Simon, D. (2002). *Elite deviance* (7th ed.). Boston: Allyn & Bacon.

Simon, R. (1975). *Women and crime.* Lexington, MA: Lexington Books.

Simonsen, C., & Spindlove, J. (2004). *Terrorism today: The past, the players, the future.* Upper Saddle River, NJ: Prentice Hall.

Smith, A. (1953). *The wealth of nations.* Cambridge, MA: Harvard University Press. (Original work published 1776)

Smith, B. (1994). *Terrorism in America: Pipe bombs and pipe dreams.* Albany: State University of New York Press.

Smith, E., & Farole, D. (2009). *Profile of intimate partner violence cases in large urban counties* (Report # NCJ 228193). Washington, DC: Bureau of Justice Statistics.

Smith, H., & Bohm, R. (2008). Beyond anomie: Alienation and crime. *Critical Criminology, 16,* 1–15.

Sowell, T. (1987). *A conflict of visions: Ideological origins of political struggles.* New York: William Morrow.

Spear, L. (2000). Neurobehavioral changes in adolescence. *Current Directions in Psychological Science, 9,* 111–114.

Spelman, W. (2000). The limited importance of prison expansion. In A. Blumstein & J. Wallman (Eds.), *The crime drop in America* (pp. 97–129). Cambridge, UK: Cambridge University Press.

Spergel, I. (1995). *The youth gang problem: A community approach.* New York: Oxford University Press.

Sprinzak, E. (1991). The process of delegitimization: Towards a linkage theory of political terrorism. In C. McCauley (Ed.), *Terrorism research and public policy* (pp. 50–68). London: Frank Cass.

Steffensmeier, D., & Haynie, D. (2000). Gender, structural disadvantage, and urban crime: Do macrosocial variables also explain female offending rates? *Criminology 38,* 403–438.

Steffensmeier, D., Zhong, H., Ackerman, J., Schwartz, J., & Agha, S. (2006). Gender gap trends for violent crimes, 1980 to 2003: A UCR–NCVS comparison. *Feminist Criminology, 1,* 72–98.

Stiles, B., Liu, X., & Kaplan, H. (2000). Relative deprivation and deviant adaptations: The mediating effects of negative self-feelings. *Journal of Research in Crime and Delinquency, 37,* 64–90.

Survey: Shoplifting losses mount. (2005). *Orlando Business Journal.* Available at http://orlando.bizjournals.com/orlando/stories/2005/12/05/daily.html

Sutherland, E. (1939). *Principles of criminology.* Philadelphia: J.B. Lippincott.

Sutherland, E. (1940). White collar criminality. *American Sociological Review, 5,* 1–20.

Sutherland, E. (1956). *The Sutherland papers* (A. Cohen, A. Lindesmith, & K. Schuessler, Eds.). Bloomington: Indiana University Press.

Sutherland, E., & Cressey, D. (1974). *Criminology* (9th ed.). Philadelphia: J.B. Lippincott.

Sykes, G., & Matza, D. (2002). Techniques of neutralization: A theory of delinquency. In S. Cote (Ed.), *Criminological theories: Bridging the past to the future* (pp. 144–150). Thousand Oaks, CA: Sage.

Talbot, T., Gilligan, L., Carter, M., & Matson, S. (2002). *An overview of sex offender management.* Washington, DC: Center for Sex Offender Management.

Tang, T., Chen, Y., & Sutarso, T. (2008). Bad apples in bad (business) barrels: The love of money, Machiavellianism, risk tolerance, and unethical behavior. *Management Decision, 46,* 243–263.

Tannenbaum, F. (1938). *Crime and community.* New York: Columbia University Press.

Tappan, P. (1947). Who is the criminal? *American Sociological Review, 12,* 96–112.

Taylor, I. (1999). Crime and social criticism. *Social Justice, 26,* 150–168.

Taylor, I., Walton, P., & Young, J. (1973). *The new criminology.* New York: Harper & Row.

Taylor, S. (2006). Tend and befriend: Biobehavioral bases of affiliation under stress. *Current Directions in Psychological Science, 15,* 273–277.

Texas v. Johnson, 491 U.S. 397 (1989)

Thorburn, K. (2004). Corporate governance and financial distress. In H. Sjogren & G. Skogh (Eds.), *New perspectives on economic crime* (pp. 76–94). Cheltenham, UK: Edward Elgar.

Thornberry, T., Huizinga, D., & Loeber, R. (2004). The causes and correlates studies: Findings and policy implication. *Juvenile Justice, 9,* 3–19.

Thornhill, R., & Palmer, C. (2000). *A natural history of rape: Biological bases of sexual coercion.* Cambridge: MIT Press.

Tibbett, S., & Hemmens, C. (2010). *Criminological theory: A text/reader.* Thousand Oaks, CA: Sage.

Tittle, C. (1983). Social class and criminal behavior: A critique of the theoretical foundation. *Social Forces, 62,* 334–358.

Tittle, C. (2000). Theoretical developments in criminology. *National Institute of Justice 2000, Vol. 1: The nature of crime: Continuity and change.* Washington, DC: National Institute of Justice.

Tolan, P., Gorman-Smith, D., & Henry, D. (2006). Family violence. *Annual Review of Psychology, 57,* 557–583.

Tonglet, M. (2001). Consumer misbehaviour: An exploratory study of shoplifting. *Journal of Consumer Behaviour, 1,* 336–354.

Topali, V. (2005). When being good is bad: An extension of neutralization theory. *Criminology, 43,* 797–837.

Transparency International. (2009). *Corruption perceptions index 2009.* Available at http://www.transparency.org/policy_research/surveys_indices/cpi/2009/cpi_2009_table

Trevino, L., & Youngblood, S. (1990). Bad apples in bad barrels: A causal analysis of ethical decision-making behavior. *Journal of Applied Psychology, 78,* 378–385.

Tseloni, A., & Pease, K. (2003). Repeat personal victimization. *British Journal of Criminology, 43,* 196–212.

Turner, H., Finkelhor, D., & Ormrod, R. (2006). The effects of lifetime victimization on the mental health of children and adolescents. *Social Science and Medicine, 62,* 13–27.

Tutty, L., & Nixon, K. (2003). Selling sex? It's really like selling your soul. Vulnerability to and the experience of exploitation through child prostitution. In K. Gorkoff & J. Runner (Eds.), *Being heard: The experience of young women in prostitution* (pp. 29–45). Black Point, Nova Scotia: Fernwood.

Udry, J. R. (2003). *The National Longitudinal Study of Adolescent Health (Add Health).* Chapel Hill: Carolina Population Center, University of North Carolina.

Unnever, J., Cullen, F., & Pratt, T. (2003). Parental management, ADHD, and delinquent involvement: Reassessing Gottfredson and Hirschi's general theory. *Justice Quarterly, 20,* 471–500.

U.S. Bureau of Justice Statistics. (2005). *Homicide trends in the U.S.* Available at http://www.popcenter.org/problems/domestic_violence/PDFs/Fox&Zawitz_2002.pdf

U.S. Bureau of Justice Statistics. (2010). *Victims.* Available at http://bjs.ojp.usdoj.gov/index.cfm?ty=tp&tid=9

United States Census Bureau. (2004). *Statistical abstracts of the United States.* Available at http//:www.census.gov/prod/2004pubs/03statab/vitstat.pdf

U.S. Department of Health and Human Services. (2002). *Results from the 2001 National Household Survey on Drug Abuse.* Washington, DC: Author.

U.S. Department of Health and Human Services. (2009). *Results from the 2008 National Household Survey on Drug Abuse.* Washington, DC: Author.

U.S. Department of Justice. (2004). *Nineteen individuals indicted in Internet "carding" conspiracy.* Available at http://www.cybercrime.gov/montovaniIndict.html

U.S. Department of State. (1995). *Patterns of global terrorism: 1994.* Washington, DC: Author.

U.S. Department of State. (2004). *Patterns of global terrorism: 2003.* Washington, DC: Author.

U.S. Department of State. (2005). *International narcotics control strategy report.* Washington, DC: Author.

U.S. Fire Administration. (2010). *Intentionally set structure fires.* Available at http://www.usfa.dhs.gov/

Umbreit, M. (1994). *Victim meets offender: The impact of restorative justice and mediation.* Monsey, NY: Criminal Justice Press.

Van Berlo, W., & Ensink, B. (2000). Problems with sexuality after sexual assault. *Annual Review of Sex Research, 11,* 235–257.

Van Dijk, J. (2008). *The world of crime.* Thousand Oaks, CA: Sage.

Van Honk, J., Harmon-Jones, E., Morgan, B., & Schutter, D. (2010). Socially explosive minds: The triple imbalance hypothesis of reactive aggression. *Journal of Personality, 78,* 67–94.

Vaughn, M., Fu, Q., DeLisi, M., Wright, J., Beaver, K., Perron, B., et al. (2010). Prevalence and correlates of fire-setting in the United States: Results from the National Epidemiological Survey on Alcohol and Related Conditions. *Comprehensive Psychiatry, 51,* 217–223.

Vetter, H., & Perlstein, G. (1991). *Perspectives on terrorism.* Pacific Grove, CA: Brooks/Cole.

Victoroff, J., & Kruglanski, A. (2009). *Psychology of terrorism: Classic and contemporary insights.* New York: Psychology Press.

Vila, B. (1994). A general paradigm for understanding criminal behavior: Extending evolutionary ecological theory. *Criminology, 32,* 311–358.

Vila, B. (1997). Human nature and crime control: Improving the feasibility of nurturant strategies. *Politics and the Life Sciences, 16,* 3–21.

Voiskounsky, A., & Smyslova, O. (2003). Flow-based model of computer hackers' motivation. *CyberPsychology & Behavior, 6,* 171–180.

Vold, G., & Bernard, T. (1986). *Theoretical criminology.* New York: Oxford University Press.

Vold, G., Bernard, T., & Snipes, J. (1998). *Theoretical criminology.* New York: Oxford University Press.

Von Hentig, H. (1941). Remarks on the interaction of perpetrator and victim. *Journal of Criminal Law, Criminology, and Police Science, 31,* 303–309.

Wakschlag, L., Pickett, K., Cook, E., Benowitz, N., & Leventhal, B. (2002). Maternal smoking during pregnancy and severe antisocial behavior in offspring: A review. *American Journal of Public Health, 92,* 966–974.

Walsh, A. (1988). Lessons and concerns from a case study of a "scientific" molester. *Corrective and Social Psychiatry, 34,* 18–23.

Walsh, A. (1994). Homosexual and heterosexual child molestation: Case characteristics and sentencing differentials. *International Journal of Offender Therapy and Comparative Criminology, 38,* 339–353.

Walsh, A. (2000). Evolutionary psychology and the origins of justice. *Justice Quarterly, 17,* 841–864.

Walsh, A. (2002). *Biosocial criminology: Introduction and integration.* Cincinnati, OH: Anderson.

Walsh, A. (2003). Intelligence and antisocial behavior. In A. Walsh & L. Ellis (Eds.), *Biosocial criminology: Challenging environmentalism's supremacy* (pp. 105–124). Hauppauge, NY: Nova Science.

Walsh, A. (2005). African Americans and serial killing in the media: The myth and the reality. *Homicide Studies, 9,* 271–291.

Walsh, A. (2006). Evolutionary psychology and criminal behavior. In J. Barkow (Ed.), *Missing the revolution: Darwinism for social scientists* (pp. 225–268). Oxford, UK: Oxford University Press.

Walsh, A. (2009a). *Biology and criminology: The biosocial synthesis.* New York: Routledge.

Walsh, A. (2009b). Criminal behavior from heritability to epigenetics: How genetics clarifies the role of the environment. In A. Walsh & K. Beaver (Eds.), *Biosocial criminology: New directions in theory and research* (pp. 29–49). New York: Routledge.

Walsh, A. (2011a). *Feminist criminology through a biosocial lens.* Durham, NC: Carolina Academic Press.

Walsh, A. (2011b). *Social class and crime: A biosocial approach.* New York: Routledge.

Walsh, A., & Ellis, L. (2004). Ideology: Criminology's Achilles' heel? *Quarterly Journal of Ideology, 27,* 1–25.

Walsh, A., & Ellis, L. (2007). *Criminology: An interdisciplinary approach.* Thousand Oaks, CA: Sage.

Walsh, A., & Hemmens, C. (2000). *From law to order: The theory and practice of law and justice.* Lanham, MD: American Correctional Association.

Walsh, A., & Hemmens, C. (2008). *Law, justice, and society: A sociolegal introduction.* New York: Oxford University Press.

Walsh, A., & Hemmens, C. (2011). *Law, justice, and society: A sociolegal introduction* (2nd ed.). New York: Oxford University Press.

Walsh, A., & Stohr, M. (2010). *Correctional assessment, casework, and counseling* (5th ed.). Lanham, MD: American Correctional Association.

Walsh, A., & Wu, H.-H. (2008). Differentiating antisocial personality disorder, psychopathy, and sociopathy: Evolutionary, genetic, neurological, and sociological considerations. *Criminal Justice Studies, 21,* 135–152.

Walsh, A., & Yun, I. (2011). Developmental neurobiology from embryonic neuron migration to adolescent synaptic pruning: Relevance for antisocial behavior. In M. DeLisi & K. Beaver (Eds.), *Criminological theory: A life-course approach* (pp. 69–84). Boston: Jones & Bartlett.

Walters, G. (1990). *The criminal lifestyle.* Newbury Park, CA: Sage.

Walters, G., & Geyer, M. (2004). Criminal thinking and identity in male white-collar offenders. *Criminal Justice and Behavior, 31,* 263–281.

Walters, G., & White, T. (1989). The thinking criminal: A cognitive model of lifestyle criminality. *Criminal Justice Research Bulletin.* Huntsville, TX: Sam Houston State University.

Wanberg, K., & Milkman, H. (1998). *Criminal conduct and substance abuse treatment: Strategies for self-improvement.* Thousand Oaks, CA: Sage.

Warchol, G. (1998). *Workplace violence, 1992–1996.* Bureau of Justice Statistics special report. Washington, DC: U.S. Department of Justice.

Ward, D., & Tittle, C. (1994). IQ and delinquency: A test of two competing explanations. *Journal of Quantitative Criminology, 10,* 189–212.

Warr, M. (2000). *Companions in crime: The social aspects of criminal conduct.* New York: Cambridge University Press.

Warr, M. (2002). *Companions in crime: The social aspects of criminal conduct.* Cambridge, UK: Cambridge University Press.

Weber, J. (1990). Managers' moral reasoning: Assessing their responses to three moral dilemmas. *Human Relations, 43,* 687–702.

Weber, M. (1978). *Economy and society: An outline of interpretative sociology, Vol. 2* (G. Roth & C. Wittich, Eds.). Berkeley: University of California Press.

Webster, C., MacDonald, R., & Simpson, M. (2006). Predicting criminality? Risk factors, neighborhood influence, and desistance. *Youth Justice, 6,* 7–22.

Weisburd, D., Wheeler, S., Waring, E., & Bode, N. (1991). *Crimes of the middle classes: White-collar offenders in the federal courts.* New Haven, CT: Yale University Press.

Weitzer, R. (1999). Prostitution control in America. *Crime, Law, and Social Change, 32,* 83–102.

Wells, R. (1995, June 16). Study finds fines don't deter Wall St. cheating. *Idaho Statesman,* pp. 1e–2e.

Wheeler, E. (1991). Terrorism and military theory: An historical perspective. In C. McCauley (Ed.), *Terrorism research and public policy* (pp. 6–33). London: Frank Cass.

White, A. (2004). *Substance use and the adolescent brain: An overview with the focus on alcohol.* Durham, NC: Duke University Medical Center.

White, J. (1998). *Terrorism: An introduction* (2nd ed.). Belmont, CA: West/Wadsworth.

White House, The. (2001). *The Office of Homeland Security* [Home page]. Available at http://www.whitehouse.gov/news/release/2001/10/print

Widom, C., & Brzustowicz, L. (2006). MAOA and the "cycle of violence": Childhood abuse and neglect, MAOA genotype, and the risk for violent and antisocial behavior. *Biological Psychiatry, 60,* 684–689.

Wiebe, R. (2004). Psychopathy and sexual coercion: A Darwinian analysis. *Counseling and Clinical Psychology Journal, 1,* 23–41.

Williams, F., & McShane, M. (2004). *Criminological theory* (4th ed.). Upper Saddle River, NJ: Prentice Hall.

Willoughby, M. (2003). Developmental course of ADHD symptomology during the transition from childhood to adolescence: A review with recommendations. *Journal of Child Psychology and Psychiatry, 43,* 609–621.

Wilson, C. (1984). *A criminal history of mankind.* London: Panther Books.

Wilson, J. (1987). *The truly disadvantaged.* Chicago: University of Chicago Press.

Wilson, J., & Kelling, G. (1982). *Broken windows.* Available at http://www.theatlantic.com/ideastour/archive/windows.html

Wilson, M., & Daly, M. (1997). Life expectancy, economic inequality, homicide and reproductive timing in Chicago neighborhoods. *British Medical Journal, 314,* 1271–1274.

Wodak, A. (2007). Ethics and drug policy. *Psychiatry, 6,* 59–62.

Wolfgang, M., & Ferracuti, F. (1967). *The subculture of violence: Towards an integrated theory in criminology.* London: Tavistock.

Wright, J. (2011). Prenatal insults and the development of persistent criminal behavior. In M. DeLisi & K. Beaver (Eds.), *Criminological theory: A life-course approach* (pp 51–67). Boston: Jones & Bartlett.

Wright, J., & Beaver, K. (2005). Do parents matter in creating self-control in their children? A genetically informed test of Gottfredson and Hirschi's theory of low self-control. *Criminology, 43,* 1169–1202.

Wright, J., Beaver, K., Delisi, M., & Vaughn, M. (2008). Evidence of negligible parenting influence on self-control, delinquent peers, and delinquency in a sample of twins. *Justice Quarterly, 25,* 544–569.

Wright, J., Dietrich, K., Ris, M., Hornung, R., Wessel, S., Lanphear, B., et al. (2008). Association of prenatal and childhood blood lead concentrations with criminal arrests in early childhood. *PLoS Medicine, 5,* 732–740.

Wright, R., & Decker, S. (1994). *Burglars on the job: Streetlife and residential break-ins.* Boston: Northeastern University Press.

Wright, R., & Decker, S. (1997). *Armed robbers in action.* Boston: Northeastern University Press.

Young, J. (2003). Merton with energy, Katz with structure: The sociology of vindictiveness and the criminology of transgression. *Theoretical Criminology, 7,* 389–414.

Zechel, J., Gamboa, J., Peterson, A., Puchowicz, M., Selman, W., & Lust, D. (2005). Neuronal migration is transiently delayed by prenatal exposure to intermittent hypoxia. *Birth Defects Research, 74,* 287–299.

Zhang, Z. (2004). *Drug and alcohol use and related matters among arrestees, 2003.* Washington, DC: National Institute of Justice.

Zuckerman, M. (1990). The psychophysiology of sensation-seeking. *Journal of Personality, 58,* 314–345.

Photo Credits

Index

About the Author

Anthony Walsh received his PhD in criminology and statistics from Bowling Green State University, Ohio, and is currently a professor at Boise State University where he teaches criminology, criminal law, statistics, and correctional counseling. He has field experience in both corrections and law enforcement. He is the author or coauthor of 27 books, including *Correctional Assessment, Casework & Counseling* with Mary Stohr, and over 100 articles. He also has a drop-dead gorgeous wife whom he loves with all his being.

SAGE Research Methods Online

The essential tool for researchers

An expert research tool

- An **expertly designed taxonomy** with more than 1,400 unique terms for social and behavioral science research methods
- **Visual and hierarchical search tools** to help you discover material and link to related methods

- Easy-to-use navigation tools
- Content organized by complexity
- Tools for citing, printing, and downloading content with ease
- Regularly updated content and features

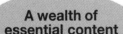

A wealth of essential content

- The most comprehensive picture of quantitative, qualitative, and mixed methods available today
- More than **100,000 pages of SAGE book and reference material** on research methods as well as editorially selected material from SAGE journals
- More than **600 books** available in their entirety online

Launching 2011!

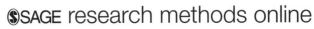

$SAGE research methods online